MARKETING LIVESTOCK AND MEAT
William H. Lesser, PhD

SOME ADVANCE REVIEWS

"A very readable text about a dynamic, important sector of the U.S. economy. Adding to the easy writing style are numerous figures and tables interspersed throughout the text. Several chapters contain an interesting historical perspective of livestock and meat marketing in the U.S. . . . Professor Lesser explains several economic concepts in concrete, easy-to-understand terminology and with practical examples. The approach to discussing consumer demand is particularly appealing."

Clement E. Ward
Professor and Extension Economist
Oklahoma State University

"*Marketing Livestock and Meat* . . . is presented at an appropriate level and with a clarity that is above average for a text. . . . The overwhelming advantage for the teacher is an up-to-date, scholarly, and comprehensive coverage of the entire field. Lesser has assimilated much of the available literature in a readable and well-balanced text that can also serve as a highly useful reference for anyone in the field."

V. James Rhodes, PhD
Professor
Department of Agricultural Economics
College of Agriculture
University of Missouri-Columbia

Marketing Livestock and Meat

FOOD PRODUCTS PRESS
Agricultural Commodity Economics,
Distribution, and Marketing
Andrew Desmond O'Rourke, PhD
Senior Editor

New, Recent, and Forthcoming Titles:

Marketing Livestock and Meat by William H. Lesser

*Understanding the Japanese Food and Agrimarket:
A Multifaceted Opportunity* edited by A. Desmond O'Rourke

The World Apple Market by A. Desmond O'Rourke

Marketing Livestock and Meat

William H. Lesser, PhD

Food Products Press
New York • London • Norwood (Australia)

Published by

Food Products Press, an imprint of The Haworth Press, Inc., 10 Alice Street, Binghamton, NY
13904-1580

Library of Congress Cataloging-in-Publication Data

Lesser, William.
  Marketing livestock and meat / William Lesser.
    p.   cm.
  Includes bibliographical references (p.   ) and index.
  ISBN 1-56022-016-3 (acid free paper)–ISBN 1-56022-017-1 (pbk.: acid free).
    1. Meat industry and trade–United States. 2. Animal industry–United States. 3. Meat–United
States–Marketing. 4. Livestock–United States–Marketing. I. Title.
HD9415.L47   1992
636´.0068´8–dc20
                                                                          91-32989
                                                                             CIP

Dedicated to Richard Vilstrup, who introduced me
to livestock marketing at a time when I considered
fish marketing to be the only possible subject of interest.

ABOUT THE AUTHOR

William H. Lesser, PhD, is Professor of Marketing and Director of the Western Societies Program at Cornell University in Ithaca, New York, where he has been a faculty member in the Department of Agricultural Economics since 1978. Dr. Lesser's research responsibilities have covered the full range of agricultural marketing issues, from the economic feasibility of kosher slaughter plants, to a strategy for the development of a feeder calf industry in New York State, and to efficiency in food distribution and retailing. Currently, his research focuses on the implications of biotechnology in agriculture, and he has written on the potential impacts of patents for livestock. Dr. Lesser, who has a specialization in industrial organization economics, was a forerunner in work on microcomputer applications for the food retail sector and offers one of the few courses on the "how-to" of export marketing. He is a member of the International Association for the Advancement of Teaching and Research in Intellectual Property, the American Agricultural Economics Association, the Food Distribution Research Society, the North Central Agricultural Economics Association, and the Northeastern Agricultural and Resource Economics Association. Dr. Lesser, a Kellogg National Fellow, received his doctorate in Agricultural Economics from the University of Wisconsin.

CONTENTS

List of Tables

List of Figures

Chapter 5

Chapter 16

Chapter 17

Chapter 18

Preface

There was a time when the livestock producer seemed to operate in a relatively stable environment. Prices and margins were moderately stable and multiple market outlets existed. During the 1980s much of that situation eroded. Product and feed prices became less stable, cycles lost much of their century-old pattern, both competition and trade barriers seemed to rise, and market outlets shrank in number and ownership diversity. At the same time, the demographic transformation of the U.S. population continued in a slow march to an older nation while new and confusing health concerns arose. Not all the changes were negative — new developments promised reductions in production costs and the industry began to take an active role in promoting the merits of its product.

This volume, which is intended for undergraduate and beginning graduate students, attempts to assimilate these numerous changes within one cover. Not every issue can be covered in detail, or even included. Undoubtedly, some topics that will prove to be of future significance have been neglected while other trivial matters receive too much attention. For this lack of foresight, I apologize. However, I have attempted to include relevant issues not likely to become dated in the near future so that this text can serve as a reference after the course is completed. Additional references are included where appropriate, to direct those readers with a deeper interest in the subjects. The material is presented in a way that is of principal value to producers needing to understand the marketing system, but it also provides other functionaries in the system with an overall concept of how the market functions.

A work like this is never solely one person's effort. In the main, what I have done is to assemble, digest, and interpret the scholarly work of livestock economists throughout the world. A disproportionate share of the material, however, draws upon the publications and staff expertise of the U.S. Department of Agriculture's various divisions. To that group I, and the livestock sector, owe a great deal. Presentation of the material was greatly assisted by my secretary, Jennifer Wyse, who went far beyond what I could reasonably expect during preparation of the manuscript for publication. I shall long remain in her debt. Numerous colleagues from Cornell assisted me in specific areas but to name them risks implying joint responsibility for any errors. Errors of fact and interpretation are, rather, solely of my creation.

William Lesser
Ithaca, NY

PART I:
BACKGROUND MATERIAL

Chapter 1

Introduction

On January 1, 1990, the United States had a standing herd of 164.6 million head of the three principal red meat species—cattle, hogs, and sheep (USDA, "Situation and Outlook," various issues).[1] Such numbers are sufficient to give Americans a level of red meat consumption of 135 pounds (retail weight) in 1987 (USDA, ERS 1989, Table 107), a volume placing us near the top of the world scale. Even in an era of light foods for an aging population, the United States remains a meat-and-potatoes nation, and the red meats remain first with consumers.

Supplying such vast amounts of meat is an amazing success story for the livestock production sector. The achievement is even more overwhelming considering that none of our principal meat species is indigenous to the hemisphere. Cattle, hogs, and sheep were either brought over from Europe with the early settlers or moved up from Mexico where they were introduced by the Spanish conquistadors and missionaries. The variety of names tells us of the ancestry—Angus and Hereford cattle, Yorkshire and Hampshire hogs—all are from the British Isles. The enormous red meats sector, the largest agricultural enterprise in terms of sales dollars, has developed from nothing over the past three centuries. And much

---

1. Hog inventories are taken December 1 of the preceding year.

of the growth, as we shall see, is far more recent, dating only to the end of World War II.

THE ROLE OF MARKETING

As much as the growth of the livestock sector is a success story for agronomists, animal scientists, and countless hard-working farmers and ranchers, it is also a testament to the marketing system. Without the simultaneous development of an effective and efficient marketing system, livestock would have remained on or near the farm. The large metropolitan areas could not have expanded as they did, and the world would be a much different place.

Despite the evident importance of the livestock and meat marketing system, it has a bad name with livestock producers. Many use the same vocabulary for marketing that is reserved for hoof-and-mouth disease and similar pestilence. In taking this view of marketing, the farmer travels in the rarefied company of Plato and Aristotle. Plato visualized marketing as little more than a necessary evil carried out by those whose "business is to remain on the spot in the market, and give money for goods to those who want to sell, and goods for money to those who want to buy." For him, marketers "are, generally speaking, persons of excessive physical weakness, who are of no use in other kinds of labor." (Figure 1-1).

While Plato's views are surprisingly modern, Aristotle made the point even more bluntly. He regarded marketers as "useless profiteering parasites" and condemned their efforts as "unnatural, mercenary, exploitative and corrupting." It would seem that many livestock producers are disciples of the Greek philosophers. For them, the societal and economic contribution is production. They view themselves, quite literally, as livestock producers. Exchange necessarily follows the production process. In earlier times, exchange involved barter; today, the universal exchange medium, money, is paramount.[2]

The idea of the exchange nevertheless remains unchanged. It is simply a transfer of ownership from seller to buyer, a process known as *selling*. Viewed in this way, selling is a purely mechanical activity, an exchange of money for whatever documents of legal ownership the system or the law requires. In such a world, the buyer is a necessary, if rather insignifi-

---

2. The quotes from Plato and Aristotle and views of the marketers' role were drawn from the talk "What is Marketing?" presented before the National Institute on Cooperative Education at Purdue University on August 4, 1982, by Professor Max E. Brunk of Cornell University.

FIGURE 1-1. Plato [For Plato, marketers were persons of excessive physical weakness.] (Source: The Louvre)

cant, intermediary who warrants a return only in proportion to this limited role. The producer is the center of this system, the source of wealth and happiness.

As comforting as the livestock producer may find this view, it is as wrong and as antiquated as the notion that the sun revolves around the earth. What drives the capitalistic, free-enterprise system is not the producer but the consumer. The consumer ultimately decides who and what will be produced just as the vast gravitational power of the sun keeps the planets suspended in space. The producer who does not satisfy the dictates of the consumer disappears from the marketplace much like an errant "shooting star" in the heavens on a hot, late summer night.

Marketing

Serving the needs of consumers is the role of marketing. In the words of Professor Levitt's classic article "Marketing Myopia," the difference between marketing and selling is more than semantic:

> Selling focuses on the needs of the seller, marketing on the needs of the buyer. Selling is preoccupied with the seller's need to convert his product into cash, marketing with the idea of satisfying the needs of the customer by means of the product and the whole cluster of things associated with creating, delivering, and finally consuming it. (1960, p. 50)

Livestock producers share their myopic world view with other mass-production industries. Where technological change and lower unit costs for larger operations are prevalent, the urge to produce, produce, produce proves irresistible. Firms become product- rather than consumer-oriented, substituting their judgement of what the consumer wants for the comsumer's true needs and desires. All too often the result is financial disaster; livestock producers talk with black humor of losing money per head but making it up on volume. Even the mighty, like General Motors, are not immune from market forces, as declining profits during the late 1980s demonstrated.

The resolution of this dilemma is marketing—the provision of services for which the consumer is willing to pay and on which a profit can be made. The remainder of this book describes marketing and its applications that help ensure the continued profitability of the red meats sector. Emphasis is placed on understanding the operations of the system with the idea that through knowledge comes the recognition of opportunities. To identify opportunities, it is necessary for each participant, each player, to understand his or her role and relationship with other participants and with the final decision maker, the consumer. In a system as large and complex as the red meats sector, this is a major undertaking.

Definition of Marketing

As intuitive as the concept may be, there is no universally accepted definition of the term, as the following passage makes clear.

> It has been described by one person or another as a business activity; as a group of related business activities; as a trade phenomenon; as a frame of mind; as a coordinative, integrative function in policy making; as a sense of business purpose; as an economic process; as a

structure of institutions; as the process of exchanging or transferring ownership of products; as a process of concentration, equalization, and dispersion; as the creation of time, place, and possession utilities, as a process of demand and supply adjustment; and as many other things. (Ohio State University 1965, p. 43)

From these wide-ranging definitions, two interrelated insights can be generated. First, marketing is a very broad concept. Selling is but one small part of it, and some would argue that production itself can be subsumed as one aspect of marketing. This latter view is extreme perhaps, although it is clear that production must be attuned to the market. Volume should be high when demand is strong, and vice versa. Second, the product type must fulfill market needs. "Lardtype" hogs were in demand when animal fats were popular as cooking oils. Now that the oil market has largely been taken over by vegetable oils, the fatty hog is penalized in the market. Lard, rather than being a valuable product in its own right, is now largely a liability which must be cut away. In response to the changing market need, animal scientists have bred a leaner hog (as shown in Figure 1-2).

The notion that marketing is on approximately equal terms with production is a particularly galling one for livestock producers; however, their view does not recognize economic realities. In 1988, consumer expenditures for foods (excluding alcoholic beverages) originating on farms totaled over $460 billion. Of that sum, only about 25 percent went to farmers (USDA, ERS, Marketing Review, p. 1, *Chartbook*, Chart 128). The remainder, or 75 percent of food expenditures, went to pay for other services such as processing, transportation, and retailing. In total, this sum is known as the farm-to-retail price spread or, simply, the marketing margin.

A marketing margin of 75 percent means that, on average, for all products, consumers value marketing services by a ratio of three to one over the value of the raw product at the farm gate. That is, for every dollar spent on food, consumers pay an average of three times as much for the marketing service component of the product as they do for the basic ingredient. Of course, the proportions vary widely from product to product. The producer receives a large share of the retail price of lamb chops, a relatively unaltered product, while the value of wheat in a loaf of bread is said to be but a few cents. Similarly, away-from-home food purchases have higher margins than food prepared at home.

One of the basic tenets of a market or capitalistic economy is that in selecting among competitive products, the consumer makes rational

FIGURE 1-2. Hogs Have Grown Leaner Over the Past 20 Years in Response to Changing Consumer Needs (Source: Duewer 1989, Fig. 1)

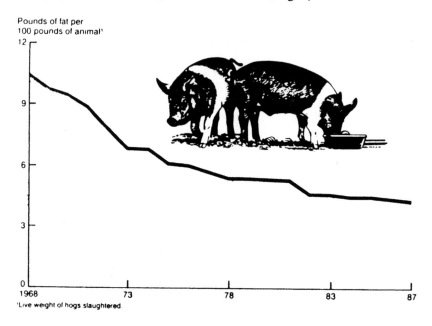

Pounds of fat per
100 pounds of animal¹

'Live weight of hogs slaughtered.

choices, concerning prices, income, and the strength of preferences. Without considering all implications of this concept, known as consumer sovereignty, it can be applied to livestock marketing. In its basic form, what consumer sovereignty means is that consumers pay for livestock marketing services because they are valued. This brings us to a second aspect of marketing — the concept of value added.

Value Added

A live steer in Colorado is worthless to a hungry vacationer in Miami. As trivial as this example may be, it does identify several major contributions of marketing. A product must be in the right place (Miami), form (rare steak), and time (6:00 p.m. today) to be useful to our traveler. Several components of the marketing system are involved in providing the necessary services: processing, aging, transportation, distribution, and preparation. Each of these activities adds value to the steer, just as the mining, forging, shipping, and assembling of a car add value to iron ore from the Minnesota taconite mines. Each of the contributors receives payment roughly in proportion to his contribution and costs.

However, these mechanical functions do not completely describe the contributions of marketing. Few would question that the packer contributes (adds value) to the carcass and needs to be compensated. Other marketing functionaries, however, provide more ephemeral yet necessary services which also add value. These services include pricing, financing, grading, and risk taking. Each facilitates the marketing process.

Financial executives recognize that the money invested in a carcass in the cooler has an opportunity cost — the interest not earned in a bank account or money-market fund. The packer must be paid for the cost and trouble of his financing arrangements. Grading allows a product to be bought unseen, thus providing a savings for the buyer but a reimbursable cost for the seller. The owner of the product accepts the risk of theft as well as deteriorating quality or falling prices. Some monetary reward is needed to provide an incentive for such risk taking. Finally, even the making of a market, the pricing, is a service. Buyers and sellers observe trends and trades as they arrive at mutually agreeable exchange prices. This function takes time, judgement, and diligence, and requires compensation.

In many cases, it is difficult to recognize what services are performed and how each is rewarded because a single participant typically combines several functions. In addition to providing the obvious slaughter and carcass-breaking functions, a packer also serves as financier, risk taker, and market observer. The packing margin, in part, covers these activities as well as the physical operations; in many cases, however, the functions are separate and participants specialize. The government provides graders (for a fee), while institutions such as the futures markets specialize in accepting risk.

Coordination

Thus far we have discussed those marketing functions that directly or indirectly benefit the consumer, but the marketing system performs a key additional market-wide function, that of coordination. The principal tool of coordination in the livestock sector is price.

The role of price can be seen most clearly by drawing a contrast with a centrally controlled economy typified by the former U.S.S.R. (at least in past years). In Russia, production decisions have largely been based on successive "Five-Year Plans," with bureaucrats allocating production quotas among the livestock collectives and other production centers. (Of course, the quotas are not always met.) The capitalistic economy operates differently; the aggregate purchase decisions of buyers dictate production needs. Industries dominated by one or a few firms function slightly differently. The large firm (relative to total industry production) interprets sales trends and establishes production levels, recognizing that the level of its output will affect price. These so-called concentrated industries recognize their pivotal roles and act to determine price rather than simply respond to it.

The red meats sector is fundamentally distinct from a concentrated industry. Particularly at the producer level, no individual operator or firm is large enough for the output to have a measurable effect on prices. With 89 million head of hogs slaughtered in 1989 (USDA, Livestock Slaughter 1990, Table 2), even a very large producer of 100,000 head could raise or lower total output by only .0011 percent by alternatively doubling or ceasing production. Thus, the producer views price as a given (a "parameter" in economic jargon) and adjusts production according to the current and expected levels of profitability. How then, if each producer acts as an independent decision maker, is some approximate sector-wide equilibrium achieved?

The coordinating mechanism is, of course, price. Lower prices suppress aggregate output by reducing the profitability of feeder production and finishing. This may take place through a total withdrawal by some producers, a cutting back by some or feeding to lower weight by others, or a combination of the three; the net effect is reduced output. Higher prices have the opposite effect, with the speed of the change influenced by lags in either case.

Income Allocation

Because of the overriding importance of price in dictating the operation of the system, its formation is a key issue to producers, packers, and consumers alike. When the mechanism works imperfectly, as with livestock cycles (Chapter 8), the costs are significant. However, there is an additional reason why prices are so sensitive. They dictate the economic rewards to each participant. Prices net of costs are the entrepreneurial rewards in a capitalistic economy. Thus, when there is a perceived problem with the way prices are determined or when they are not efficiently transmitted through the system, and in particular when they are seen as inequitable or not "fair," there is substantial discontent. Hence, market problems or "failures" are a key issue.

MARKET FAILURES

The perfect competition model used by economists leads to some very desirable outcomes. Principal among these are production at minimum achievable costs and at profit levels just adequate to keep resources committed to that enterprise. Moreover, each input to the process receives compensation determined by its relative contribution. And finally, adjustments in output are treated as if they were made instantaneously, and exchanges in vertical systems are considered to be costless.

With this setting, market failure can now be identified. Basically, it exists whenever any of the above conditions do not hold — whenever operations are inefficient, participants are not compensated equitably (as would happen if one segment were monopolized, for example), or creaky markets impede information transmission. Of course, market failure is a matter of degree; perfect competition never has and never will exist, so some extent of market failure will always be present. However, we shall sidestep that issue by considering only those instances in which market failure is "substantial." Where market failure is present, it may be classified as either technical inefficiency or pricing inefficiency, or both.

Technical and Operational Inefficiency

As the name implies, technical inefficiency refers to the technical or physical aspects of marketing. For example, using two straight chassis trucks to transport livestock to market when a semi would suffice, at lower costs, would be technically inefficient. Similarly, the old bed system in packing plants is technically inefficient compared to the lower cost overhead rail method which prevails today.

Technical inefficiency can be partitioned into operational inefficiency and sub-optimum operating levels. The distinction can be seen in a study of livestock auction markets by Lesser and Greene (1980). Using a statistical technique known as frontier function analysis, they estimated the minimum achievable costs for all 1,596 U.S. auction markets for 1976. As would be expected, costs are partially related to the size of the operation, with larger markets operating at lower cost per unit than smaller ones. For the smallest market size category, the unit costs were found to be 57 percent higher than the attainable level at the most efficient size (or over one dollar per head). That estimate, however, assumes markets are operating at 100 percent efficiency given their size category. In practice, such efficiency of operation is seldom achievable; the regular auctioneer may be ill, pen layout may slow the sorting operation, or poor maintenance may allow an animal to fall through a loading chute. Each of these and many other management, labor scheduling, and layout factors can reduce productivity and raise costs. Conditions for which costs are found to be above minimum achievable levels for protracted periods are known as operational inefficiencies. Lesser and Greene found their sample ranged in degree of operational inefficiency from essentially zero to over 400 percent. Overall, they concluded that the U.S. livestock auction market system was not a good example of technical efficiency.

Operational efficiency is also known by the more colorful name of "x-efficiency," coined by Professor Leibenstein. Leibenstein cited examples of several industrial firms able to reduce costs in the range of 25-30 percent (1966, pp. 398-99). Although figures of this magnitude are much in dispute, there is evidence of substantial cost "slack" in the system, something Leibenstein attributed to a lack of competitive pressures and "[in concentrated industries] a variety of [other] reasons people and organizations normally work neither as hard nor as effectively as they could" (p. 413).

Pricing Inefficiency

Pricing inefficiency has a particular meaning in economics related to the association of production, costs, and profits and the way they are shared.[3] Definitionally, within the confines of economic theory, prices that do not equal average and marginal cost are inefficient. Furthermore, if labor receives less than its economic contribution while capital receives more, then the prices (e.g., factor shares) of these inputs are inefficient.

A monopoly is a good example of pricing inefficiency. As monopolists earn excess profits (that is, price exceeds average costs plus competitive profits over a sustained period), monopoly prices are inefficient. What is the practical effect of this? Income could be transferred from shoppers and/or producers to the monopolist. Due to the higher prices, less will be produced under a monopoly and, hence, less of the raw material supplied to the monopolist is needed. Finally, there is a net loss to society (known as dead weight loss) caused by the monopoly.[4]

However, pricing inefficiency is not limited to monopolies. If a sector slowly adjusts prices to reflect changing market conditions, this is a form of pricing inefficiency. Supermarkets are often charged by livestock and

---

3. A third term, allocative ineffiency, is also used. This concept is closely related to pricing inefficiency.

4. For a simplified explanation, see F.M. Scherer, *Industrial Market Structure and Economic Performance*. Chicago: Rand McNally College Publishing Co., Second Edition, 1980, pp. 17-18.

other agricultural producers with adjusting prices too slowly, especially during periods of declining prices.[5] Under these conditions, for example:

> If retail lettuce prices decline little or only with a long lag when farm prices fall sharply as supply increases, consumers are insufficiently encouraged to buy and growers are too much discouraged from producing next year. If retail prices also respond too little to decreased supplies, price variability and risk are increased for producers, while consumers lose from unwarranted stability. (Brandow 1977, p. 92)

The price responsiveness issue described here is but one example of possible coordination problems in systems (such as red meats) where prices serve the principal coordinating role. There are others. A favorite example of the problems with grades and standards is fatty bacon. In a smoothly functioning system, the housewife preferring lean bacon will select it even at a slightly higher price. Carrying the incentive of higher prices, this preference will be transferred through the marketing system to the hog producer who will expand lean pork production while curtailing the output of fat animals. Although this system sounds so simple as to be nearly foolproof, it is anything but. Suppose, for example, that all brands of bacon include both fat and lean selections at the same price. The distributor reviewing gross sales data will have no way to identify a relative preference for lean hogs. Hence, no signal is identified or transmitted and fat hogs will continue to be overproduced relative to lean ones. A similar problem would arise if the grading system for slaughter hogs did not recognize fatness as a pricing variable. Packers would select the leaner animals; however, producers of lean hogs might not associate the choice with the degree of fatness and, therefore, would not alter feeding and breeding practices. When this problem occurred with feeder cattle, it led, in part, to recent changes in grade standards (see Chapter 9). Completing the list of possible coordination problems are spatial and temporal misalignments.

Technical and pricing inefficiency often coexist. For example, as Leibenstein observed, monopoly (a pricing efficiency problem) can create the slack that encourages inefficiencies (technical efficiency problems). Moreover, by charging higher prices, monopolies enlarge the marketing margin. Broader marketing margins make production less responsive to retail price changes, thereby reducing the communication power of prices.

---

5. Typically, these charges have not been substantiated by empirical analysis. See, e.g., L. L. Hall, W. G. Tomek, N. L. Ruther, and S. S. Kyereme, *Case Studies in the Transmission of Farm Prices*, Dept. of Agricultural Economics, Cornell University, A. E. Res. 81-12, August 1981.

However, livestock producers' concerns about bargaining power and lack of buyers are principally pricing efficiency issues, and they are given the greatest attention in the following chapters.

SYNOPSIS

This chapter introduces the concept of *marketing*, emphasizing the importance of the focus on the consumer. In that respect, marketing differs from *selling* which emphasizes the exchange process. In a free enterprise economy, marketing is essential.

Since marketing is more than exchange, it involves more components, each of which *add value*. In total, value added (or marketing services) adds 75 cents for every 25 cents worth of food at the farm gate. Some value added, like slaughter, are obvious physical tasks, while others are more ephemeral activities such as risk taking and price setting.

In an unconcentrated idustry like livestock production, the principal coordination — the determination of what is produced, when, and where — is accomplished by price. Hence, *market failure* is anything that impedes price determination and dissemination, upsets the sector, and is costly. The major sources of market failure are *technical inefficiencies* (say, the use of outdated equipment) and *pricing inefficiency* (for example, under a monopoly, when too little is produced for a price that is too high compared to a competitive system).

This book examines the components of the system which lead to marketing efficiencies (components such as grading systems, the packing industry, and risk shifting through futures markets) discusses their role and use and, where appropriate, provides commentary on the apparent level of efficiency. Effective use by producers of efficient systems is the goal to which the system, and this book, are directed.

Study Questions

1. What are the key differences between selling and marketing? Why is marketing important to livestock producers?
2. What is the role of the consumer in a market economy?
3. What functions do prices play?
4. What is meant by market failures, and what effects can these failures have?
5. Why do livestock producers typically take such a dim view of marketing? Are they justified in that view?

REFERENCES

Brandow, G. E. "Appraising the Economic Performance of the Food Industry," *Lectures in Agricultural Economics*, USDA, June 1977, p. 92.

*Bressler, R. G., Jr. and R. A. King. *Markets, Prices and Interregional Trade*. New York: John Wiley & Sons, Inc., 1970, (particularly Chapter 21).

Brunk, M. "What is Marketing?" Speech before the National Institute on Cooperative Education at Purdue University on August 4, 1982.

Duewer, L. A. "Changes in the Beef and Pork Industries." 12 NFR, March 1989, pp. 5-8.

*Fowler, S. H. *The Marketing of Livestock and Meat*. Danville, IL: The Interstate Printers and Publishers, Inc., Second Edition, 1961.

Hall, L. L., W. G. Tomek, N. L. Ruther, and S. S. Kyereme. "Case Studies in the Transmission of Farm Prices." Dept.of Agricultural Economics, Cornell University, A.E. Res. 81-12, August 1981.

*Kotler, D. *Marketing Management: Analysis, Planning and Control*. Englewood Cliffs, NJ: Prentice-Hall, Inc., Fourth Edition, 1980 (particularly Chapter 1).

Leibenstein, H. "Allocative Efficiency vs. X-Efficiency," *American Economic Review* LVI(1966):392-415.

Lesser, W. H. and W. H. Greene. "Economies of Size and Operating Efficiency of Livestock Markets: A Frontier Function Approach," *Journal of Northeastern Agricultural Economics Council* IX(1980):37-40.

Levitt, T. "Marketing Myopia," *Harvard Business Review* July-August (1960): 45-56.

*McCoy, J. H. *Livestock and Meat Marketing*. Westport, CT: AVI Publishing Co., Second Edition, 1979.

Ohio State University, Marketing Staff. "A Statement of Marketing Philosophy," *Journal of Marketing*, 29 (1965):43-44.

Scherer, F. M. *Industrial Market Structure and Economic Performance*. Chicago: Rand McNally College Publishing Co., Second Edition, 1980 (particularly pp. 12-21).

USDA, Economic Research Service. "Livestock and Poultry: Situation and Outlook Report." Various issues.

USDA, Economic Research Service. "Livestock and Meat Statistics, 1984-88." Stat. Bull. No. 784, September 1989.

USDA, Economic Research Service. "Food Marketing Review, 1988." Ag. Econ. Rpt. No. 614, August 1989.

USDA, National Agricultural Statistics Service, Ag. Stat. Board. "Livestock Slaughter, 1989 Summary." MtAn 1-2-1(90), March 1990.

USDA, "1989 Agricultural Chartbook." Ag. Handbook No. 684, March 1989.

*Indicates supplementary references.

Chapter 2

Economic Concepts

While livestock marketing is very much a practical pursuit, a foundation in economic principles is nevertheless helpful for an understanding of the economic environment in which the system operates. Readers without any background in microeconomic theory are referred to one of the introductory texts listed at the end of the chapter (or one of the many others available). The material included here is limited to two topics of particular importance to livestock marketing: elasticities and market equilibrium, both of which are often troublesome to grasp completely.

ELASTICITIES

Consumers are the basic decision makers in a market economy. The consumer operates in a marketplace of choices and constraints, chief of which is income. Few of us have the resources to buy everything we want. Operating under this limitation, the consumer typically chooses to buy fewer of relatively costly products, but another option is available – selecting other products. As the price of round steak rises, the income-limited shopper may switch to less costly chuck steak, ground chuck, or across product lines to pork or chicken or cheese.

The readiness with which consumers make these adjustments is the "elasticity." Although this term is commonly employed, there are actually *three* forms which should be differentiated: the so-called price (or "own"), cross, and income elasticities of demand. Of these, the price elasticity is the most used and quoted. Supply curves also allow the calculation of an elasticity known as the price elasticity of supply. Supply elasticities shall not be discussed in detail here.

Price Elasticities

The price elasticity of demand is a measure of the relative responsiveness of demand to changes in product prices. More specifically, it describes how demand changes in response to a price increase or decrease.

The range of the elasticity can be blocked out easily. When the demand curve is perfectly vertical (Figure 2-1), consumers will pay any amount to secure their "Q." Their demand elasticity is zero. Alternatively, when the curve is perfectly horizontal, an increase in price of even a few cents will lead to none being sold. This demand is known as perfectly elastic. Thus, price elasticities of demand vary from zero to infinity. This discussion can be demonstrated in a very straightforward fashion by examining the slope of the curves in Figure 2-1.

The slope of a line is defined as the run over the rise: the change in Q over that in P. From Figure 2-1, this is 0/250 on curve (B). The elasticity is zero when the curve is vertical. In (A), the ratio is 100/0, which mathematicians identify with the symbol ∞ (meaning infinity). When the curve is horizontal, the price elasticity is infinite. The reader will note that since demand curves are typically downward sloping, the slope should be designated with a negative sign. This is absolutely correct, but for convenience price elasticities are usually expressed with the positive sign implied (i.e., the absolute value).

What about all the in-between lines? Can they be described as slope? Unfortunately, no; it is more involved than that. To see why, consider the demand curves in Figure 2-2 where (A) represents the demand for chewing gum, and (B) that for Duesenberg boat-tailed speedsters, probably the most sought-after American-built collector's car. By construction, the two curves have the same slope, but the units on the axes are clearly different. What happens with a 10-cent price change? Figure 2-2 (A) tells us that demand changes (moves along the demand curve) substantially, but for Duesenberg owners (Figure 2-2 (B)) there is no discernible change. (They probably would not stoop to pick up a loose $50 bill.) Clearly, then, the slope of the demand curve does not tell us all we need to know when comparing products. What is required is an adjustment that allows for differences in interproduct prices and quantities.

Definition: The required adjustment is made by dividing the slope by the point on the demand curve, (Q/P) where the estimate is being made. (In other cases, the estimate is made for the mean point, designated by $\bar{Q}/\bar{P}$.) Then the equation for calculating the price elasticity of demand is:

$$\text{Price Elasticity of Demand: } E_D = \frac{\text{slope}}{Q/P}$$

Defining the slope over an interval as: $\dfrac{Q_2 - Q_1}{P_2 - P_1}$

gives us $\qquad \dfrac{[Q_2 - Q_1]/Q}{[P_2 - P_1]/P}$

or inverting and multiplying $= (P/Q) \times \dfrac{[Q_2 - Q_1]}{[P_2 - P_1]} = \dfrac{P}{Q} \cdot \dfrac{dQ}{dP}$

A shorthand way of writing $Q_2 - Q_1$ is as dQ.[1] Along a linear demand curve (such as in Figure 2-2), any "d" will suffice — the slope is the same everywhere. However, with a curved demand schedule, the slope is different everywhere. Here two distinctions are made: the elasticity over an interval, which is known as an *arc*, and the elasticity at one particular spot, or *point*. The point is identified with the symbol of dQ where d stands for derivative. Students of calculus recognize that d as a Δ with the length of the interval approaching zero. Two definitions are now possible.

Arc Price Elasticity of Demand $= E_{arc} = \dfrac{\Delta O}{\Delta P} \times \dfrac{P}{Q}$

Point Price Elasticity of Demand $= E_{point} = \dfrac{dQ}{dP} \times \dfrac{P}{Q}$

It is now possible to calculate the elasticities from the demand curves in Figure 2-2. This is done in Table 2-1. As noted, in this example with its linear (i.e., straight-line) demand curves, the arc and point elasticities are the same. The demand for chewing gum is less elastic (.75 vs. 1.0) than that for Duesenbergs, meaning chewing gum buyers in this fictitious example are less price sensitive than are buyers of the other product. (Of course, they may physically be the same person, but we are concerned here with their behavior in distinct product markets.)

Elasticities at Points on a Curve: With the denominator (Q/P) changing at every point along a demand curve, the calculated elasticity will similarly be different. This leads to the useful generalization that the elasticity of a linear demand curve is one (unity) at the midpoint. Above the mean, the elasticity declines (becomes inelastic) while, below the mean, it increases and demand becomes elastic (Figure 2-3).

Interpretation and Uses of Demand Elasticities: The price elasticity of demand succinctly describes the relationship between price and quantity for a particular product. A key distinction is whether the demand is elastic or inelastic. Demand is termed elastic if the elasticity exceeds one (in absolute value), while it is called inelastic if the value lies between zero and one. A calculated value of one is, as previously noted, referred to as unitary elasticity. When describing elasticities, it is important to limit the

---

1. Note that another way to calculate elasticities is as the proportional change in quantity for the proportional change in price, or $\dfrac{dQ \,/\, dP}{Q \,/\, P} = \dfrac{dQ}{Q} \times \dfrac{P}{dP} = \dfrac{dQ}{dP} \times \dfrac{P}{Q}$

FIGURE 2-1. Price Elasticities of Demand for Perfectly Elastic (A) and Inelastic (B) Demand Curves

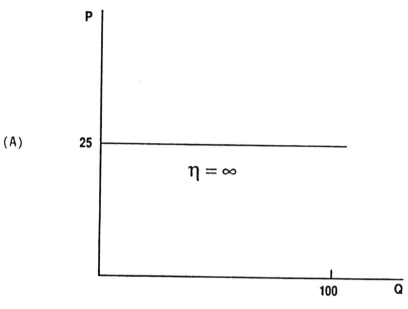

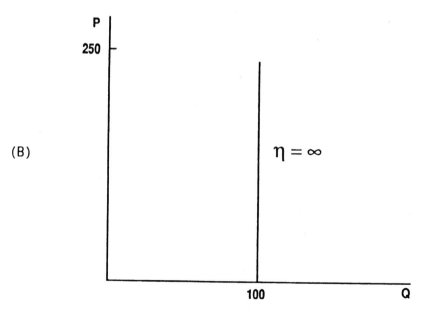

FIGURE 2-2. Equal Sloped Demand Curves for Dissimilar Products: (A) = Chewing Gum; (B) = Duesenberg Boat-tailed Speedsters

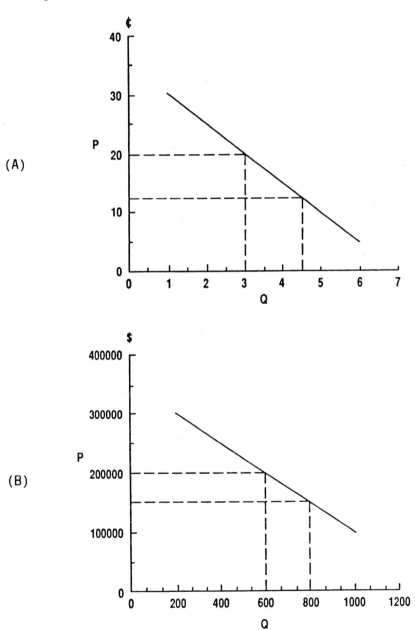

discussion to a small area around the estimation points; large movements along the curve may cause the value to change. (Alternatively, the mean elasticity, more correctly known as the elasticity calculated at the mean point, should be used.)

The price elasticity of demand may be thought of as a percentage change in quantity resulting from a one-percent change in price. This facile interpretation is made possible by our division of the slope by the price quantity relationship at that point — a manipulation which also converts the value to a percentage. Assumed in this distinction is that all other factors, such as changes in prices of other products and incomes, remain unchanged.

An inelastic demand (remember the importance of limiting the description to a small area around the point) then means that the quantity demanded does not change as much as the price. Hence, noting that total sales or revenue is price times quantity, a price increase raises total revenue when the price elasticity of demand is inelastic. This comes about because the quantity demanded falls, but not as rapidly as price is increased, so the product of the two increases. Conversely, when the price elasticity of demand is inelastic, a reduction in price increases the quantity demanded less than proportionally. The total revenue falls.

Another way to examine the effects of inelastic demand is to consider what happens when the quantity sold varies. (For all non-storable products, such as live animals and fresh meat, once produced, they are consumed, therefore the quantity sold is typically determined by the amount supplied.) Higher quantities decrease prices more than proportionally, while lower quantities increase prices more than proportionally. The sig-

TABLE 2-1. Computed Demand Elasticities for Chewing Gum and Duesenbergs

Chewing gum: $.10 price charge

$$E_D = \frac{dQ}{dP} \times \frac{P}{Q} = \frac{4.5 - 3}{.22 - .12} \times \frac{(.30 - 0)/2}{(6 - 0)/2} = \frac{1.5}{.10} \times \frac{.15}{3} = 15 \times .05 = .75$$

Duesenbergs: $50,000 price charge

$$E_D = \frac{dQ}{dP} \times \frac{P}{Q} = \frac{800 - 600}{270,000 - 220,000} \times \frac{(300,000 - 0)/2}{(1200 - 0)/2}$$

$$= \frac{200}{50,000} \times \frac{150,000}{600} = .004 \times 250 = 1.0$$

FIGURE 2-3. Demand Elasticities Along a Linear Demand Curve

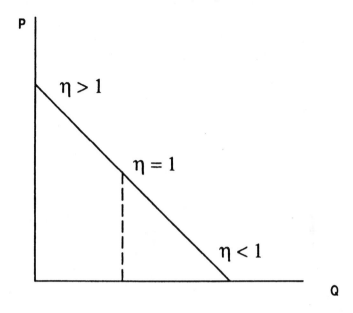

nificance of products with inelastic price elasticities of demand and variable supplies is clear: prices will be very unstable.

Elastic price elasticity of demand has the opposite interpretation: a one-percent increase in price reduces quantity demanded by more than one percent. When demand is price elastic, total revenue will decline following a price increase. Described in terms of quantity supplied, an increase in supply will lead to a less-than-proportional reduction in price. Hence, products with elastic price elasticities of demand have a "built-in" stabilizing influence.

Cross Elasticities

If a rise in the price of lamb causes consumers to shift to pork or beef, then the meats are substitutes for each other. The degree of substitutability can be described by another elasticity called the cross (for cross product) elasticity of demand. The cross elasticity of demand is defined in a similar way to the price elasticity except that it is now necessary to identify the two products. Referring to the cross elasticity of demand of product X (e.g., beef) in response to a change in the price of Y (e.g., lamb), this elasticity may be defined as:

Cross Elasticity of Demand $E_{xy} = \dfrac{dQ_x}{dP_y} \cdot \dfrac{P_y}{Q_x}$

The subscripts x and y identify the two products.

In some sense, all products are related in the marketplace. Many, however, have strictly limited cross effects and are known as independent goods. Rain hats and veal are highly unrelated products, and the cross elasticity of demand among them would be expected to be zero. Other products complement each other, as horseradish does beef, mint jelly for lamb, and applesauce for pork. These complementary goods are typically negatively related: an increase in the price of Y leads to a reduction in Y purchased, and hence the need to buy X. Complements, then, have negative cross demand elasticities.

The most significant use of cross demand elasticities is for substitute goods. The relationship between these goods is positive: the increase in the price of Y leads to increased purchase of X. Throughout this text, when the cross elasticity of demand is discussed it will refer to substitute goods unless otherwise noted. The cross elasticity of demand may then be interpreted as the percentage increase in the quantity of a product demanded as the result of a one-percent increase in the price of a substitute good.

Income Elasticity

A second major factor which causes demand to shift is income. Typically, rising incomes will lead to more of a product being purchased. Goods for which this is true are called normal goods, as opposed to inferior goods which are characterized by a decline in purchases with higher incomes. A positive income effect is consistent with the rationale that more is generally preferred to less, and consumption is constrained by means. Rising incomes ease this restraint and consumption advances. Sharply higher incomes following World War II have been a significant factor in the demand for red meats in both the U.S. and the world. In the 1980s, trends toward static and even declining incomes (after adjustments for taxes and inflation) in some income categories may also be having a distinct effect on the livestock sector, but in the opposite way.

Following customary usage, the income elasticity of demand is defined as:

Income Elasticity of Demand $\eta_y = \dfrac{dQ}{dY} \cdot \dfrac{Y}{Q}$

where Y represents income and Q, quantity. Interpreted further, this means that a one-percent change in income leads to a given percentage change in quantity demanded. As with the other elasticities, income elasticity can be described as elastic, inelastic, or of unitary elasticity. Additionally, the elasticity assumes that all other factors other than income and quantity are held constant.

Price Flexibility

Demand elasticities are calculated assuming price is determined independently and quantity demanded is a function of that price. Sometimes, however, the quantity will be the choice variable, and price is a function of that volume. An importer, for example, may find the volume he/she can supply to be limited by government quotas. His/her supply is thereby fixed and the sales responsibility involves selling that volume for the highest price.

"Elasticities" estimated from relationships in which price is the dependent variable are known as price *flexibilities*. The price flexibility of demand is defined as follows:

$$\text{Price Flexibility of Demand } F = \frac{dP}{P} \div \frac{dQ}{Q} = \frac{dP}{P} \times \frac{Q}{dQ} = \frac{dP}{dQ} \times \frac{Q}{P}$$

Cross price flexibilities and income flexibilities are defined in an analogous manner. The price flexibility is defined as the percentage change in price caused by a one-percent change in quantity.

Mathematically, a flexibility is just the inverse of an elasticity. In fact, in some cases, an estimated flexibility may be converted into an elasticity by taking the inverse. However, care must be taken, as this is not always the case (Tomek and Robinson 1990). Generally, if the price elasticity of demand for a product is inelastic, then the price flexibility will be greater than one, and vice versa.

Total Elasticity and Temporal Effect

Elasticities were defined earlier as the interactive effects between two variables, *all other factors held constant*. While this definition is adequate for heuristic purposes, it is not as valuable in the real world because many of these factors are linked. This is especially true among substitute products. Consider pork and lamb. A rise in lamb prices will shift some consumers to pork in the first period. Subsequently, the increased demand for pork will cause its price to rise, mitigating the benefit of substituting. Moreover, changing relative prices will affect the supply functions for both products. When all these effects are worked through, the total elasticity may be computed.

Other factors change elasticities as they are allowed to vary. The significant factors are tastes and substitute products. Tastes rooted in habit and tradition change slowly, but change they do. Broilers were essentially a new product after World War II when new feeds and breeds reduced production costs radically. In subsequent years, they have become a major source of animal protein, providing about 36 percent by retail weight in 1987 (USDA, ERS 1989, Table 107). Thus, typically, the longer the time frame, the more elastic demand becomes. The broiler example also points out a component of this temporal factor—substitutes. Longer periods allow for the development of more substitutes which makes demand more elastic.

Empirical economists have devised means of estimating the short- and long-run elasticities. This is done by using two distinct data sources. One is based on replicated family expenditure records across a wide area with different prices, tastes, and incomes. According to the arguments, these data provide a view of long-term consumption adjustments. These are known as *cross-sectional* estimates and may be compared with *time series* analysis. Time series data are taken from price, consumption, and income records for the same area over a period of weeks, months, or years. Changes in these factors over these periods do not measure the new equilibrium levels but, rather, initial responses as consumers adjust to changing prices. Thus, elasticities calculated from time series data are considered short-term values. Following our expectations, short-term elasticities will be smaller than long-term ones.

MAINTAINING MARKET EQUILIBRIUM

Readers with even a passing familiarity with basic economic theory will know that market equilibrium (e.g., supply and demand are balanced at the prevailing price) under perfect competition is maintained through the "instantaneous" entry and exit of firms. While no one really expects these changes in firm numbers to be made so rapidly, there are industries in which supply can be changed quickly or surpluses placed in long-term storage. Those conditions are not applicable to livestock products that have a production lag, and once produced must be consumed promptly. That is, once the breeding decision has been made, it takes almost exactly a year to produce a market hog, and once the production decision has been made, it is generally completed unless market conditions dramatically turn bad. At the same time, once the hog reaches slaughter weight, it must be processed and consumed in a short time. (Belly, hams, and a few other parts may be stored, but that does not change the fundamental relationships.)

What we have described are the conditions for the so-called cobweb model. The workings of this model, as well as the reason for the name, can be seen by following a series of adjustments through time. It is convenient to begin with a relatively high price like P_0 (Figure 2-4). At a favorable price like P_0, hog feeders are encouraged to expand production. Assuming expectations are simplistic ("naive" is the term used by economists), their future price expectations are based on current prices leading to a total production decision of Q_0 head. However, Q_0 can only be sold at a sharply lower price of P_1. The sharply lower P_1 price leads producers, again assuming they view P_1 as their best estimate of market price in the future period, to reduce output to Q_1. Low output, in turn, leads to higher prices of P_2 and so on, tracing out a cobweb-like pattern until equilibrium is achieved. At any point in the process, some external force like disease or a short corn crop or a demand shift could disrupt the process and cause it to begin again. Therefore, in the real world, equilibrium may actually never be achieved or, if achieved, may not be stable.

FIGURE 2-4. Achieving Equilibrium in a Cobweb Model

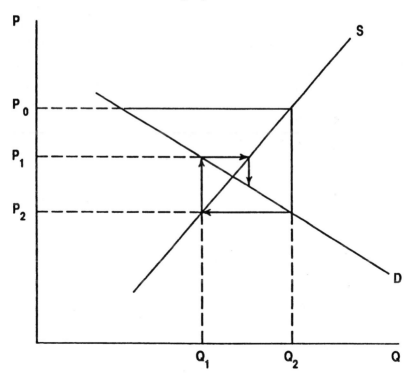

In the example shown here (Figure 2-4) the oscillations are damped and move toward the equilibrium. This is because the supply curve is steeper than the demand curve. If the demand curve were steeper than the supply curve, the opposite would happen: the supply/demand balance would move progressively farther away from equilibrium. And if the two were of equal slopes, then the system would be caught in a perpetual pattern of movement around the origin.

With the appropriate supply and demand relationships, the cobweb model does explain how a perfectly competitive market with production lags and non-storable products can achieve equilibrium. In these respects, it is descriptive of the livestock sector but not of products such as milk where the market is regulated by federal and state marketing order legislation. The cobweb further provides a theoretical justification for the production cycles seen in the livestock sector. In the case of our pork example, the cycle is typically four years.

In the simplest form presented here, this model does have some key limitations when applied to livestock. It suggests that production cycles tend to be dampened over time as equilibrium is approached. Yet, casual observation reveals that livestock cycles occur regularly and repeatedly. A partial explanation for this phenomenon is that exogenous factors, including weather, intervene, and production intended is not always production realized. More fundamentally, the model assumes producers have limited foresight in formulating production decisions. More recently, economists have recognized that farmers are shrewd decision makers who include a number of factors along with current prices when making production decisions. Once these factors are incorporated, the model loses its distinctive cobweb shape. Nevertheless, at this stage of our analysis, the cobweb model does provide an understandable and plausible (if not entirely convincing) explanation for equilibrium in competitive markets with lagged production response.

SYNOPSIS

Elasticities (consisting of own price, cross price and income) are derived from the slopes of these curves, adjusted for the position on the curve. Price elasticities for meat products are typically inelastic (e.g., less than one in absolute value), meaning that total revenues are increased as prices rise because demand declines less than proportionally to the price increase (and vice versa). Very price inelastic products can, however, be quite unstable as prices or quantities change. The existence of substitute products is important to consumer choices and, hence, to elasticities for individual products.

The cobweb model is an effective means of conceptualizing why supplies (and, therefore, prices) can fluctuate for products, like livestock, which have production lags. The reality of supply adjustments is more complex.

Study Questions

1. The price elasticity of red meats is generally described as inelastic. What are the ramifications of this market situation for the industry?
2. Distinguish among the major types of elasticities and the proper interpretation of each.
3. How does the time period enter into the determination of elasticities?
4. Identify another product, other than livestock, for which the cobweb model might be appropriate.

REFERENCES

**Baumol, W. J. *Economic Theory and Operations Analysis*. Englewood Cliffs, NJ: Prentice-Hall, Inc., Fourth Edition, 1977.

**Gould, J. P. and C. E. Lazear. *Microeconomic Theory*. Homewood, IL: Richard D. Irwin, Inc., Sixth Edition, 1989.

*Samuelson, P. A. and W. D. Nordhaus. *Economics*. New York: McGraw-Hill, Thirteenth edition, 1989.

**Tomek, W. G. and K. L. Robinson. *Agricultural Product Prices*. Ithaca, NY: Cornell University Press, Third Edition, 1990.

U.S. Department of Agriculture, Economic Research Service, "Livestock and Meat Statistics, 1984-1988." Stat. Bul. No. 784, September 1989.

References in economic theory: Almost any of the numerous introductory macroeconomic theory or price theory texts is appropriate. At the most basic level, Samuelson's *Economics* contains a great deal of information and is very readable, but massive. Also suggested are Tomek and Robinson, and Shepherd.

*Introductory theory text.
**Intermediate level theory.

PART II:
ECONOMICS OF THE DEMAND
AND SUPPLY FOR RED MEATS

Chapter 3

Historical Overview

The livestock sector is one of the most important to the world economy. According to the USDA's Foreign Agricultural Service estimates, world livestock numbers in 1988 were 2.48 billion. Of this total, U.S. production contributed 66 percent (USDA, *Ag. Stat.* 1989, Tables 389, 406, 423). This mammoth size is even more impressive when we recognize how much of the output occurred following World War II. In the post-war period, worldwide cattle numbers were up 30 percent to 948.3 million head while in the U.S., the increase was a dramatic 39 percent from 1936-46 to 1986-88 (USDA, *Ag. Stat.* 1947, Table 410 and 1988, Table 385).

With any sector of this magnitude and growth rate, instability is likely to be a factor. And so it is with livestock and the hog-corn price ratio, one indicator of profitability, varying by over 100 percent from 1970-1989 (Figure 3-1). In a generally competitive sector like livestock, much of the short- and long-term variability is explained by changes in supply and demand. The purpose of this section is to review the underlying supply and demand factors over time and to summarize empirical studies useful to understanding current changes and making projections for the future.

The chapter contains a brief historical sketch which will put the existing

FIGURE 3-1. Hog-Corn Price Ratio, 1970-1989 (Source: USDA, *Agricultural Chartbook*, 1989, T35)

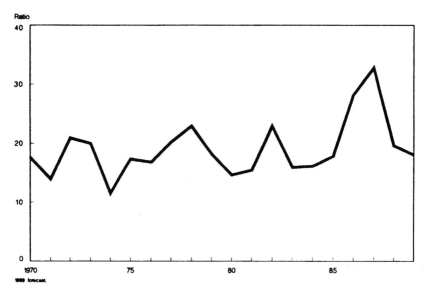

system into a broader perspective. In the following chapters are considered world production, trade, and consumption, followed by an evaluation of prices and the causes and effects of livestock cycles.

HISTORICAL SKETCH

Dodge City, Abilene, Chicago, and Fort Worth: sod-busters and the longhorn; sheepherders and the cowboy: the round-up and trail-drive; coffee around the chuck wagon; sleeping on the saddle blanket and whooping it up after payoff—all symbols of a brief period many consider to be the golden era of livestock production in North America.[1] It was a period so fleeting it was straddled by the life of one of its legends, "Buffalo Bill" Cody, who, by the time of his death in 1917, had outlived most of the wild in the West and earned his living by packaging and parading the myth, Broadway-style, in his show.

---

1. This section is based largely on Walsh (1982); Fowler (1952); Weld (1924); Ives (1966); and Thompson (1942).

We are getting ahead of ourselves; our interest begins with the dawn of man, probably in central Africa at least 2 million years ago. Domestication of livestock came much later, about 10,000 B.C. for cattle and sheep and more recently for pigs (*Encyclopedia Britannica* 1983). To this primitive hunting and gathering society, domesticated livestock provided four key functions: (1) a controllable supply of high-quality protein essential for the diet, (2) the ability to utilize and store foodstuffs not directly consumable by humans, (3) hides and wool for clothing and shoes, and (4) in the case of cattle, motive power. Over the millennia, there may have been a change in the ordering of these advantages, but they continue to explain the enduring dependence of mankind on the domesticated meat animal (Figure 3-2).

Despite the importance of stock to the agrarian world preceding the Industrial Revolution, production techniques remained primitive until the early 18th century. Animals were typically kept without shelter and were required to forage for themselves on hillsides and in forests. Winter was a particularly difficult time and the carry-over of breeding stock from year to year was limited, thereby holding down total herd sizes. For this reason, wild game remained an important and closely guarded source of meat

FIGURE 3-2. Oxen Are Still Used As Draft Animals in Many Parts of the World — Here Delivering Fish in Guatemala (Source: W. Lesser)

throughout much of this period. In fact, Robin Hood is a fictitious representation of the numerous real individuals who spent a large share of their time protecting themselves and other poachers of the King's deer from the Sheriff of Nottingham. So valuable was game in the Middle Ages that the penalty for poaching was often death.

Animal husbandry practices began to improve by the reign of Charles I (1625-49), and the history of livestock in North America from this period gives a good view of changes throughout the Western world. In North America, livestock production can be divided into four periods: the Colonial Era (1620-1794), the New West (1795-1848), Development (1849-1922), and Maturity (1923-present). The following sketch covers the first three of these periods.

The Colonial Era: 1620-1794

East

The earliest appearance of domesticated livestock in the New World was associated with exploration. Columbus introduced cattle and sheep to Hispaniola on his second voyage. Horses and cattle either stolen or lost from the Coronado exploration of the Southwest in 1540-42 explain the existence of the mustang and black Spanish cattle on the southern plains when settlement pushed out from the East three centuries later. However, by and large, livestock was associated with settlement — cattle, sheep, and hogs arrived with the Mayflower in 1620, and even earlier at Jamestown, Virginia. Because conditions during passage and the first winters there were so difficult, several shipments were required before a self-sufficient population of animals was established.

Animal husbandry on Manhattan Island gives an indication of predominant practices during that time. Cattle were allowed to shift for themselves, feeding primarily on shoots and forest grasses. The present Wall Street was the site of a wall separating the inhabited southern tip from the largely forested upper reaches. Hogs, too, ran unattended in the woods, but were better suited to this than were cattle. The rigors of survival, combined with local breeding efforts, led to the development of indigenous breeds including Poland China, Doroc-Jersey, and Chester White (Thompson 1942, p. 10).

Of the three principal species, sheep presented the greatest problems to early settlers. They were poor foragers who had few natural defenses against wolves and bear. As a result, sheep production was limited until a major effort began in the mid-1600s to develop protected public sheep pastures — the origin of the commons in North America. During this period, sheep were raised primarily for meat. The short, thick wool was considered inferior by the wealthy who preferred English clothing, leaving the native wool for the poorer farmers and servants.

Close ties with England were interrupted by the Puritan War of 1640. The loss of the market in Great Britain was compensated for by the opening of the West Indian trade. The New England colonies shipped meat, both live and preserved, and horses to the sugar plantations in the Caribbean. The northern monopoly on this trade weakened following the settlement of southern Pennsylvania by English yeomen beginning about 1681. Unlike the earliest settlers who sought religious freedom, the yeoman was an experienced farmer who brought knowledge and capital to agriculture in the New World. Their scientific farming methods were much encouraged by William Penn, an early supporter of improved pasture and shelters for livestock. Advanced use of barns for shelter came only with the impoverished German immigrants beginning in the 1700s. These frugal and hard-working peoples typically arrived as indentured servants, but quickly established themselves as important livestock producers. Their other principal innovation was planting turnips for feed.

The middle colonies were soon winning the competition with New England as a source of meat for the West Indies. Thus, the settling of Pennsylvania and the subsequent expansion into the rich Shenandoah Valley marked the decline of agriculture in New England. Certainly, the greater productivity of the mid-continental soils and the milder climates were major factors in New England's decline, but land costs were also important. By now, the northern colonies were becoming relatively industrialized and populated, and land prices were much above those of newly opened areas. Then, as now, livestock production was predominantly a frontier activity.

Succeeding years showed continued, if slow, westward expansion. The Catskills and Mohawk Valley of New York were opened in the early 1700s. Earlier settlement of these regions had been prevented by the Indians, but their power was broken during King Phillip's War (1675-76). Somewhat later, settlers moved into the western reaches of the Carolinas. There, rich native grasses and legumes provided excellent forage. Cattle towns soon grew—towns more associated with the modern perception of the Plains than with the Southeast. Hogs flourished in the eastern woodlands, but sheep, following the pattern of New England, proved difficult to feed and protect.

Modern breeding practices are also traceable to this period, and particularly to the efforts of Robert Bakewell in England. He questioned the popular concept that a large frame was required to hold the flesh, and he bred for smaller frame sizes in cattle. It was not until many years later that animal scientists were able to prove what he had recognized by observation — that meatier, small frame cattle provide greater yield percentages than rangy, large animals. Although he is credited with this observation, it had been put into practice on a small scale in many New England towns where runty stallions were prohibited from running with the mares on the village greens. The effects of the early breeding efforts were seen quickly. One observer recorded that the average market weight of English "beeves" was 800 pounds in 1795, up from 370 pounds in 1710 (quoted in Thompson 1942). Nevertheless, it would be many years before many producers recognized that fewer, meatier animals were more profitable than a larger number of poorly finished ones.

Marketing: The early industrialization of the nation by the mid-1700s meant stock had to travel greater distances to market. The earliest references to specialized marketing efforts date to John Pynchon of Springfield, Massachusetts, in 1655. His earliest recorded activities included driving cattle to market in Boston and subsequently fattening them during the winter months. At this point in history, fresh meat was typically sold in the established public markets. At first they operated only a few days a week, later expanding to six days. These outlets were similar to today's "farmers' markets" with producers selling directly to consumers in the absence of middlemen. Distance and specialization, however, made these direct exchanges unwieldy, and by 1742, Faneuil Hall in Boston opened as a permanent market operated by middlemen. It quickly grew to be the major outlet for meat and meat products, and producers and drovers went there seeking sales. As a result, an active live animal market grew up nearby. Buyers were the shopowners who combined slaughter with retailing. Butchers formed a powerful guild which, at times, was responsible for controlling the supply and, hence, the price of meat.

Meat that was not consumed immediately had to be preserved, a process limited at that time to drying and smoking or pickling. Much preserved or "packed" meat was shipped to the West Indies or used as provisions for cargo, passenger, and fishing vessels. Limited to the cold months because of spoilage problems, meat packing was done on farms removed from market or by specialized packers such as William Pynchon (who apparently changed from droving to packing later in his career).

Regulation of livestock marketing and slaughtering also has roots deep in the history of settlement. To a large extent, regulation was possible only when sales and marketing were centralized, as at Faneuil Hall. Coinciding with the opening of this market was the passage of laws controlling the quality of meats sold. In New Amsterdam, the process was slightly different. There, animals required a slaughter certificate from the "slaughter farmer" after a fee, or tax, based on the value of the animal was paid. Because this law was so difficult to enforce when slaughter was dispersed, public slaughterhouses were built, and by 1749 slaughter outside these facilities was prohibited.

The packing industry remained small scale until about 1756 when Jonathan Winship, who had a military contract to fulfill, opened a slaughterhouse on his land in Brighton. His needs attracted sellers who, in turn, attracted other competing buyers. As a result, a local slaughter industry was established, and the Brighton Cattle Market was founded. This market was the forerunner of the great terminal markets of the 1800s.

The livestock producer in the middle colonies who was farther from market had a long drive to Philadelphia, Baltimore, Norfolk, or Charleston. Droving was done by the producer or by specialized drovers who took title to the animals as does a modern-day livestock dealer. Both cattle and hogs were driven to market despite the obvious problems of moving half-wild hogs through wilderness areas. As yet, sheep were unimportant throughout the region.

Southwest

Although the initial settlement of the Southwest occurred at about the same time as the East, the early communities were unsuccessful and permanent habitation did not begin until the refounding of Santa Fe in 1693. There, as well as throughout the Spanish Southwest, the locus of settlement was the mission, each a self-contained and self-supporting community. Due to the aridity of the area and that of the subsequent settlement of San Diego (1769), stock, and particularly sheep, were a mainstay. The original movement up from Mexico City was difficult, but once in place livestock thrived on the natural hay cured by the dry conditions.

Marketing per se was limited because most animals belonged to a mission that decided its disposition. Wool and surplus meat were sent south to Mexico and provided a source of income. This essentially feudal system remained largely unchanged until the aftermath of the Mexican Revolution.

Revolutionary War and Its Aftermath

The American Revolution is popularly perceived as a general uprising of the population against tyranny. In truth, the War was quite restricted, involving only a small part of the population in limited areas. Those who were spared direct involvement were only slightly affected and often prospered, particularly if they sold supplies to the British forces.

Thus, during the Revolution, livestock production continued to expand. Settlement also moved ahead into Kentucky where the limestone soils supported bluegrass and corn, an ideal combination for cattle. In uncleared areas, switch cane nourished the cattle, while hogs thrived in the woods. The remoteness from market and the difficulty of the route, however, limited the potential of this enterprise. Fermenting and distilling the grain to make whiskey was far more profitable. The equivalent of many bushels could fit into a small keg and be taken to the coastal areas at great profit.

Then, in 1794, two events occurred that had a major influence on livestock production in North America. The first was the placement of an excise tax on distilled liquors. Farmers of western Pennsylvania, deprived of the right to produce homemade whiskey, started the Whiskey Rebellion, but did not prevail in the face of troops sent by President Washington. A new market for corn would have to be found. Also in 1794, the Indians were defeated at the Battle of the Fallen Timbers, thereby opening Ohio and Indiana for settlement.

The New West: 1795-1848

By 1795, conditions were set for a major expansion of livestock production. New productive areas, particularly in Ohio, were safe for widespread settlement. The collapse of the whiskey empire led to stock feeding as the only outlet for grain and forage, and transportation improvements connected the new production areas with major national and international markets. In particular, steamboat service on the inland rivers began about 1810, and the Erie Canal opened in 1825.

The immediate response of Pennsylvania, Kentucky, and Tennessee producers to the loss of the whiskey market was to ship live animals, bacon, and hams down the Ohio and Mississippi Rivers to New Orleans. When flatboats were used, the journey, while long, was very inexpensive. Because of this, as well as the growth of production in the rapidly settled plains of Ohio, the southern market was soon glutted. The oversupply in New Orleans occurred despite the fact that livestock production in the Southeast declined after the Louisiana Purchase of 1803 when large plantations concentrated on cotton rather than stock and feed grains. Even in South Carolina, the frontier cattle towns were quickly replaced by large cotton plantations.

While the market potential to the south diminished, that to the east improved. Increased population concentrations in Boston, New York, Baltimore, and Philadelphia put pressure on Eastern production and prices. First to respond to this opportunity was George Renick who drove cattle from Ohio to Baltimore in 1805. Hogs and cattle were both driven, sometimes together. New York City, at nearly one thousand miles, was more distant and the first drive did not occur until 1817. By 1820, long-distance droving had become commonplace, reaching its peak in the 1830s and 1840s. Railroads led to its eventual demise in the 1850s, but the substitution was slow, in part because of the high rail rates compared to the costs of overland transport.

Sheep

Frontier conditions had held back the production of sheep through the eighteenth century. Additionally, the light breeds available in North America with their short coats were economically unattractive. This latter problem was overcome about 1810 when the Napoleonic Wars made the jealously guarded Merino sheep, which were raised primarily for wool, available to North America. The timing was propitious because the blocking of English woolen imports during the War of 1812 caused prices to rise by four times from 1811-1813. Merino sheep, able to thrive on rocky hillsides, led to the revival of animal husbandry in New England, especially in Vermont which supplied the Massachusetts and Rhode Island woolen mills. Ohio, also was caught in the Merino craze, and both production and processing expanded greatly there.

The speculative spree was broken in 1815 when the war ended and competition with English imports again became a dominant market factor. Not until the tariffs of 1828 provided some protection did sheep production begin to recover.

Marketing

The size of production units, distance from market, and road conditions made specialization a factor from the earliest days. Although the feeder could drive his own cattle to market, this was often impractical and ill-advised, so the task was typically left to the drover. Drovers existed in three types: the hired man, the agent, and the professional drover. Agents acted much as the packer-buyer or order-buyer of modern times who works for a user such as a butcher or exporter (see Chapter 9). The drover would buy on his account or operate on wages. Unscrupulous dealings were commonplace, especially because information on distant market prices was poor. Sometimes speculators, seeing the telltale dust cloud, would ride out from market areas and buy cattle and hogs at low prices from unsuspecting farmers and drovers. As we shall see, abuses like these contributed to the creation of the Packers and Stockyards Commission in 1921.

Specialization in production separated the activities of calf production, range feeding, and fattening in much the same fashion as is seen today. Feeder pig and hog production, however, typically remained combined as the enterprise took little attention in the pre-selective breeding period.

Both specialization in product and geographic separation necessitate an exchange mechanism that brings sellers and buyers together. Filling this need was the monthly or quarterly sale ("market fair") typified by that of Madison County, Ohio. Sales there were made privately between breeder and feeder or feeder and butcher, or impromptu auctions ran simultaneously for those unwilling or unable to negotiate directly with buyers. Although informal and largely lacking in physical facilities beyond pens, these early markets carried out the same basic functions as the great terminal markets of the post-Civil War era. However, with increasing production and the coming of railroads, a more formalized market was required, and the market fairs declined.

Livestock naturally assembled, and exchanges logically followed, at transshipment points. For livestock in the pre-mechanized days, these points were navigable rivers. While there were a number of suitable riverfront locations, only one—Cincinnati, Ohio—combined river transport with access to the major overland trails to the east and southeast. Largely because of its advantageous location, Cincinnati became and remained the center of the livestock empire from about 1830 to 1850.

The role of Cincinnati in the meat trade varied substantially by species. For sheep, it was little known. For cattle, it was an assembly and embarkation point for drives over the mountains to eastern markets. Hogs, however, were different because they lost more weight and quality during the long overland drives. Moreover, prepared pork products were more palatable than those made from beef or mutton, a condition to which our present-day diet of bacon, ham, sauerbraten, and sausage attests. Hence, it made good sense to slaughter and process hogs near the production points but to send cattle and sheep live to market. And pack hogs Cincinnati did, earning for its efforts the name Porkopolis and credit for the creation of the modern meat packing industry.

This period was short-lived, however. Due to the extension of major trunk and feeder lines to Chicago around 1850, Missouri and Illinois cattle feeders could ship their stock directly to eastern markets rather than curing it locally. Hogs fed out in the expanding corn belt were also accessible to the Chicago and neighboring packers. Hence, this area's importance grew in production and packing while Ohio became industrialized, thereby displacing livestock enterprises. Opened in 1848, the first of the Chicago area stockyards, "Bull's Head Market," provides a convenient departure date from which to measure the great Chicago packing empire.

Southwest

The agricultural potential of Texas was not long concealed from the enterprising pioneers. Eastern Texas provided exceptional forage land for cattle and hogs while the western reaches, including New Mexico, were ideally suited for sheep. Mexico, for obvious reasons, resisted settlement of its territories, and it was not until 1821 when Connecticut-born Moses Austin received permission to build a colony. This colony shipped livestock, particularly the Mexican black cattle, overland to New Orleans or to the coast and by sea to Cuba and the West Indies. After about 1824, the major trade route became the famous Santa Fe Trail which ran from Independence to Santa Fe via Fort Dodge. Manufactured goods traveled southwest and unprocessed products, including hides and livestock, returned. Austin used this trail to import superior American breeds of sheep. Subsequently, a branch opened to Van Buren and was favored by drovers.

A principal reason Austin gained permission for colonization of Spanish territory in 1821 was because his bid coincided with the Mexican Revolution and the collapse of Spanish power in North America. Independent Mexico was far weaker, and the Texas revolt of 1835 led to the 1846 Mexican War. The war culminated in the grant of the Spanish Southwest to the United States with the Treaty of Guadalupe Hidalgo in 1848. The gold rush of '49 led to the occupation of California and to the creation of a new livestock market.

The Southwest mission system declined soon after the Mexican Revolution. The division of the country into sectors was a goal of the revolt and followed soon after (1834) with the passage of law that the mission land was to be seized and distributed equally to the Indian inhabitants and to the government. Needless to say, this law was not implemented smoothly nor equitably. The ensuing rush between government officials and the friars to profit from the turmoil led to a wholesale destruction of the large cattle herds. The principal salable products from these herds were the hides and tallow sold to traders, primarily for use in the New England shoe factories. Merchant vessels visited the California coast for as long as two years, acquiring and preparing a cargo of hides.[2] With the decline of the missions, more cattle were slaughtered at a younger age and the hides were handled poorly to be disposed of for very low prices. In less than a decade, a herd once numbering in the hundreds of thousands and built up over centuries was reduced to less than 30,000.

Development: 1849-1922

By 1849 the United States had achieved most of its current size. Industrialization had begun, railroads were expanding rapidly, and the nation was poised for enormous growth. A major contributor to and supporter of this growth was the expansion of the livestock sector and the marketing and packing industries which permitted the sector to exist.

Perhaps the most obvious change throughout this period was the expansion of railroad service (Table 3-1). Nearby railheads made it possible for remote producers to ship to large markets rather than rely on local exchange. The large concentrations of stock at these points distant from producers required both bigger facilities and improved marketing arrangements that would allow the seller better representation. For buyers, the concentrations of numbers permitted the construction of larger plants. Additionally, as the need for the long trail drive began to decline, breeding could emphasize conformation rather than stamina.

TABLE 3-1. Rail Miles in the United States, Selected Years, 1830-1910 (Source: Ives, 1966)

YEAR	MILES*	YEAR	MILES*
1830	32	1860	30,626
1835	1,098	1970	52,922
1840	2,818	1880	93,267
1845	4,633	1890	163,597
1850	9,021	1900	193,346
1855	18,374	1910	240,439

*Point-to-point distances excluding sidings and parallel track.

---

2. For an interesting account of this era, see Richard Henry Dana, *Two Years Before the Mast*.

Terminal Markets and Commission Agents

The first major Chicago market, the Bull's Head Market, was not actually a terminal market since it did not have direct rail access. That distinction belonged to John Sherman's market, a 30-acre yard built in 1856 and situated on the Michigan Central and Illinois Central railroads, with rail linkages to the other lines. Soon, numerous other yards were opened, principally by the railroads who recognized the importance of livestock as a source of trade and built the yards to facilitate and stimulate this use. However, multiple competing yards created their own problems. Buyers moved toward the largest yards, while settlers often had to move their animals through the streets to place their animals in these markets. Clearly, the system was inefficient. The response was the merger of all the yards into the Chicago Union Stock Yard and Transit Company (commonly known as the Union Stock Yards) which opened on Christmas Day, 1865. The growth of these yards was phenomenal. Volume increased from 0.4 million cattle, one million hogs, and 0.2 million sheep in 1866 to 2.7, 8.1, and 3.5 million head, respectively, by the turn of the century (*Drovers Journal*). Clearly, this was the way to sell stock, particularly slaughter animals, during the boom period of the American livestock sector, and Chicago was the place to sell. Times nevertheless change, and this entire complex closed just over one hundred years later.

One factor that contributed greatly to the growth of these, and other stockyards, was the appearance of the commission agent. Previously, as noted, the producer often took full marketing responsibility and accompanied his animals to market and sold them. Alternatively, they may have been sold locally to the drover or livestock dealer. Here, the producer was at a disadvantage because he did not know market conditions. Moreover, transit times were long, and the cash-poor drovers frequently dealt on credit and did not pay the producer until after receiving payment.

Appearing in 1857, the commission agent resolved these difficulties. Agents would receive shipments and sell for a fee. Their regular presence in the market assured a good knowledge of prices and helped the seller make a more favorable exchange. Thus, the agent served as a knowledgeable representative of the seller in the market. Payments were more prompt, thereby allowing the seller quick access to his capital through the banking system. Although the idea of an agent may seem simple, his role is invaluable, as anyone familiar with areas lacking such specialized marketing services would recognize.

MEAT PACKING

The other factor that allowed concentrations of slaughter animals in Chicago and elsewhere was the presence of a large packing industry. In order to understand better the significance of a large, centralized packing industry, it is helpful to understand what that sector consisted of up to mid-century. At that point, packing was seasonal (depending on cold weather to slow spoilage) and speculation was an inescapable part of the operation. Speculation was related to the uncertainty of supply from numerous very small producers, the vicissitudes of the market, and transportation difficulties.

As a result, no one was willing to invest heavily in the enterprise; rather, slaughter was often carried out in a crude building in the corner of a field. The salable hide and meat were sold locally while all other products were dumped into a stream or fed to hogs living below the structure. Alternatively, many general dry goods merchants engaged in packing during the winter months when waterways were impassable. Entry was easy since any partially vacant warehouse could be converted to a slaughter and packing facility by moving in some crude tables. For the merchant, meat packing was an attractive extension of other marketing activities. It provided an opportunity to trade stock for goods with the farmer and, ideally, make a profit on both parts of the exchange.

Although such decentralized packing arrangements benefited the individuals involved, the system had several key limitations. The small size and transient nature of the participants made it difficult to fill large, uniform orders. The small, non-specialized operations lacked the economies associated with specialization of labor and equipment. The by-products generated in small quantities at disparate locations had to be discarded rather than salvaged. And finally, inspection was sporadic to nonexistent, so that many slaughter operations were health hazards.

Beginning in the 1860s, the giants of the trade stepped into this void, destined to become household names, including Nels Morris, P. D. Armour, Gustavus F. Swift, and John Morrell (Figure 3-3). To the small, decentralized industry, they brought integration. This amounted to the development of the modern "disassembly-line" concept, employing mechanized equipment and division of labor. Equally important was the integration of by-product manufacturing into finished products such as soap, lard, candles, sausage casings, glue, brushes, combs, and buttons. To make the increased capital investment practical, ice rooms were added which extended the packing season from four to ten months. Typical of such plants was the 1871 Plankington & Armour facility in Kansas City

FIGURE 3-3. The Men Who Revolutionized the Meat Packing Industry

GUSTAVUS F. SWIFT   PHILLIP D. ARMOUR
THOMAS E. WILSON
MICHAEL CUDAHY   EDWARD CUDAHY

which had a daily capacity of 1,200 hogs or 1,000 cattle (Walsh 1982, p. 81). True, not all of these innovations were entirely new, but the growth era of the great meat packing empires brought them together on a scale never seen before.

The extension of telegraph lines throughout the Midwest in the 1850s and 1860s facilitated the development of a large-scale packing industry with its requirements for large numbers of similar animals. The telegraph was used to transmit price and volume information, particularly among the several terminal markets in existence by the 1870s. The result was a more efficient and better coordinated marketing system.

All these changes reduced the cost of the packing system. In order to justify the investments, demand was necessary. Here the Civil War benefited the industry, while damaging much of the rest of the country. The Northern army of 1.5 million men consumed over two-thirds of a billion pounds of cured meat during the War and an additional 160 million pounds of fresh (Walsh 1982, p. 57). This meat sold at highly favorable prices, allowing a large pool of profits to be re-invested in the sector. But Chicago and other leading northern packing centers benefitted in another way from the war between the states. Normal product flows were interrupted, resulting in major southern supply points like Louisville losing major markets and supply points. Largely due to these disruptions, Chicago jumped from third in pork packing in the Midwest in the pre-war era to a leading first-place position with a 25-percent share during the conflict. Louisville, for its part, fell to a distant fourth with only a three-percent share (Walsh 1982, p. 59).

Despite its leading position, Chicago was by no means the only packing center in the latter half of the nineteenth century. Cincinnati retained a significant position with hogs, while new yards and plants were constructed in Milwaukee, Cincinnati, St.Louis, Omaha, and later in Kansas City, Oklahoma City, and Fort Worth. Even second-tier areas like Cedar Rapids, Iowa, and Peoria joined the trend.

Obviously, this concentration of packing capacity was the major supply point for the expanding and industrializing nation. Supplying was facilitated by the extension of natural (e.g., ice) refrigeration from the plant to insulated rail cars, beginning in the early 1870s. Suddenly *fresh* western beef, pork, and lamb could be sold in major eastern cities for less than locally killed stock. It took nearly a century for virtually all the large northeastern slaughter plants to close, but the patterns were set.

Although natural ice refrigeration was a major innovation at the time, it was nevertheless cumbersome, costly, and, above all, subject to the vicissitudes of the weather. These obstacles were overcome within 20 years when mechanical refrigeration was quickly adapted for the chilling and shipping fresh meat. With this advance, prepared meats began losing popularity to fresh. Also, as we shall see, mechanical refrigeration ultimately contributed to the demise of the large, integrated, nineteenth-century packing plant. However, that was not realized at the time the benefits of mechanical refrigeration and other uses of electricity for lighting and mechanical power were first realized.

OVERPRODUCTION AND THE END OF AN ERA

The demands of the packing houses for more head, especially for cattle during a period of territorial expansion into the grasslands of the High Plains, created an irresistible draw for cattlemen to produce. And produce they did, with the Texas drive doubling to 600,000 from 1870 to 1871. That winter, severe weather decimated herds on the Plains (Figure 3-4) causing the 1872 drive to drop to 200,000, but the worst was saved for 1873 when the depression, which began two years earlier in the East,

FIGURE 3-4. Waiting for a Chinook—C. M. Russell's Depiction of Range Conditions in 1886 [C. M. Russell's drawing expressed, better than any words, the losses caused by the severe winter of 1872.] (Source: Montana Stockgrowers Association, Helena)

finally set in. With supplies high and prices falling rapidly, the cattleman could normally hold stock over on grass for the following season. In 1873, that was not an option because much of the recent expansion was financed with bank credit. And banks had such poor liquidity that no extensions could be offered. Many head had to be sold at whatever price the depressed market would bear, a price which all too often did not cover costs, and many of the large cattle ranches went out of business.

The lure of immense profits nevertheless continued, and a renewed cycle of land speculation (at that time, using foreign capital), overgrazing, depression, and severe weather continued through 1884-87. At this point, however, it was evident that primitive herd management was inadequate. Some scientific management of grazing was needed, as was the systematic production of hay for winter feed. The great era of livestock production in the West was ebbing.

In truth, the end was predestined back in 1874 with the development of a practical barbed wire. Heretofore, there was no economical way to enclose the great prairies, and they were managed as one immense range. Barbed wire changed that. New settlers could protect their holdings and sheep grazing became possible. Fencing, of course, can also be used to prevent entry, and many cattlemen hastened to erect fencing, much of it illegally, on public land. In 1885, this was outlawed, and over the next several years the unauthorized fencing came down. In one move, Congress had removed much of the effective control of the cattle companies.

Fencing is an effective means of halting the spread of disease, real or imaginary. Imaginary (at least partially imaginary) infectious diseases were used as an excuse for quarantining certain central states. This meant that cattle had to be driven around states like Kansas. Fenced land, quarantined states, and the expansion of railroad service made the trail drive both impractical and unnecessary. By 1890, the trail drive all but ceased.

The introduction of sheep to the Plains created a bloody conflict with cattle ranchers. An early source was the Mormon expedition of 1847, but sheep numbers remained small until after the Civil War and in particular, until the introduction of barbed wire. Sheepherders had as much right to public grazing lands as did the cattlemen, but might and tradition prevailed. The introduction of sheep was vigorously opposed because they were believed to destroy rangeland and foul water tanks for cattle. Despite these efforts, the number of sheep in Idaho, Utah, and Colorado rose steadily until the turn of the century.

Hog production developed naturally with more scientific management as midwestern forests were replaced by croplands. The animals were now fattened for slaughter, following a rediscovery of the colonial era observation that fewer, higher quality animals were more profitable than many poorly framed and finished ones. Competing with hogs for grain were an increasing number of cattle who were fattened on the way from the range to the packing plant. An increasingly wealthy nation was becoming more sophisticated in its meat expectations.

RELATIONS WITH THE PUBLIC

An enormous and significant sector like livestock could not operate forever outside the public domain. Public attention and an increasingly activist government first responded to perceptions of monopoly and land grabbing by the large cattle companies during the 1870s. By 1889, attention turned to the packing sector. Two issues were involved: safety and competition. Beginning in 1891, Congress addressed the safety factor with a series of more encompassing laws which culminated in the Meat Inspection Act of 1906. (For further information, see Chapter 11.)

Monopoly issues, as might be expected, proved to be a more complex matter for public policy. This attention to meat packing paralleled an increased public awareness of "trusts" which was instrumental in the passage of the first antitrust act, the Sherman Act of 1890. The first specific attention to meat packing dates to 1902 when the major packers were widely seen as manipulating prices. In 1912, the major firms were acquitted by a jury, of a Sherman Act price-fixing case. This did not satisfy the cry for action and a controversial investigation by the Federal Trade Commission (newly created in 1914) led to "The Consent Decree" of 1922 and — in a related action — to the passage of the Packer and Stockyards Act of 1921. (For further details, see Chapter 11.) The new regulatory environment marked a loss of innocence and the entry of livestock production into the modern era.

SYNOPSIS

In three centuries, the United States livestock sector grew from essentially nothing to the largest of its kind in the world. Much of this development was, in fact, limited to the period between the Civil War and World War I. During that period, fortunes and legends were made, but the era was short-lived. The open lands were quickly exploited and a growing social awareness and resource scarcity led to bounds on the once free-wheeling sector. It is in this more limited, yet efficient, environment that the sector presently operates, and it is to that which we turn our attention now.

Study Questions

1. What events separate the major periods in U.S. livestock production?
2. When and why did specialization begin to enter production, marketing, and packing?
3. Identify the livestock marketing innovations that appeared from prerevolutionary times to the 1920s. What emerging needs led to their development?
4. in numbers, sheep have long been a weak third compared to cattle and hogs. Why is that the case?

REFERENCES

Drovers Journal, Annual Statistical Issue, various years.

The New Encyclopedia Britannica. Fifteenth Edition, 1983.

Fowler, Bertram B., et al. *Man, Meat, and Miracles*. New York: Julian Messner, Inc., 1952.

Ives, J. R. *The Livestock and Meat Economy of the United States*. American Meat Institute, Center for Continuing Education, 1966.

Thompson, J. W. *A History of Livestock Raising in the United States, 1607-1860*. USDA, Ag. History Series 5, 1942.

USDA. *Agricultural Statistics*. Various years.

USDA. *Agricultural Chartbook*, Ag. Handbook No. 684, March 1989.

Walsh, M. *The Rise of the Midwestern Meat Packing Industry*. Lexington, KY: The University Press of Kentucky, 1982.

Weld, L. D. H. "The Packing Industry: Its History and General Economics." In *The Packing Industry* by The Institute of American Meat Packers. Chicago: The University of Chicago Press, 1924.

Chapter 4

World Livestock and Meat Production

A myriad of livestock production data are available on a world, national and — in some instances — regional basis. This attention to data is explained, in large part, by the importance of livestock to the world economy; with livestock providing a large share of the world's wealth and income, it is important for planners to document this sector. Others involved in the industry on a daily basis (such as producers, packers, retailers, and import/exporters) need current and regionalized information to carry out their activities efficiently. Much of the data generated by public and private sources are for their use and benefit. (For a discussion of price information uses, see Chapter 10.)

For our purposes here, industry data serve a different role. We shall first use data to document the magnitude of the industry and establish the major nations and regions producing and consuming red meats as well as the domestic and international trade that developed to equilibrate supply and demand. Subsequently, we will use price data, both directly and indirectly, through a review of empirical studies, to understand the patterns and causes of fluctuations in prices across time and space. The intent of this material is to help the reader understand past actions as an aid to anticipating and responding to future change.

WORLD HISTORICAL DATA

The most complete source of data (following World War I) on livestock numbers and consumption is available from the United Nations and its predecessor organization, the League of Nations. When using these figures, which are compiled rather than generated by the U.N., it is important to keep in mind the difficult nature of assembling these numbers and the likely resultant errors. Accuracy problems are particularly acute for developing areas where census procedures are primitive or nonexistent and for tightly controlled countries where reliable data may not be released for security reasons. Nevertheless, the U.N.series is the best available and is useful, at the minimum, for tracking the major changes over time.

Major Producers

A review of on-farm numbers from 1925 to the present (Table 4-1) reveals several significant characteristics of the world livestock economy. First, it is extremely widespread. With the exception of the poles (and a few tiny, industrialized nations like Monaco and Vatican City), virtually every area produces at least one of the three leading red meat species. Despite the near ubiquity of the enterprise, there is specialization among countries. As befits its size and population, The People's Republic of China is the leading producer in terms of total head. Considering individual species, the United States is now third to the former U.S.S.R. (Commonwealth of Independent States) and Brazil in the production of cattle for meat. India leads in size of standing herd, but religious custom prohibits most of these animals (the count also includes water buffalo) from being eaten. Hog production is led by Mainland China, with the Commonwealth in second place, and the U.S. a distant third. However, because of low productivity in the Commonwealth, the U.S. is second in pork production (Shagam 1989). Australia is the standout in sheep production; the U.S., although a larger contributor to supply, is not among the top group.

Cattle production is the most widely dispersed livestock enterprise in the world. U.N. data show cattle raised in virtually every area and climate from Iceland to Mali. The few exceptions are the small land masses such as Hong Kong and the Falkland Islands. The widespread nature of this species is attributable to its ability to forage and consume otherwise unusable feed stuffs, its genetic adaptation to different climates (Figure 4-1), and its range of products (including motive power, milk, and meat).

Sheep share some desirable characteristics with cattle: the multiple product output (wool and meat) and a forage capacity that makes them suited to many areas with limited agricultural potential. Nevertheless, the high labor demands during the lambing season and their poor natural defense system restrict their geographic range. Certainly, the predator situation is a major concern in the United States as opposed to New Zealand and the Falkland Islands.

Pigs are the most limited of the three major species, both in terms of numbers and worldwide distribution. However, that perspective is somewhat misleading because the high productivity of hogs, combined with their single use as meat producers, allows them to provide the greatest amount of red meat production worldwide of any of the major species. Production is concentrated, with the five leading regions (P. R. C., European Community, Eastern Europe, U.S., and former U.S.S.R.) providing 85 percent of total supply (Shagam 1989). This is partially due to feed supplies and partially to religion. Islam, with the largest following of any religion on earth, prohibits the consumption of this meat. Orthodox Judaism has similar dietary laws. This religious prescription is largely responsible for the virtual absence of hog production throughout a wide region in the Middle East from Afghanistan to Syria to Morocco.

Changes Over Time

World livestock production, as measured by the number of head, approximately doubled from 1925 to 1969/70, and increased another 15 percent over the next 15-plus years. The gains were greatest with cattle and smallest for hogs although the growth rate differences are not marked. Of course, there is always a limitation to discussing livestock numbers based on individual years because considerable variability exists due to livestock cycles, particularly for cattle. Nevertheless, it is evident that the trend has moved upward since the mid-1920s, especially for cattle, but there is a distinct flattening of that trend in recent years.

Much of the growth has indeed taken place in the years following the end of World War II, a period of enormous economic advancement in the Western world. In fact, as is shown below, data on numbers do not fully reflect the transition because of a concomitant increase in slaughter weights. Thus, while a livestock sector goes back to the dawn of civilization, the form and magnitude is really quite recent, with its origins in the post-war boom which may now have run its course in the Western world.

While trends have been quite predictable, there have been some realignments among nations in their importance as producers of particular species. Cattle numbers have risen particularly rapidly in Africa and Brazil, while the trend has been much flatter in Europe, Canada, and the former U.S.S.R. Lagging production in Russia has undoubtedly been caused by an inept agricultural policy and limited feed grain imports (in response to foreign exchange shortages and other political and economic considerations). At the other extreme, Brazil has followed an expansionist agricultural policy including the clearing of new lands and introduction of soybean production. Argentina, which has long had a pricing/taxation policy burdensome to agriculture, has shown only modest growth in each of its livestock enterprises.

TABLE 4-1. Livestock Numbers for Areas and Major Producing Countries, Selected Years, 1925-1986§ (Source: United Nations, various years)

	1925			Pre-War¶			1953/54		
	Cattle	Hogs	Sheep	Cattle	Hogs	Sheep	Cattle	Hogs	Sheep
WORLD	523261	167753	560194	694000	296600	744000	814000°	349000°	855000°
AFRICA	39420	1872	63484	64352	2513	85590	60359*	1687*	85685*
South Africa	9738	801	35570	11852	1037	38289	11655*	537*	35992*
Algeria	892	89	6171	886	60	6406	864	88	6014
Morocco	1955	67	9278	1871	52	10798	2284*	64*	13556*
AMERICAS	165547	76999	107051	202365	92426	155643	716307°	109638°	168438°
Canada	9307	4426	2756	8247	4779	2742	9379	4723	1179
U.S.	59122	52148	39730	66029	50012	51595	94787	48560	31218
Argentina	37065	1437	36209	33207	3966	43880	45263*	3989*	54684*
Brazil	34271	16169	7933	40745	21763	10745	57626	32721	16800
ASIA	162761	13058	50363	201996	12367	96001	34972°	10051°	50264°
India	149184		36909	137933	2702	41506	NA	NA	NA
Pakistan	NA	NA	NA	24444		5941	NA	NA	NA
Turkey			11444	9311		25221	10759	NI	27287
PRC	NI	NI	NI	NI	NI	NI	NI	NI	NI
EUROPE	138749	74256	211185	157086	107171	192879	187742°	129458°	297054°
FRG	17202	16200	4753	12187	12280	2097	11641	12435	1352
Spain	3794	5267	20067	3899	5613	24237	4500	5000	20000
France	14373	5793	10537	15622	7127	9875	16889	7328	7826
Poland		1519	668	9924	9684	1940	7385*	9730*	3330*
U.S.S.R.	59630	20939	107031	59700	32200	62600	63000	47600	112700
OCEANIA§§	16784	1568	128111	17265	1873	140258	21550	1907	164911
Australia	13280	1128	103563	12862	1156	111058	15601	1197	126944
New Zealand	3504	440	24548	4254	676	29200	5782	656	38011

Notes:
§Count or estimate nearest to January 1 of each year
¶1937, 1938 or (most often) 1939
§§Includes Hawaii after 1925
*1952/53
°Includes some values for 1952/53
NA - Not Available   NI - Not Included

Sources:
Societe Des Nations, Annuaire Statistique International 1929.
United Nations, Statistical Yearbook, 1955, 1967, 1971, 1981, 1985/86

1965/66			1969/70			1980			1986		
Cattle	Hogs	Sheep	Cattle	Hogs	Sheep	Cattle	Hogs	Sheep	Cattle	Hogs	Sheep
1073674	582019	1027132	1123627	628855	1068839	1216052	797867	1120092	1271811	822443	1145690
140675	6085	130628	155453	6529	140502	169933	9233	183562	176841	11987	191746
12380	1500	37406	12251	1230	39136	12577	1317	31641	11750	1445	29481
602	10	5726	870	6	7400	1433	4	12500	1557	5	14795
3277	29	12570	3600	15	17000	3680	10	16100	2570	8	12100
349498	145848	159604	366179	163331	153411	390950	152953	129701	423568	144544	128269
11651	5108	783	11836	6460	616	12403	9688	481	11465	10721	722
108862	47414	24734	112330	56655	20288	111192	67353	12687	105468	52313	9983
48800	3100	48500	48440	4250	44000	56000	3800	33000	53000	4000	29243
90505	62534	22312	95008	65734	24333	93000	36800	18500	128918	33000	18473
343277	251506	259666	351460	267808	275586	337317	382889	324561	387727	404023	321584
176057	4975	42014	176450	4800	42600	182500	10000	41300	200000	8700	54460
40100	94	13400	43700	96	15000	15038		26239	16749		25826
13203	NA	33382	13189	NA	36351	15567	13	46026	16200	12	40400
62904	208936	68400	63203	223048	70600	64681	325123	102880	66925	338074	94210
214643	175907	263314	219152	186929	258972	134471	173715	134249	130360	179134	136027
13680	17723	797	14286	19323	841	15050	22374	1145	15627	24282	1296
3694	4681	18785	4288	6915	18729	4679	10715	14547	5084	10367	17735
20640	9239	9056	21719	10463	10037	24009	11432	11799	22896	10956	10790
10391	14251	3164	10844	13446	3199	12685	21326	4207	10919	18949	4991
93436	59576	129764	95162	56100	130665	115100	73898	143599	120888	140850	77772
25581	2673	214920	31383	3258	240368	35218	5188	204369	32427	4983	227214
17936	1747	157563	22162	2398	180079	26321	2488	135706	23451	2553	155561
7217	667	57343	8777	578	60276	8375	540	68653	8392	450	71646

FIGURE 4-1. Cattle Breeds Adapted to Different Conditions and Products [The Brahman (top) is better suited to hot weather than the European breeds (bottom).]

Brazil is the only major western hemisphere nation to show any growth in sheep output. The remaining nations were stagnant or posted decreases. This means that Asia and Oceania lead the growth of the species. Hogs are the most difficult to track because data on the major producer, The People's Republic of China, are missing for the early period. Strongest gains since the mid-1970s are shown by the P. R. C. and the European Community, perhaps because pigs are more efficient converters than fed cattle and because there are good export markets for some processed pork products. Total growth would have been faster if not for the declines in the U.S.

Meat Production

For several reasons, the ranking of countries by livestock on the hoof does not give a complete picture of actual production. Differences arise if the principal use varies by region (e.g., sheep) for wool vs. meat, and cattle for milk vs. meat. To be sure, most animals (except in special places such as predominantly vegetarian India) will eventually be slaughtered and consumed, but a milking cow will typically live five or more years compared to approximately two for a fed steer. Thus, areas emphasizing non-meat uses will have lower meat production for a given herd size than will those feeding specifically for slaughter.

A second distinction among areas is in slaughter weights. Because some breeds such as Brahmans are much larger than the small-framed Angus, a simple head count will not give a clear indication of volume. Along with breed differences, feeding regimes also vary so that the same breeds reach different market weights at different ages across areas. A key variant is finish feeding compared to foraging.

The unambiguous way to rank regions is by their actual level of meat production (Table 4-2). From these data, it is clear that the United States is the world's second leading producer, recently overtaken by the People's Republic of China. Third place goes to the Commonwealth of Independent States. This information tells us that, compared to the Commonwealth which leads in numbers, the stock in the United States is heavier and/or younger. In fourth place is former West Germany, followed at nearly the same figures by Brazil and Argentina in South America, and France in Europe.

About two-thirds of the red meat production in the U.S. is from cattle, while in the Commonwealth of Independent States it is but 50 percent. In contrast to this dominant position, mutton and goat meat production in the U.S. is a distant ninth in the world, with Australia and New Zealand a close one and two (U.N. 1987, Table 99).

TABLE 4-2. Total Red Meat Production in World and Selected Countries, Average 1979-1981, Selected Years 1975-1986 (Source: U.N. *Statistical Yearbook* 1986/87, Table 97)

Country or Area	1979-81	1975	1980	1984	1986
			- Thousands of Metric Tons -		
World	101,112	89,801	101,594	108,891	114,941
Africa	4,024	3,413	4,023	4,256	4,488
Ethiopia	216	269	297	301	303
Kenya	206	151	214	129	174
Morocco	169	125	160	168	182
Nigeria	332	241	333	329	350
South Africa	810	721	804	872	878
America, N.	22,097	20,426	22,318	22,891	23,036
Canada	1,795	1,638	1,832	1,868	1,903
Mexico	2,007	1,400	2,014	2,401	2,369
U.S.	17,484	16,675	17,680	17,819	17,831
America, S.	8,827	7,942	8,703	8,411	8,827
Argentina	3,310	2,817	3,214	2,892	3,136
Brazil	3,087	2,961	3,093	2,970	3,029
Venezuela	420	349	425	417	473

Asia	19,043	13,850	19,301	23,757	28,006
Afghanistan	197	155	195	201	201
PRC	12,118	8,145	12,374	15,746	19,538
India	321	285	320	366	393
Japan	1,864	1,392	1,893	1,960	2,109
Philippines	503	414	508	627	554
Turkey	462	543	440	532	550
Europe	30,756	27,688	31,112	32,322	32,835
Denmark	1,201	970	1,218	1,282	1,389
France	3,812	3,463	3,815	3,984	3,912
FRG	4,722	4,133	4,759	4,864	5,053
Poland	2,256	2,478	2,397	1,975	2,461
UK	2,248	2,290	2,305	2,387	2,329
Oceania	3,587	3,322	3,467	3,253	3,364
Australia	2,439	2,249	2,332	2,063	2,232
New Zealand	1,103	1,033	1,090	1,143	1,080
USSR	12,777	13,162	12,668	14,001	14,385

Production by country further emphasizes the disparity between production regions and areas of major population. To a large extent, the system is equilibrated by varying consumption patterns. Sometimes, the difference is religious as is the case with heavily Hindu India; in others, price and income differences vary the per capita consumption from a high of 228 pounds in Argentina to under 10 pounds in Zaire during 1980 (Table 6-1). Final, rather modest adjustments are made through world trade of red meats and derivative products as noted below.

DISTRIBUTION OF PRODUCTION IN THE U.S.

The center of livestock production in the U.S. has moved westward since Colonial days. To a good measure, this westward movement is the result of virgin lands opening to agriculture and the displacement of livestock production from industrialized regions. These factors were discussed in detail within the preceding historical sketch (Chapter 3). Other factors specific to the livestock sector have also had profound effects on the location and level of production.

Principal among these factors has been the genetic advances in feed grain production. Beginning in the 1930s, hybridization, particularly of corn, allowed increased output at lower cost and gave a decided advantage to the "corn belt" states in livestock production. This comparative advantage was reinforced when a taste for grain-fed pork and beef developed in the later 1800s. Since stock can be moved to feed more cheaply than the other way around, the Midwest excelled in finishing operations. Developments in meat packing, especially mechanical refrigeration, also made it possible for slaughter and processing to be done more cheaply near production as opposed to consumption points. (The role of the meat packing industry is explained in depth in Chapter 15.)

More recently, grain sorghum hybridization has restored the competitive position of the Plains states in cattle feeding, in particular. Many of the advances in feed production were made possible by large-scale irrigation projects. When irrigation becomes less cost-effective as the result of both higher energy costs and receding supplies, we can again expect a rearrangement of production within the U.S. and between the U.S. and other producing areas.

Cattle Trends

Trends for cattle numbers and meat production have both moved upward since the post-Depression days, with cattle numbers reaching their peak about 1975 (Figure 4-2). The trends have, of course, not been smooth due to factors such as production cycles and external influences ranging from war to price controls. There is an apparent discontinuity in the meat production trend about 1952: it is notably higher in the post-1952 period than before. This is primarily the result of the increase in grain finishing which began in the prosperous aftermath of World War II and continues through the present. An aberration took place in the mid-to-later 1970s when high grain prices, combined with price controls, led producers to grass feed rather than finish a large number of steers and heifers (Figure 4-3). Since 1975, cattle numbers have moved below previous trends due to a combination of factors including price competition and changes in tastes (see Chapters 6, 7, and 8).

FIGURE 4-2. Cattle Numbers and Beef and Veal Production, 1950-1988 (Source: USDA, *1989 Agricultural Chartbook*, Chart 191)

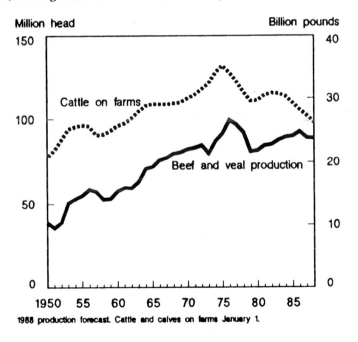

FIGURE 4-3. U.S. Cattle Slaughter by Class, 1970-1988 (Source: USDA, ERS 1989(a), p. 13)

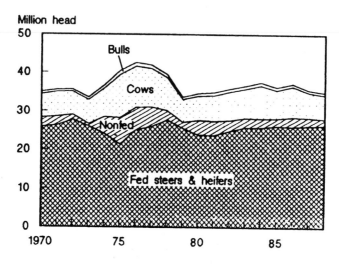

Increases in slaughter weights as a result of heavier grain feeding and other genetic and management factors can be seen in several ways. The average live weight of slaughter cattle rose to 1,104 pounds in 1987 from 956 pounds in 1950, an increase of 15 percent (USDA, *Ag. Stat.* 1980; ERS 1989, Table 84). Dressing yields—the proportion of the live weight remaining after removal of the head, viscera, blood, and hide—have also increased with changes in genetics and feeding regimes. The percentage increased from 54.9 percent in 1950-1953 to an average of 59.3 percent over 1985-1987 (USDA, ERS 1989, Table 101 and previous years). Thus, there has been a trend to more meat per pound liveweight over time.

Beef production has been affected by a changing balance between beef and dairy breeds. In the 1930s, dairy cows dominated beef animals by a ratio of over two to one. By 1950, the ratio was 1.5 to one and, by 1980, the relationship had reversed so that beef brood cows outnumbered their dairy cousins by nearly four to one (Figure 4-4 (A)). Two reasons explain this change. First, advances in dairy practices and breed genetics have proceeded to a point where, over 20 years, the herd has declined by about half, although milk production has remained generally constant (Figure 4-4 (B)). Furthermore, it is clear that per capita consumption of dairy products has lagged, thereby causing the sector to suffer both a relative and absolute decline in relation to the remainder of the livestock industry.

The decline of the dairy sector has the direct effect of raising the aver-

FIGURE 4-4. (A) U.S. Dairy and Beef Cattle Herd, 1930-1989 (Source: USDA, ERS 1989, p. 7); (B) Milk Production, Number of Cows, and Milk per Cow, 1974-1988 (Source: USDA, *1989 Agricultural Chartbook*, Chart 198)

(A)

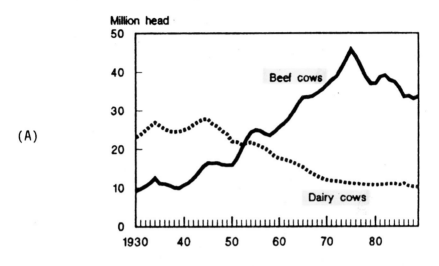

(B)

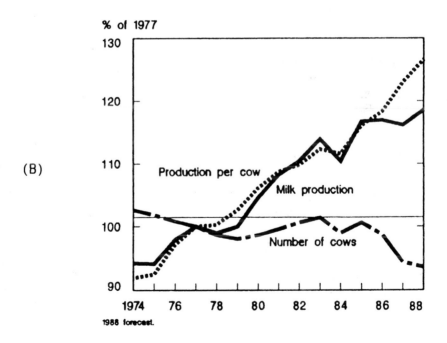

age dressing yield for cattle. This happens because dairy breeds have greater intestinal capacities than beef breeds, so the dairy animal loses more weight in the "drop" during dressing. More fundamentally, the lean dairy stock provided much of the trimmings (hamburger) and sandwich (bull) meat consumed domestically. The rapid decline of these supplies at a time when taste and lifestyles favor these products has stimulated a significant imported beef market.

Hog Trends

Hogs showed a steady increase in numbers up through World War I. Subsequently, numbers have fluctuated around 60 million head. Year-to-year changes are greater than for cattle, an attribute of the multiple head pig litters which allow a more rapid buildup of numbers. The greater supply response compared to cattle allows pork producers to take advantage of rapidly expanding export markets, especially those of wartime. The end of these markets has a dramatic impact that can be seen in 1945 (after World War II) and 1954 (following the Korean Conflict) (Figure 4-5). Most recently, hog producers suffered during the price-cost squeeze of the mid-1970s, but prices and profitability have subsequently stabilized at below-record levels.

Unlike cattle, slaughter hogs have remained quite constant in weight over the past three decades with an average weight of about 240 pounds but edging up slightly to 247 pounds in 1987 (USDA, *Ag. Stat.*, 1950, 1980; ERS 1989, Table 84). Carcass characteristics, however, have changed: a 1930 hog yielded an average of 33 pounds of lard, whereas the 1987 counterpart supplied but 11 pounds (USDA, *Ag. Stat.* 1930, 1988, Tables 411 and 417). As a result of the reduced lard yield and other genetic improvements in confirmation, the dressed weights have increased while the live weights have remained stable. Over the 15-year period from 1973-1987, average dressed weights grew by 13 pounds, or 8 percent (USDA, *Ag. Stat.* 1988, Table 413). Overall, pork production has increased since the 1930s. Increases are all due to breeding improvements. This means that the total increase lags behind beef production which grew in two dimensions—numbers and yield per head.

Lamb and Sheep Trends

Running counter to trends for cattle and hogs—and indeed trends throughout much of the world—lamb production in the United States has been in a long and steady decline for the past 30 years, although the most recent years show some flattening (Figure 4-6). For explanation, one can

FIGURE 4-5. Hog and Pig Numbers, 1866-1987 (Sources: USDA, SRS 1966; USDA, *Ag. Stat.* 1988)

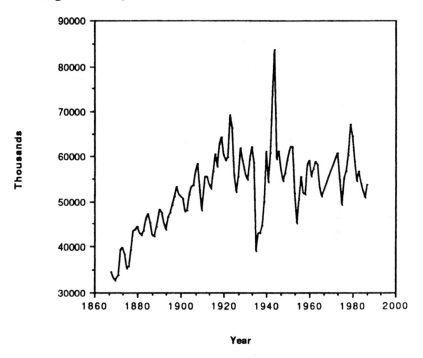

look toward production costs and wool prices. Production has encountered several special problems attracting labor that will accept the long, lonely hours required for sheepherding. Susceptibility to parasites and disease further raises the labor requirements. For a time, some producers even relied on Basques (traditional sheepherders from northern Spain) and other non-resident aliens.

Predators, an overwhelming problem during the settlement period, continue to be a problem in the western areas. For a long time, coyotes were held in check by a wide-ranging poison program, but that was severely limited as a result of environmentalist complaints that many other animals, including the bald eagle, were also being killed. In many of the same areas, large prairie dog populations destroy pasture and lead to animal and equipment damage. Restrictions on poison programs have also allowed these animals to enlarge their range at the expense of the sheepherder.

Production will continue to be focused on the 17 western states where

FIGURE 4-6. Sheep Numbers, Lamb and Mutton Production, U.S., Annual, 1950-1988 (Source: USDA, *Agricultural Chartbook 1989*, Chart 192)

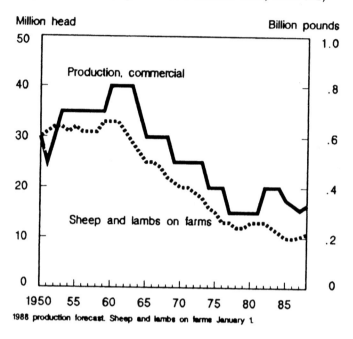

there are few alternative enterprises available in the arid environment. Aridity is itself a threat as additional dryness reduces the carrying capacity of the rangeland and prevents expansion, as happened in 1983. Reductions in lamb numbers have been partially compensated by feeding to heavier weights. An estimated 65 percent of lambs pass through feed lots (Stillman 1989).

The price of wool is supported by the government in somewhat the same procedure as grains and milk (program descriptions in Glaser 1986). Essentially, if the price falls below some target level announced by the Secretary of Agriculture, the government effectively purchases the surplus to keep the price at the target level. In 1987, almost $180 million was paid out (USDA, *Ag. Stat.* 1988, Table 434). Thus, wool market prices have not been falling as fast as they would in an open market given the widespread transference to artificial fibers and a general movement of textile production out of the nation. Practically, the effect of this program on lamb production has not been great because in recent years only about 15 percent of receipts have come from wool sales as compared to meat production (USDA, *Ag. Stat.* 1988).

Overall, the decline of the sheep industry in the United States must be attributed to a notable disinclination to consume lamb compared to competing red meat and poultry products. (See the section on regional tastes in Chapter 6.) In the face of this weak demand, production has declined.

Geographic Distribution of Cattle in the U.S.

In its annual January 1 inventory, the Crop Reporting Board of the USDA reports some cattle in every state in the union. Of course, distribution is highly uneven around the country (largely due to cost factors). The costs also vary depending on what stage of the production cycle is considered, and different phases of the system have different location characteristics.

Beef cows—the basis of the cow-calf system that produces young, weaned animals for subsequent feeding and finishing—are heavily concentrated in the heartland of the country (Figure 4-7). In terms of regions, this production area encompasses much of what is designated as the Southwest and North Central.

Those areas are characterized by abundant rangeland, some with a long growing season. Since grazing is a major feed source for cows, the range has been the traditional home of cow-calf operations. In more recent years, available rangeland has declined due to a variety of factors including changes in government access arrangements for publicly held lands, expansion of irrigation that made more intensive use possible, and alternate uses including mining and residential/commercial. Cow-calf operations have also expanded geographically because of their suitability to part-time producers. Thus, cow-calf production frequently supplements other farming operations such as cash grains and dairy, or provides additional income for non-farm workers. Low, non-land investment requirements and tax advantages have made cow-calf operations especially attractive to the higher income, non-farm operator (for analysis, see Milligan et al. 1983). However, the current income tax law, which dropped a distinction between earned income and capital gains, removed much of that inducement. It is one reason why beef cow numbers declined over the past two agricultural census years (Figure 4-8).

Cow-calf operations remain quite small when size is measured in terms of cow numbers. From a special survey in 1976, it was estimated that 47 percent of the operations had fewer than 100 cows. This figure would have been larger except that farms with less than 20 cows were excluded

FIGURE 4-7. Cows and Heifers That Had Calved, Inventory 1982 (Source: U.S. Dept. of Commerce, Bureau Census 1984)

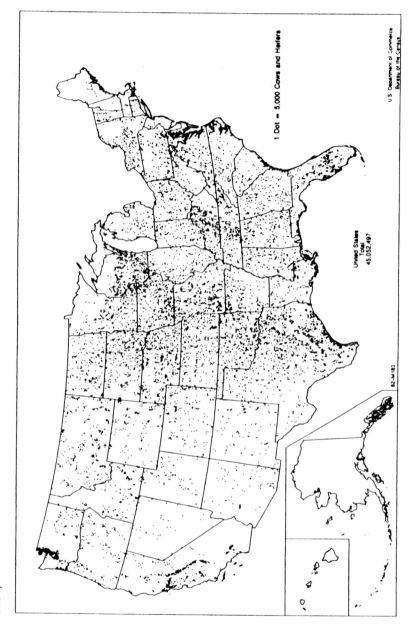

66

FIGURE 4-8. Beef Cows, Inventory: 1982 and Changes: 1978-1982 (Source: U.S. Dept. of Commerce, Bureau Census 1984)

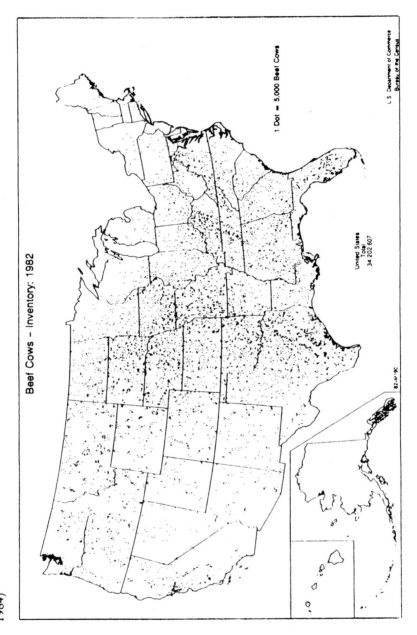

Beef Cows – Inventory: 1982

1 Dot = 5,000 Beef Cows

United States
Total
34,202,607

U S Department of Commerce
Bureau of the Census

67

FIGURE 4-8 (continued)

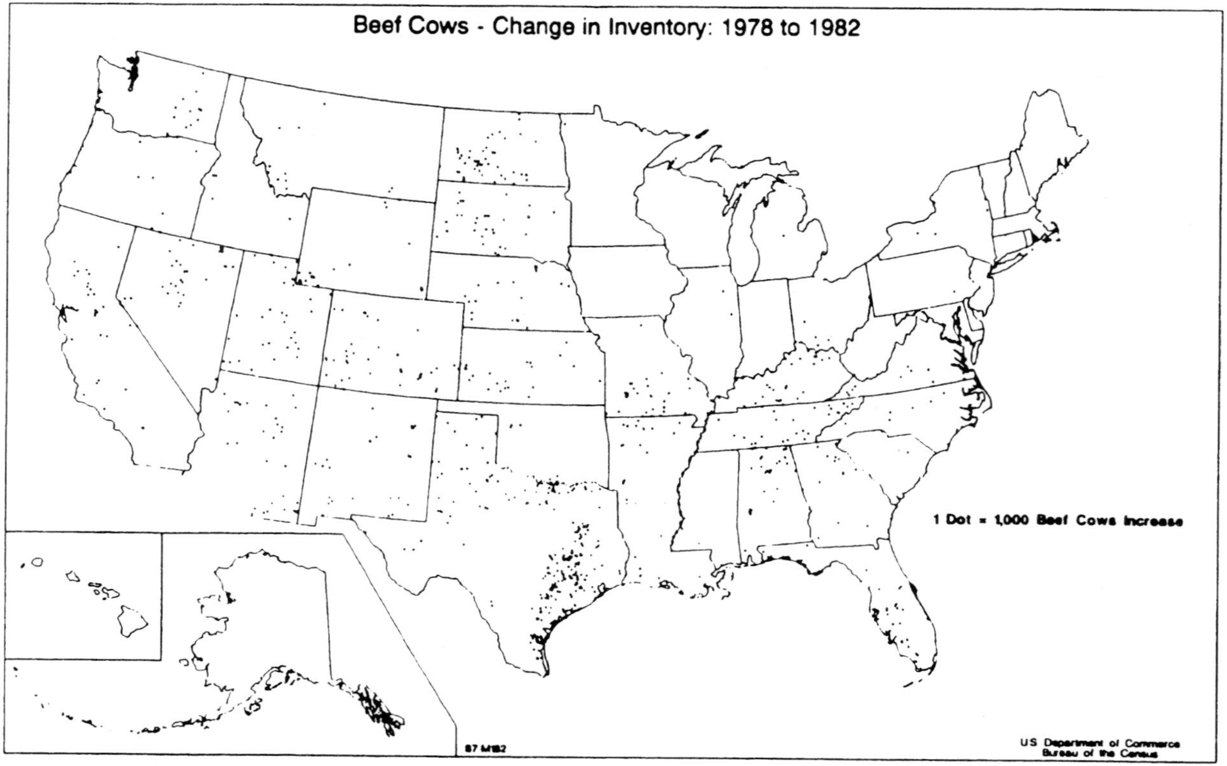

Beef Cows - Change in Inventory: 1978 to 1982

1 Dot = 1,000 Beef Cows Increase

US Department of Commerce
Bureau of the Census

87 M82

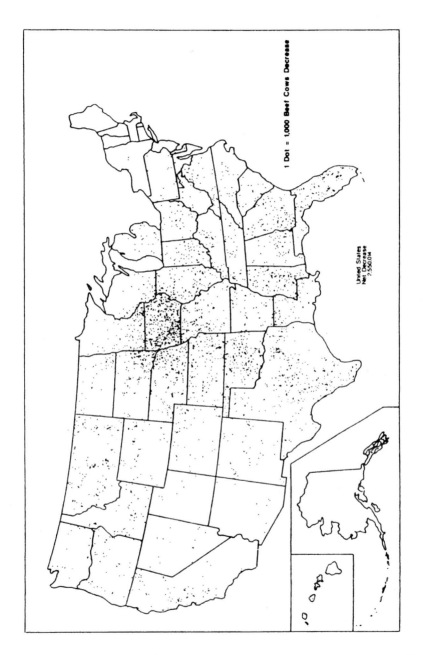

1 Dot = 1,000 Beef Cows Decrease

United States
Net Decrease
7,550,014

69

from the survey. The large number of small cow-calf farms is instrumental in keeping the sector highly competitive; it presents costs to the system in the form of high assembly expenses and limited coordination, contributing to the cattle cycle. The importance of the very small cow-calf operation is waning as 47 percent of the firms owned only 18 percent of the cows in 1978. The median number of cows was between 200 and 500, suggesting the predominance of larger farms in the future (Boykin, Gilliam, and Gustafson 1980, Table 6).

Cattle feeding is far more highly concentrated geographically than is calf rearing (Table 4-3). Indeed, the USDA Crop Reporting Board publishes a monthly report titled "Cattle on Feed—7 States"[1] Those states have an estimated 80 percent of U.S. Placements for the 7 states surveyed.[2] Feeding locations are determined by the availability of feed and other cost factors.

From Table 4-3, small feedlots (1,000 herd or less capacity, referred to as farm feedlots) made up 97 percent of numbers in 1984 but finished only 19 percent of all fed cattle that year. Conversely, the 32 large feedlots in Texas and Oklahoma (less than 1 percent total) produced nearly 12 percent of total marketings. Indeed, the share of marketings from farm feedlots is in precipitous decline, losing nearly two-thirds of its total volume from 1970-84 (Crom 1988, p. 57). Not surprisingly, much of this change is due to cost considerations and studies indicate that lots in the Southern Plains are lowest cost overall (Clarey, Deitrich, and Farris 1984). However, the "corn belt" continues to dominate in terms of farm numbers (Figure 4-9).

Geographic Distribution of Hogs and Pigs in the U.S.

Hog and pig operations remain an activity of the grain belt states, particularly the west North-Central area, whence it moved prior to the Civil War (Figure 4-10). That region leads in all areas of the sector including inventory, numbers of litters, and sales of slaughter hogs and feeder pigs. By far the leading state is Iowa followed by Illinois in the east North-Central region. Other important regions and states are the South Atlantic, and especially North Carolina (U.S. Dept. of Commerce, Bureau Census 1984, Table 18). The leading fattening states are also principal producers

---

1. "Cattle on Feed—13 States" is part of this series.
2. The states are Arizona, California, Colorado, Iowa, Kansas, Nebraska, and Texas.

TABLE 4-3. Cattle Feeding by Region and Size of Enterprise, 1984 (Source: Crom 1988, Table 31)

Region	Feedlot capacity							
	Fewer than 1,000 head		1,000-7,999 head		8,000-32,000 head		More than 32,000 head	
	Lots	Marketings	Lots	Marketings	Lots	Marketings	Lots	Marketings
	No.	Pct.	No.	Pct.	No.	Pct.	No.	Pct.
West1/	384 5/	0.8 5/	228 5/	3.7 5/	91 5/	10.2 5/	19	6.9
Southern Plains 2/	1,107 5/	.3 5/	55	1.6	81	11.7	32	11.9
Northern Plains 3/	17,470	8.0	443 5/	9.5 5/	123	14.2	14	6.5
Corn Belt 4/	38,888	10.0	603	4.3	9 5/	.7 5/	6/	6/
13-State total	57,835	18.8	1,383	18.3	310	35.7	69	27.2

1/ Arizona, California, Colorado, Idaho, Washington
2/ Oklahoma, Texas
3/ Kansas, Nebraska, South Dakota
4/ Illinois, Iowa, Minnesota
5/ Lots and marketings from other size groups are included to avoid disclosing individual operations
6/ Combined into previous size capacity

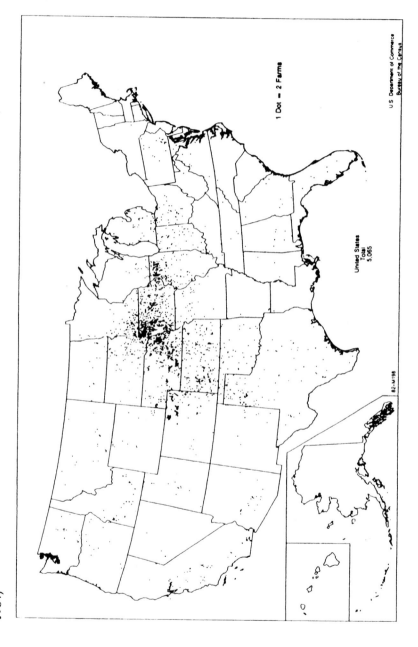

FIGURE 4-9. Farms with 500 Fattened Cattle or More Sold in 1982 (Source: U.S. Dept. of Commerce, Bureau Census 1984)

72

FIGURE 4-10. Hogs and Pigs Inventory by State, 1982 (Source: U.S. Dept. of Commerce, Bureau Census 1984, p. 125)

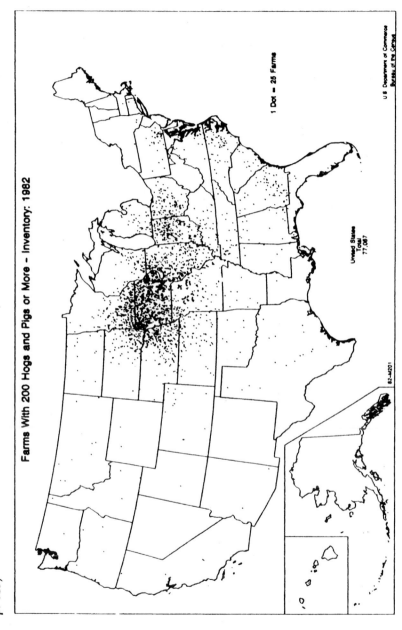

Farms With 200 Hogs and Pigs or More – Inventory: 1982

1 Dot = 25 Farms

United States
Total
77,087

U S Department of Commerce
Bureau of the Census

73

of feeder pigs, because many operations are integrated from farrowing through finishing. Thus, overall, the sector is more geographically concentrated than for cattle.

Concentration has also been increasing at the production level. Traditionally a supplementary enterprise with substantial year-to-year variability, hog feeding has become a more specialized and concentrated industry in the last decades. Presently, of the operations that produce 500 head or more annually, 16 percent provide 70 percent of the total supply, and the half-percent of farms with capacities over 5,000 provided 12 percent of marketing in 1984 (U.S. Dept. of Commerce, Bureau Census 1984). Much of the investment in these operations has been made since the late 1970s. The west North-Central region has a greater proportion of larger operations than the Southeast (USDA, Nat. *Ag. Stat.* Service, various issues).

Geographic Distribution of Sheep and Lambs in the U.S.

Sheep have benefited from a recent resurgence of interest throughout the Northeast. In number terms, this has not amounted to anything of significance. The mountain states (Montana and Idaho south to Arizona), plus Texas and California, are still the dominant producing regions. Together the ten states have over two-thirds of the U.S. flock (Table 4-4). Sheep were successfully introduced there by the Spanish, and the underlying conditions have not changed. Operations are small, mostly in the 10-99 head range. Only two percent have over 1,000 (U.S. Dept. of Commerce, Bureau Census 1984, Table 21).

SYNOPSIS

Red meat production is a worldwide enterprise. Specifically, this is true of the ruminant species. The U.S. is a leading player in cattle and hogs, but not sheep. The position of the U.S. is more evident when examining production trends than it is for herd size because most other countries get lower productivity out of their stock.

U.S. cattle and hog numbers, rising since the post-Depression era, have been declining following the mid-1970s. Price competition and changes in eating habits are implicated. Sheep numbers began the decline earlier— about 1960—but seem to have stabilized.

Regionally, the U.S. situation has been quite stable, especially for hogs

TABLE 4-4. Sheep and Lamb Inventories by State, 1988 (Source: USDA, "Livestock Meat Statistics" Table 9)

- 1,000 head -

Major States	Number
Arizona	284
California	1,015
Colorado	860
Idaho	324
Iowa	405
Montana	538
New Mexico	451
Oregon	490
South Dakota	610
Texas	1,960
Utah	478
Wyoming	865
Other States - by Region	
Northeast	241
Southeast	340
Central	1,731
West	264

in the midwest and sheep in the west. Feeder cattle production is more widespread but feeding is concentrated in the feed-rich areas of the midwest and the southern plains.

Study Questions

1. What are the principal factors influencing country-by-country trends in livestock production?
2. How and why have trends in numbers of head and meat production diverged?
3. In the U.S., cattle, hogs, and sheep each have different production areas and typical enterprise sizes. What impact does that diversity have on the marketing systems for these species?

REFERENCES

Boykin, C. C., H. C. Gilliam, and R. A. Gustafson. "Structural Characteristics of Beef Cattle Raising in the United States." USDA, ESCS, Ag. Econ. Report 450, March 1980.

Clarey, G. M., R. A. Dietrich, and D. E. Farris. "Interregional Competition in the U.S. Cattle Feeding Fed-Beef Economy." Texas Ag. Exp. St., College Station, Bull. 1487, November 1984.

Crom, R. J. *Economics of the U.S. Meat Industry.* USDA, Economic Research Service, Ag. Info. Bull. 545, November 1988.

Glaser, L. K. "Provisions of the Food Security Act of 1985." USDA, Economic Research Service, Ag. Info. Bull. 498, April 1986.

Milligan, R. A., C. J. Nowak, W. A. Knoblauch, and D. G. Fox. "Economic Viability of Investing in Alternative Part-Time Cow-Calf Forms in the Northeastern United States." Cornell University, Department of Agricultural Economics, A. E. Staff Paper 83-10, June 1983.

Shagam, S. D. "Trends in World Pork Production and Trade." In USDA, Economic Research Service, "Livestock and Poultry: Situation and Outlook Report." LPS-36, July 1989.

Stillman, R. "Sheep Industry Trends." In USDA, Economic Research Service, "Livestock and Poultry: Situation and Outlook Report." LPS-37, August 1989.

United Nations. *Statistical Yearbook.* New York, various years.

USDA. *1989 Agricultural Chartbook.* Ag. Handbook No. 684, March 1989.

USDA. *Agricultural Statistics*, Washington, DC, various years.

U.S. Department of Commerce, Bureau Census. *1982 Census of Agriculture.* Vol. 2, Subject Series, Part 1: Graphic Summary, 1984.

U.S. Department of Commerce, Bureau Census. *Census of Agriculture 1982 and 1987.* Vol. 1, Part 51: Summary and State Data.

USDA, Economic Research Service. "Livestock and Meat Statistics, 1984-88," Stat. Bull. No. 784, September 1989.

USDA, National Ag. Stat. Service. "Hogs and Pigs." Various issues.

USDA, Statistical Research Service. "A Century of Agriculture in Charts and Tables," Ag. Handbook No. 318, July 1966.

USDA, Statistical Research Service, Crop Reporting Board. "Cattle on Feed — 7 States." Various dates.

Chapter 5

Trade in Livestock and Meat Products

Compared to the enormous worldwide volume of livestock production, the amount that passes intercountry is minuscule. In 1982, of some 81.6 million metric tons (MT) of red meat produced in the major nations, only 6.3 million MT (7.6 percent) were traded internationally (USDA, FAS 1982). Trade in live animals, as might be expected, is even more limited. From the United States, $330.6 million worth of live animal exports (excluding poultry) were made in 1987, while imports were valued at $609.7 million (out of total agricultural imports that year of over $20.6 billion) (USDA, *Ag. Stat.* 1988, Tables 696, 697). Internationally, 1989 live animal imports for consumption came to $7.8 billion out of total imports exceeding 3.1 trillion (U.N., *Yearbook* 1989, Vol. I, p. 53; 1991, Vol. II, p. 3).

The expense and perishability of live animal trade is an obvious limiting factor. These imports exist only when special needs and opportunities exist. The United States, for example, typically ships and receives feeder cattle across the border with Canada and Mexico and imports hogs from Canada. Other live exports are dairy cows for breeding to Eastern Europe and hogs for breeding to the People's Republic of China. Some contiguous nations, like Guatemala and Mexico, trade live cattle. Orthodox Muslim countries such as Saudi Arabia have a preference for importing live sheep over meat, primarily so that proper religious slaughter and handling procedures can be followed. Because trade in meat does not face the same physical difficulties, it is instructive to examine the principal factors determining trade levels. These factors shall command most of our attention in this chapter. Prior to the analysis, however, it is important to review past and current trends.

HISTORICAL TRENDS

Livestock travelled on the earliest sea voyages, serving as both a source of food on route and breeding stock for the new settlements. Costs at that time were too high to ship animals for immediate consumption, and large volume trade in live animals required technological advances in shipbuilding and power. Processed (packed) meats were nonetheless important trade products, especially when used as sailors' rations and military provisions. Major volume movements of meat, however, awaited the advent of refrigeration, especially mechanical refrigeration.

United States

U.S. foreign trade statistics go back to 1790 and, over the ensuing 200 years, show a considerable growth in current dollars and in real (adjusted for inflation) dollars (Table 5-1). The year-to-year instability in trade is evident from these beginnings. In terms of gross national product (GNP), the figures are more stable and, initially, always less than 10 percent. Currently (1987), foreign trade (imports plus exports) accounts for about 22 percent of U.S. GNP (U.S. Dept. of Commerce, Bureau Census, *Stat. Abstract*, Table 685).

During the post-Civil War period, pork exports predominated. From that period, up to the 1920s, the U.S. exported much more pork than it

TABLE 5-1. U.S. General Merchandise Imports and Exports, Selected Years, 1790-1930 (Source: U.S. Dept. of Commerce, Bureau Census, *Historical Statistics of the United States:* Series U1-14, E1-12, E13-24, and F1-5)

| | - $ Millions - | | | | | |
| | Imports | | | Exports | | |
	Current $	Real $²/	% GNP	Current $	Real $²/	% GNP
1790	23	25.5	NA	NA	NA	NA
1800	91	70.5	NA	32	24.8	NA
1810	85	64.9	NA	42	32.1	NA
1820	74	69.8	NA	52	49.0	NA
1830	63	69.2	NA	59	64.8	NA
1840	98	103.1	NA	112	117.9	NA
1850	174	207.1	NA	135	160.7	NA
1860	354	380.6	NA	316	339.8	NA
1870	436	333.0	6.5	377	279.2	5.0
1880	668	668.0	7.3	824	824.0	9.0
1890	789	962.2	5.8	858	1,046.3	6.3
1900	850	1,036.6	4.9	1,371	1,671.9	7.9
1910	1,577	1,538.5	5.0	1,710	1,668.3	5.4
1920	5,278	2,334.4	5.9	8,080	3,573.6	9.1
1930	3,061	2,419.8	3.4	3,781	2,988.9	4.1

NA: Not Available
²/Adjusted by the Wholesale Price Index

currently does, both in terms of absolute amount and proportion of production (Figure 5-1). The shift to a net importing nation came after World War II (Figure 5-2). Wars are obviously major spurs to meat product trade. Presently, fresh beef is the principal import item followed distantly by canned pork products (Table 5-2).

Examining the more recent period (beginning in 1930) in more detail, exports can only be described as modest (Table 5-3). In terms of value, two percent of U.S. meat production was exported in 1987 (USDA, *Ag. Statistics* 1988, Tables 696 and 443). Because lower value products predominate in exports, the proportion of weight would be greater. Total meat imports that year, by value, were 4.2 percent of domestic production (USDA, *Ag. Statistics* 1988, Tables 670 and 443). On balance, U.S. trade of meat products affects only four percent of domestic supply.

Examining U.S. red meat exports only, however, leads to a substantial understatement of the sector's value as a generator of exchange earnings. This is because most U.S. exports are not meat and meat products, but rather by-products, especially hides and skins, fats and oils, and variety meats. In 1989, some $2.75 billion of these products were exported, amounting to over half the value of all livestock and product sales outside the country that year. Fats and oils are the major items, followed closely by hides and skins (Table 5-4).

It is reasonable to expect that the U.S., as the major producer of fed animals, would be a leading exporter of fats in their several forms. Egypt, Korea, Japan, Mexico, Pakistan, and the former Soviet Union are among the major importers of the inedible products. Hides are exported largely because high domestic labor costs, combined with pollution standards, have often priced this nation out of the tanning and shoe manufacture businesses. Japan and Korea are the major importers of these products. The situation with varietal meats (entrails) is different because they are not traditionally consumed domestically. Thus, they are exported to nations where they are either prized (e.g., kidneys in France) or consumed as lower-priced protein sources.

International

In total, world trade of market economies has been growing markedly over the past three decades. Beef trade increased six-fold to 3.1 million tons from 1952-1956 to 1976-1980 (GATT 1983, p. 4) The increase is evident in both nominal dollar amounts and in volume indexes (Figure 5-3). Imports reached the first trillion dollars in 1976, taking only four years to pass two trillion. The volume, however, is inflated by price increases, especially for petroleum and petroleum products. The more re-

FIGURE 5-1. U.S. Exports of Beef and Pork, 1865-1964 (Source: U.S. Dept. of Commerce, Bureau Census, *Historical Statistics of the United States:* Series U1-14, E1-12, E13-24, and F1-5)

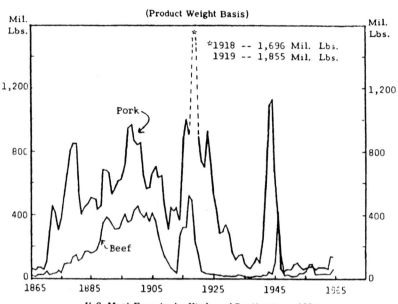

U.S. Meat Exports, by Kinds and Destinations, 1964

Item	Million Pounds (Product Weight)					
	Canada & Mexico	U.K.	Other W. Europe	C. & S.[1] America	Other Countries	Total
Beef & Veal:						
Fresh, Chilled & Fr. . . .	3.3	1.2	3.2	4.5	23.1	35.3
Canned & Other.	14.1	.2	.5	5.4	1.7	21.9
Pork:						
Fresh or Frozen	47.1	*	24.7	10.3	14.1	96.2
Canned	.4	*	.7	1.2	1.0	3.3
Cured & Other	11.5	.1	4.8	16.7	.5	33.6
Other Meats:						
Lamb & Mutton	.5	.1	*	.5	.2	*1.3
Sausage	.4	.7	.3	2.0	1.4	4.8
Other Canned	.6	.1	.2	.6	1.3	2.8
Variety Meats.	6.0	52.9	165.3	1.6	5.6	231.4
Total.	83.9	55.3	199.7	42.8	48.9	430.6
Lard.	22.1	549.7	23.5	53.7	33.0	682.0
Hides & Skins (Thous.)[2]. .	2,160	448	4,928	75	6,286	13,894
Tallow & Greases	39.7	62.7	808.0	248.0	1,245.6	2,404.0

*Less than 50,000 pounds.
[1]Includes Caribbean Islands.
[2]Includes Cattle, Calf and Kip Hides and Skins.

FIGURE 5-2. U.S. Imports of Beef and Pork, 1915-1964 (Source: U.S. Dept. of Commerce)

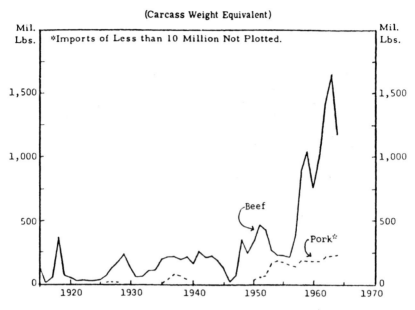

(Carcass Weight Equivalent)

U.S. Meat Imports, by Kinds and Country Sources, 1964

Item	Million Pounds (Product Weight)					
	Canada & Mexico	Europe	C. & S.[1] America	Oceania	Other Countries	Total
Beef & Veal:						
Fresh & Froz., Bone-in........	16.6	—	2.8	15.3	*	34.7
Fresh & Froz., Boneless	60.4	19.9	61.6	528.7	.3	670.9
Canned..........	*	.1	78.5	.1	*	78.7
Other...........	.6	.4	8.9	1.3	—	11.2
Pork:						
Canned Hams & Sh...	5.5	135.0	*	—	—	140.5
Other Pork	44.6	23.5	—	*	*	68.1
Other Meats:						
Lamb & Mutton	.7	*	.3	43.2	.5	44.7
Sausage	.6	5.9	4.9	—	*	11.4
All Other	.7	6.3	.4	1.1	.2	8.7
Total	129.7	191.1	157.4	589.7	1.0	1,068.9

*Less than 50,000 pounds.
[1]Includes Caribbean Islands.

flective volume index shows inequality among developed and undeveloped market economies (Figure 5-4). Developing economies showed particularly strong gains in exports up to 1972 when momentum was lost to the developed economies. After 1975, the developing economies have posted a volume trade deficit. That is especially critical as those countries tend to export low unit value commodities and import higher valued manufactured goods.

Beef was the major product traded internationally in the 1980s, and it constituted over half of total exports in terms of volume (Tables 5-5 and

TABLE 5-2. United States Meat Imports, by Type of Product, 1987 (Source: USDA, *Ag. Stat.* 1988, Table 697)

- Millions of Dollars -

Beef and Veal		Lamb1/		Pork	
Fresh/ Frozen	Processed	Fresh	Processed	Fresh/ Frozen	Processed
1,361	214	31	1	596	529

1/Includes mutton and goat meat

TABLE 5-3. U.S. Meat and Meat Product Exports, Selected Years, 1930-1987 (Source: U.S. Dept. of Commerce, Bureau Census, *Statistical Abstract of the U.S.*, 1988, Table 696)

- $ Millions -

Year	$ Current	$ Real*
1930	66	143.8
1940	22	62.3
1950	114	153.0
1960	125	142.0
1965	170	180.1
1970	132	118.4
1975	432	246.3
1978	743	351.5
1980	890	349.6
1983	926	318.2
1985	906	293.2
1986	1,012	317.2
1987	1,300	391.6

*Adjusted by the Consumer Price Index for Food.

TABLE 5-4. U.S. Exports of Livestock, Meats and Meat Products by Major Commodity, 1989* (Source: USDA, FAS 1990, Table 1)

($1000)	
Total	4,948,957
Meat and Meat Products**	4,596,275
Red Meats	580,229
Beef and Veal	1,314,731
Pork	299,883
Lamb and Mutton	5,811
Variety Meats	329,019
Lard, Tallow and Greases	453,734
Meat, Bone, Feathermeal and Semen	271,939
Hides and Skins	1,574,517
Hair and Wool	140,689
Livestock	352,682

*Jan.-Nov. only
**Includes miscellaneous products like sausage and caseings.

5-6). Among the beef importers, the United States is the leader, consuming something around a quarter of the total movement. This is a relatively recent phenomenon for the U.S., dating back to the early 1960s. Most imports are manufacturing beef (boneless beef for hamburger) that feeds into the fast and institutional food industries (Table 5-2). U.S. imports were facilitated by the low trade barriers which existed until 1964 when a quota was imposed (see below).

MAJOR IMPORTERS AND EXPORTERS

A handful of countries account for the great bulk of red meat and by-products imported and exported over the 1970s and early 1980s. For export, these countries include the EC (particularly Denmark and the Netherlands), Argentina, Australia, and New Zealand. However, much of the EC trade was intra-Community exchange and thus did not involve world markets as they are normally defined. Leading importers include the U.S., Germany, Italy, the United Kingdom, the Commonwealth of Independent States, and Japan (Tables 5-5 and 5-6). For the three major species, the situation varies slightly and can be summarized for the 1980s as shown in Figure 5.5.

Thus, while *total* meat movement internationally may be modest, it is important for particular countries. Australia exports nearly 50 percent of its production (mostly beef), while the European Economic Community imports about 15 percent of its beef and veal consumption (USDA, FAS,

FIGURE 5-3. Exports of Market Economies in Current Dollars and Volume Indexes, Selected Years, 1970-1986 (Source: U.N. *Statistical Yearbook*, 1979/1980, Tables 12, 136)

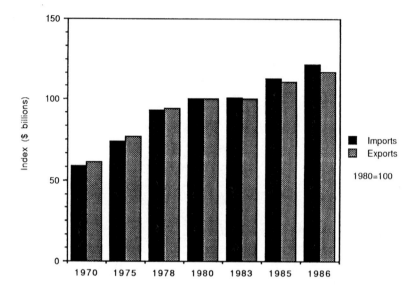

FL&P 1987). The United States, as noted, is less involved in terms of percentages, but because of the inelastic nature of meat demand, the price effect of even relatively modest imports can be substantial. (See "Prices," Chapter 7.) International meat trade, therefore, should not be discounted.

SOURCES AND LIMITATIONS OF TRADE

In 1988, per capita red meat consumption ranged from 185 pounds to less than 15 pounds (see Table 6-1). Except for a few countries where religious beliefs limit intake, given the opportunity, most people would likely choose to consume more meat. Explaining why they do not or cannot requires understanding what is produced, where it is produced and, finally, how it is paid for.

Trade Theory

Certainly the first choice of any nation is to produce domestically. While this may reduce transport costs, more importantly, it preserves jobs and promotes food independence. Employment is a perennial issue in

FIGURE 5-4. World Trade Indexes of Market Economies by Level of Development, Selected Years, 1970-1986 (Source: U.N. *Statistical Yearbook* 1985/86, Table 13)

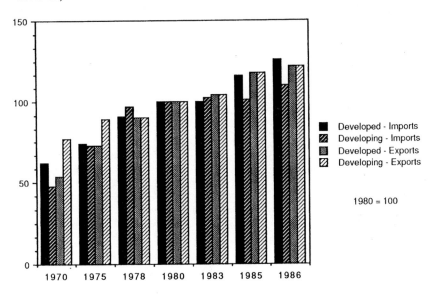

many countries, while food independence looms large among nations that have endured hunger due to war or pestilence. Yet, domestic production may not be economically unsound because of weather and/or soil conditions or because of land or labor scarcity. Then, as a second alternative, the country would choose to import. Japan is a likely meat importing nation, but why the United States with its extraordinary agricultural endowments? Explaining why the U.S. imports takes us to a trade theory concept known as *comparative advantage*.

If the U.S. is the low cost producer of computer chips, while Australia is more efficient in cattle production, a basis for trade clearly exists. Each nation benefits from trade and can actually have more of both products than if self-sufficiency were pursued.

But, what happens if the U.S. is the low cost producer for *both* products? Would we be self-sufficient with Australia importing some chips? No, not if one nation has a comparative advantage. That is, if the U.S. **produces less beef than Australia**, while Australia produces fewer electronic products than the U.S., mutually beneficial trade can exist. Why? Because nations can maximize their returns by concentrating on products

TABLE 5-5. Fresh Red Meat Trade in Selected Countries, Average 1974-1978, Selected Years 1980, 1982, and 1986 (Source: USDA, FAS, "Foreign Ag. Circular: Livestock and Poultry," FL&P, various issues)

(In thousands of metric tons)

Continent & Country	Average 1974-78		1980		1982		1986	
	Exports	Imports	Exports	Imports	Exports	Imports	Exports	Imports
NORTH AMERICA								
Canada	85	181	183	112	199	116	392	140
Costa Rica	42	-	32	-	42	-	215/	-
Dominican Rep.	5	-	3	-	4	-	--5/	78 6/
El Salvador	5	-	2	-	4	-	15/	-
Guatemala	20	-	13	1	20	1	15/	-
Honduras	24	-	36	-	33	-	96/	-
Mexico	34	7	1	12	1	21	-	128 6/
Nicaragua	34	-	23	-	13	-	95/	115/
Panama	2	-	2	-	4	-	-	-
United States	153	1,069	195	1,210	264	1,022	890	1,010
TOTAL	404	1,257	490	1,335	584	1,160	1,376	3,337
SOUTH AMERICA								
Argentina	525	-	487	5	499	-	158	19
Brazil	145	52	170	46	331	25	461	192
Chile	1	5	5	7	6	5	9	4
Colombia	18	-	10	-	16	-	11	-
Peru	-	7	-	6	-	17	-	75
Uruguay	143	-	128	-	146	-	-	151
Venezuela	-	36	-	20	-	22	-	60
TOTAL	832	100	800	84	998	69	639	501

WESTERN EUROPE

EC-10								
Belgium/ Luxembourg	276	80	350	91	356	98	448	177
Denmark	695	4	849	4	880	5	622	29
France	333	468	362	594	400	583	848	917
GDR	199	604	414	700	407	693	421	233
Ireland	330	5	440	14	338	24	419	29
Italy	42	588	109	722	130	697	-	998
Netherlands	653	108	843	167	965	142	1,352	168
UK	158	1,167	228	1,122	180	1,010	361	208
Greece	-	86	-	134	-	112	-	263
TOTAL EC-10 2/	2,686	3,110	3,595	3,548	3,656	3,364	4,471	3,022
3/	NA	NA	850	778	828	637	1,485	756
Austria	8	16	24	16	19	14	67	-
Finland	12	2	26	-	43	-	-	-
Norway	-	19	-	22	1	2	-	39
Portugal	-	43	-	15	-	10	-	156
Spain	2	74	5	26	13	11	-	-
Sweden	38	29	48	17	69	11	37	-
Switzerland	2	26	7	23	5	25	-	68
TOTAL W. EUROPE	2,748	3,320	3,705	3,667	3,806	3,437	4,575	3,285

EASTERN EUROPE

Bulgaria	52	17	48	2	49	2	NA	NA
Czechoslovakia	14	24	13	21	13	21	-	NA
GDR	118	12	247	-	270	-	NA	NA
Hungary	103	15	121	5	126	5	-	-
Poland	134	39	141	46	40	60	79	12
Yugoslavia	89	24	98	85	85	40	53	113
TOTAL E. EUROPE	509	131	668	159	583	128	132	125
TOTAL EUROPE	3,258	3,451	4,373	3,826	4,389	3,565	4,707	3,410

87

TABLE 5-5 (continued)

	56	368	30	662	30	780	14	635
Soviet Union	56	368	30	662	30	780	14	635
AFRICA								
Morocco	–	2	–	4	–	4	–	–
South Africa	21	58	14	14	2	10	NA	NA
TOTAL	21	60	14	18	2	14	NA	NA
ASIA								
ROC (Taiwan)	18	10	18	15	20	12	–	–
India	10	–	47	–	51	–	NA	NA
Israel	–	29	–	52	–	55	NA	NA
Japan	–	513	–	486	–	560	–	179
Rep. of Korea	14	24	14	15	13	48	–	–
Philippines	–	12	–	6	–	9	–	–
Turkey	6	–	6	–	23	–	41	25
TOTAL	48	588	85	574	107	684	41	204
OCEANIA								
Australia	1,079	–	1,090	1	943	1	782	–
New Zealand	726	1	795	2	821	1	754	–
TOTAL	1,805	1	1,885	3	1,764	2	1,536	–
Total Selected Countries	6,425	5,826	7,677	6,502	7,874	6,274	8,313	8,087

Note: -- denotes trade less than half the unit shown, or no trade.
1/Carcass weight equivalent basis: excludes fat, offals, and live animals.
2/Includes intra-EC trade.
3/Excludes intra-EC trade.
4/Estimate based on USSR statistics and on trading partner data. Reported on product weight basis.
NA: Not Available
Sources: USDA, FAS, "Foreign Ag. Circular: Livestock and Poultry," FL & P, various issues

TABLE 5-6. International Trade in Selected Countries for Major Red Meat Products, 1987 (Sources: USDA, FAS, "Foreign Ag. Circular: Livestock and Poultry," FL&P, various issues; USDA, FAS, "World Livestock and Poultry Situation," FL&P, various issues)

Continent and Country	Beef and Veal Exports	Beef and Veal Imports	Pork Exports	Pork Imports	Lamb and Mutton Exports	Lamb and Mutton Imports
NORTH AMERICA						
Canada	91	165	235	12	—	—
Mexico	1	6	0	2	—	2
United States	275	991	45	533	1	20
TOTAL	367	1162	280	547	1	22
SOUTH AMERICA						
Argentina	300	—	—	5	8	—
Brazil	250	—	15	0	1	—
Chile	—	90			5	—
Colombia	18	—			—	
Peru	—	4			—	2
Uruguay	110	—			11	—
Venezuela	—	21		11		—
TOTAL	678	115	15	16	25	2
WESTERN EUROPE						
EC-11						
Belgium/Luxembourg	100	31	320	32	4	14
Denmark	180	24	822	—	—	2
France	505	305	95	420	4	90
Germany, FDR	540	300	137	555	1	26
Greece	—	130	—	60	—	15
Ireland	386	7	34	21	26	—
Italy	125	495	49	466	—	19
Netherlands	375	60	915	40	5	2
Portugal	—	13	—	2	0	1
Spain	3	17	1	70	5	10
United Kingdom	190	370	61	520	37	120
TOTAL EC-11 3/	2404	1752	2434	2186	82	299
4/	60	452	302	98	6	192

TABLE 5-6 (continued)

Austria	16	1	0	4	—	1
Finland	1	—	8	—	—	—
Norway	—	13	35	6	—	3
Sweden	10	10	2	80	—	—
Switzerland	1	12	—	4	—	5
TOTAL WESTERN EUROPE	2432	1788	2479	2280	82	308
EASTERN EUROPE						
Bulgaria	3	1	10	—	30	1
Czechoslovakia	50	10	30	5	3	—
Germany, DDR	10	—	230	—	8	—
Hungary	40	3	145	—	3	—
Poland	40	2	85	20	0	—
Romania	115	—	140	—	45	—
Yugoslavia	40	30	50	15	—	—
TOTAL EASTERN EUROPE	298	46	690	40	89	1
Soviet Union 5/	7	342	6	260	1	42
AFRICA						
South Africa	2	40	2	—	—	—
Egypt	—	230	—	—	—	15
ASIA						
Tawain (ROC)	—	—	—	—	—	—
China (PRC)	—	—	—	—	—	—
India	52	—	—	230	21	—
Hong Kong	1	300	—	310	—	—
Japan	—	0	2	—	13	155
Republic of Korea	—	6	—	—	14	15
Philippines	—	—	—	1	—	—
TOTAL	53	306	2	541	48	170

MIDDLE EAST				
Israel	—	47		
Saudi Arabia	—	55		
Turkey	3	25		35
TOTAL	3	127		35
OCEANIA				
Australia	780	—	4	240
New Zealand	407	—	0	538
TOTAL	1178		4	778
TOTAL SELECTED COUNTRIES				—

Note: — denotes trade less than half the unit shown, or no trade.

1/ Carcass weight equivalent basis: excludes fat, offals, and live animals.

2/ Includes goat meat.

3/ Includes intra-EC trade.

4/ Excludes intra-EC trade.

5/ Estimate based on U.S.S.R. statistics and on trading partner data. Reported on product weight basis.

FIGURE 5-5. Leading Import and Export Nations for Red Meats, 1980s (Sources: Tables 5-5 and 5-6)

Beef and Veal

Principal Exporters: Argentina, Australia, New Zealand
Principal Importers: U.S., EC, Soviet Union

Pork

Principal Exporters: U.S., Canada, EC, East Germany
Principal Importers: U.S., EC, Japan

Lamb, Mutton and Goat Meat

Principal Exporters: Australia, New Zealand
Principal Importers: EC, Japan

that they produce more efficiently. By emphasizing production of the relatively lower cost products, costs are minimized and profits maximized. Thus, in our example, the U.S. would stress chip production. However, that would take away labor and resources that could otherwise go into beef production. Australia is relatively more efficient in beef production, and emphasizes it at the expense of chip manufacture. The nations have counterbalancing surpluses and deficits—the U.S. lacks beef, Australia lacks chips—that can be compensated for by trade. Yet, since each nation is specializing in the product for which it is *relatively* efficient, the total cost of producing all products is minimized. Both countries benefit through trade. (See any introductory trade book, e.g., Terpstra 1983, Chapter 2.)

Is this an isolated example—a matter of luck that trade between these two nations could exist? The beef producer must be endowed with a *relatively* favorable climate and productive resources, a matter of "luck," but the point is that someone is going to produce beef and that nations can benefit from specialization. In general, some country will produce products for which there is effective demand and, moreover, will specialize in that production. Exchange equilibrates supply and demand so that trade is nothing more than exchange intercountry rather than intracountry. However, trade is more than that, as its turbulent history tells us. Many nations erect barriers to trade, particularly for meat. Why, when the theory shows trade is mutually beneficial? Like any complex issue, there are numerous explanations.

Trade Impediments

Disease and Sanitation

Disease (animal health) control and food sanitation (human health) are major concerns of any nation. Among the major contagious animal diseases, the one of particular importance is foot-and-mouth disease because an outbreak requires destruction or vaccination of infected and exposed animals. This viral disease is particularly noxious because it affects cattle, swine, and sheep to nearly the same degree and can be transmitted through fresh meat. As a protection against this devastating disease, the U.S. limits imports from infected areas to cooked products like canned hams and corned beef. Other countries, such as Britain, which are more dependent on meat imports than the United States, have taken the intermediate approach of banning fresh carcass imports but permitting movement of deboned meat from infected areas.

The second health-related basis for restricting meat importation from designated areas is the condition of slaughter facilities. Most countries have strict public health criteria that must be met before a packing plant may sell its products for human consumption. (See Chapter 11.) While these criteria are typically different outside the nation's boundaries, certain standards must nonetheless be met before a plant can be certified as a point of origin. As with domestic plants, the U.S. Department of Agriculture is responsible for certifying overseas plants as well. Uncertified plants may export meat products and by-products not intended for human consumption into the U.S. if the minimal health requirements are met.

Preferences

Trade is affected by consumer preferences, and throughout much of the world there is a decided inclination toward fresh (not frozen) red meats. Yet, necessity requires that much meat in international trade be frozen. This is particularly true for the U.S. because the major supply area (one certified free of FMD) is Oceania — Australia and New Zealand — which is half a world away. Partly as a result, the emphasis on imports is for processed cuts (e.g., trimmings for hamburger or smoked bellies for bacon) rather than fresh items. The need to freeze meat for transit has been reduced in recent years as a result of cryogenic packaging developments, but the spoilage risk remains higher. (See Chapter 11.) Thus, trade will be limited until further technical breakthroughs are made in chilled product protection or until U.S. consumers become more accepting of frozen meat.

Balance of Payments

Even if the preceding conditions are satisfied, a nation must be able to pay for imports. Here, the comparison can be made to any individual's account (if you cannot pay for something, you do without) with one difference that relates to a country's ability to coin money. A nation may establish a currency that serves all domestic needs. The problem comes when payment is to be made outside national boundaries. The exporter naturally wishes to convert (exchange) the buyer's currency (let us call them doubloons) into exchange honored in the seller's country or in some third nation. The seller who is able to find someone in need of doubloons could make the trade. However, a problem emerges if there is more demand to sell doubloons than to buy them. This occurs if a country is a net importer (has an unfavorable balance of trade). Governments are the purchasers (converters) of last resort, and most nations have a pool of currency used to make such exchanges. Formerly, gold was used, but the U.S. dollar is now the international medium of exchange.

When the pool of exchange dollars shrinks as a result of net imports, a currency can no longer be converted. It becomes a soft, or nonconvertible, currency as compared to a hard currency like the Pound, Dollar, Franc, Guilder, Deutsche Mark or Yen, to name some principal ones. Countries with foreign exchange shortages cannot import readily, and trade is curtailed. During the 1980s, many developing countries found themselves in that situation largely as a result of the oil price increases of the previous decade. The centrally controlled economies also have nonconvertible currencies and rely on exports or sales of gold to generate the dollars needed for imports. Foreign exchange holdings, then, determine much of the effective level of international trade. The International Monetary Fund keeps track of countries' exchange (hard currency) holdings along with the government short- and long-term debt obligations and private overseas commitments.

Quotas and Tariffs

A final (and major) inhibitor of trade in meats and meat products is protectionism. Since most countries have a livestock sector to protect, it is not surprising to see the prevalence of barriers for those products. Barriers are erected for various reasons. Countries may feel the domestic sector needs shielding from competition for a time while the industry grows (the "infant industry" case). Or, import prices may be so low that the domestic industry withers. One source of low prices is export subsidies in the producing areas. To counter these subsidies, countervailing duties are often imposed. Or, finally, a country may have a pressing domestic employment problem and choose the temporary expedient of restricting imports, even if the long-term cost is high.

An elaborate tariff system could be seen by examining the Japanese pork import regulations as of June 1982 (USDA, FAS 1982, pp. 18-19). This system involved both a fixed or minimum tariff of 6-9 percent plus a sliding scale used to maintain a minimum price on imports. The sliding scale is keyed to domestic support prices consisting of floor prices, below which pork was purchased for government stocks, and ceiling prices, above which stocks were released for consumption. The sliding tariff scale pegs pork imports at a point about midway between these two prices. Bringing in product at a price below the midpoint causes the imposition of the sliding tariff, raising it to the target price. Importers have no incentive to increase the duty and hence choose a mix of high and low valued cuts so the mix comes near the target. Similarly, products entering at a price above the target (i.e., a mix with a greater proportion of high valued items) cause the minimum tariff of 6-9 percent to raise the final price notably, so that is discouraged. The effect of this tariff scheme is an overall reduction in import demand due to the concomitant increase in retail prices. More selectively, the system discriminates against lower-priced cuts as the lower prices are not passed along to the consumer.

Yet the Japanese system is loosening up, partially in response to pressure from the U.S. One of the self-claimed "breakthroughs" was the agreement by the Japanese government to expand the beef import quota by 60,000 metric tons annually until it is eliminated in 1991. The 1988 quota increase was slightly greater than for the preceding decade – 21 percent compared to 17 percent – and does represent a benefit. However, offsetting quota liberalization is an increase in tariffs, from 25 percent in 1988 to 70 percent in 1991, then down to 50 percent in 1993 (and negotiated thereafter) (Gorman and Mori 1990).

The net impact of declining quotas with rising tariffs is difficult to predict. Complicating the analysis is the removal of the incentive under a quota system to import only the highest quality. That opens the way for Australia whereas high quality beef imports had been virtually a U.S. monopoly. (Some believe the system was structured to favor the U.S.) At the same time, the price advantage of edible offals, which are exempt from quotas, have a lower tariff (around 15 percent). In 1987, the U.S. had an 85 percent share of that market (Gorman and Mori 1990, Table A-1; Alstan, Carter, and Jarvis 1990). Perhaps the most significant factor is the elimination of the quasi-governmental monopoly that operated through the Livestock Industry Promotion Corporation and which attached its own service charge. Now buyers and sellers can deal directly.

Quotas are easier to administer and provide more absolute protection. Quotas are also the bane of free market economists since they inhibit trade no matter how efficient the exporter may be. In general, trade restrictions raise consumer prices and mollify the incentive for the domestic producer to be efficient. Thus, open trade is preferred on public welfare grounds, but self-sufficiency, employment, and start-up considerations are also important and often take precedence over theoretical considerations of social welfare. Quotas in the U.S. and Europe constitute a complex mixture of barriers and market sharing. In the United States, beef imports were limited from 1964-1979 to about 6-7 percent of domestic production (PL 88-481). Sub-quotas were set for each authorized supply country. While the total quantity permitted under the law was not a large percent, it nonetheless grew over the years as domestic production expanded. What was particularly galling to producers was the mechanism which permitted the greatest imports to enter during years of higher supply. Since supply and price are inversely related, cattlemen saw imports as driving prices down even further when they were already depressed. As a result of these objections, the revised 1979 Meat Import Act contained a variable, "anti-cyclical" quota. Based on this formula, the 1988 quota was 1,525 million pounds with actual imports of 1,316 million pounds as of the end of October. The 1980 imports reached 1,431.0 million pounds because restrictions were suspended (USDA, Office of Governmental and Public Affairs 1983, pp. 2-9). The overall quota formula allowed by the 1979 Act is:

Annual Quota = Average annual imports (1968-1977),
or (1,204.6 million pounds)

$$\times \quad \frac{\text{3-yr. moving av. domestic prod.}}{\text{10-year (1968-1977) av. of domestic prod.}}$$

$$\times \frac{\text{5-yr. moving av. of domestic cow beef prod.}}{\text{2-yr. moving av. of domestic cow beef prod.}}$$

This law allows a minimum quota of 1.2 billion pounds with upward adjustments made as the 2-year moving average falls below the 5-year moving average, the counter-cyclical component of the measure. The Act covers 85-90 percent of imported beef, exempting sealed beef products, mainly from Brazil and Argentina.

This quota system serves the needs of U.S. producers by protecting against large imports during years of high volume and low prices. Attempts to serve consumers by stabilizing retail prices would be less effective because there is a limited supply of cattle available overseas to import during high price periods. This happens because the United States is such a major beef importer that the cattle cycle of exporting countries invariably becomes synchronized with the U.S. cycle. That is, when U.S. prices are low, foreign producers liquidate so that fewer head are available as prices rise once again (Simpson and Farris 1982, pp. 179-183).

In addition to the established quota levels (which the Secretary of Agriculture is required to enforce if imports equal or exceed the "trigger level" of 110 percent of the stated amount), "voluntary" restrictions were established beginning with the fourth quarter of 1982. In 1988, Australia and New Zealand restricted themselves to shipments of 800 and 445 million pounds, respectively (USDA, ERS "Situation and Outlook" Nov. 1988, p. 30).

The Act was amended, in 1988, principally to exempt Canada which now falls under the U.S.-Canada Free Trade Agreement. Restrictions on agricultural trade between the two countries is to be phased out over a ten-year period. In the interim, the Canadian quota will become available to other countries.

The 1979 Meat Import Act (like its predecessor) does contain an escape clause allowing the President to suspend or increase quotas when it is considered to be in the "public interest." This clause has been enacted a number of times, most notably by President Nixon in 1973 when domestic choice beef prices reached $1.35 a pound, an increase of 20 percent for the year. Producers complained about the increased competition and prices indeed dropped, but the causality was never obvious. (See below for a discussion of the effects of imports on U.S. prices.)

Under the law, the President may act when (1) national economic or security provisions dictate, (2) there is an inadequate domestic supply, or (3) special trade conditions are made with an importing country. These conditions hold if the countercyclical measure has a value of 1.0 or more. If less, the President can change quotas only in cases of a declared national emergency, or if there has been a major disruption in domestic cattle production or marketing (Petry 1982).

The European quota system may be examined by evaluating the schedule for pork. For member countries, these policies are established by the twelve-member European Economic Community (EEC or EC), some-times called the Common Market.[1] Since 1967, the EC has operated as a single market with free internal trade. One of the keystones of the Community is its Common Agricultural Policy (CAP), which is frequently a source of internal and external friction. Nevertheless, with a population of over 300 million, it is the *major* market of U.S. agricultural exports and the recipient of about 50 percent of the world's red meat imports (Table 5-5). Thus, it commands a major position in world agriculture in general, and in red meat, in particular.

The goal of the CAP is to encourage production by guaranteeing adequate foodstuffs through the maintenance of attractive prices. Since the early 1970s, this has been achieved by means of establishing a target or "guide" price that is maintained, when necessary, by mandatory public intervention in the marketplace. Protecting this system necessitates that imports be limited, and this is done through a combined fixed duty and sliding scale duty which balances world prices and domestic guide prices. The actual computation can become rather intricate as the following formula for pork shows.

Quarterly, a "sluice gate" price for imported pigs and pig meat is set to represent cost of production in non-EC countries. Variable tariffs are established to bring import prices up to the designated levels. This levy consists of two parts:

a. The difference between world market and European Community feed costs, and

b. Seven percent of the sluice gate price, thus giving domestic producers a preferential advantage. (European Communities 1977, pp. 24-25).

Japan functioned in a very similar way with a semiannual quota. Domestic prices are further honed by the Livestock Industry Promotion Corporation (LIPC) which buys and sells beef to keep prices within the government's established target range. Profits are used to improve the efficiency of the meat distribution system. Also, as in Europe, the high-price policy is followed to stimulate local production in a region still haunted by the food shortages of World War II (Ag. Policy Res. Com. 1982).

---

1. Member nations are Belgium, France, Germany, Italy, Luxembourg, Netherlands (charter) and Denmark, Ireland and the United Kingdom, Spain, Greece, Portugal.

A notable bright spot is the Republic of Korea (South) which resumed beef imports in 1988 following a three-and-a-half-year ban. The quota for 1988 was 14,500 metric tons, of which the U.S. supplied half. The announced 1989 quota was up to 35,000-39,000 metric tons (and later revised upward to 50,000). Actual imports exceeded that level, with the U.S. providing a quarter of the total (USDA, FAS 1990, pp. 10-11).

Formalized tariffs and quotas are by no means the only form these restraints take. Certain "voluntary" limitations, under which an exporting nation will offer to restrict shipments to a specified volume, are also used. Perhaps best known of these was the self-imposed restriction by Japanese auto manufacturers of sales to about two million cars and light trucks in the U.S. market annually for 1981-1985. Restrictions may be even more informal than that. Australian lamb exporters to the U.S., for example, show (in private) a keen awareness of the domestic sector and plan imports for periods of low local supplies.

As a result of the structure of quotas, tariffs, and support prices in existence around the world, domestic prices vary substantially from nation to nation (Table 5-7). Japan, with its relatively inefficient and heavily protected agricultural sector, has traditionally had the highest domestic prices.

Non-Tariff Barriers

Non-tariff barriers include the number of nuisance regulations and restrictions whose purpose is the limitation of trade. Quotas also fall under this category, but in this discussion we shall consider only non-tariff and non-quota regulations. Health regulations may be included in this heading, but since many have objectives other than inhibiting trade, they shall not be discussed here. Rather, attention is focused on indirect means of restricting trade. Identifying those cases requires a value judgment: Is the objective legitimate or not? Not wishing to make those decisions, this section shall instead describe some recent practices that have no apparent compelling justification as examples of the impediments to red meats (and other) trade which exist.

One form that impediments can take is the establishment of documentation requirements that exceed the norm. An example from the Philippines is shown in Figure 5-6.

Beyond these specifics, there are a number of "nuisance" regulations which, intentionally or not, limit trade. Many nations, for example, vary their approved food colors so that a colored ham that is legal in the U.S. may not be in the EC or Japan, and vice versa. Pack size and labeling

TABLE 5-7. Retail Prices of Selected Red Meat Products, Various Countries, October 1981 (Source: International Labor Office, Bull. Labor Stat. 1982, Part III, p. 87)

-- $ per kg --

Country	Sirloin (w/o bone)	Loin Chops	Mutton Leg
Egypt	4.00	NA	4.00
Ghana	2.76 (w/bone)	3.86	6.60
Colombia	2.62	3.36	3.19
Mexico	NA	5.35	6.60
U.S.	6.71 (w/bone)	4.89	NA
Japan	25.56	8.58 (w/o bone)	NA
France	10.79	6.10	9.37
Italy	9.08	5.79	8.79
UK	10.17	5.00	NA
Australia	4.43	5.13	3.83

NA: Not Available

requirements also differ so that products moving into Canada from the U.S., for example, can require a separate packing line.

One issue that is hardly a "nuisance" to U.S. beef producers is the ban on hormones by the EC beginning January 1, 1989. Because the USDA does not certify hormone-free meat (and state certification is of questionable acceptability), U.S. producers lost an estimated $100 million annual market. While the EC authority considers this ban to be health-related, the U.S. position is that it has no scientific basis and serves only to restrict imports. Retaliating, as permitted by the Omnibus Trade and Competitiveness Act of 1988, the U.S. placed 100 percent duties on imported EC meats and a selection of other products including prepared tomatoes, instant coffee, and pet foods to bring the value up to the estimated $100 million in lost sales. The EC has threatened to retaliate in its own right by placing a 100 percent tariff on targeted U.S. imports, while the U.S. retains the option to ban EC imports altogether (see Holder 1989).

The prevention of such requirements is a principal function of the General Agreement on Tariffs and Trade (GATT). GATT resulted from a recognition, particularly by the industrialized nations, of the bankruptcy of trade restrictions in the name of protecting domestic employment. Successively more restrictive tariffs, quotas, and non-tariff restrictions, which culminated for the U.S. in the Smoot-Hawley Tariff Act of 1930, contributed to the Great Depression of the 1930s. Determined to avoid a repetition of those destructive practices in the future, 23 nations (led by the U.S.) gathered in 1947 to negotiate tariffs and quotas and thus prevent rounds of protectionism and retaliation. GATT, with its ongoing rounds of negotiating, has been credited with spurring the enormous prosperity of the post-World War II period. GATT, however, was effective primarily for manufactured goods. Raw material exporters, most notably those in developing countries, found the terms of trade (the price of their exports compared to imports) tipping against them. A second trade group, the United Nations Conference on Trade and Development (UNCTAD), founded in 1964, attempts to reduce this problem. The principal thrust is controlling supply by regulating exports, but that has proven an elusive goal.

GATT has been effective in rationalizing formal tariff and quota agreements compared to the pre-World War II period. As a result, protectionist trade practices have partially shifted to "voluntary" quotas and to nuisance regulations as exemplified above. Considering the temptation that countries have to erect trade barriers, the record of multinational negotiations has been admirable, if not completely successful. Further pressures for protectionism arose during the worldwide recession of 1982-1983, raising concerns about the future of free trade and issues such as up to 90 percent "local content" legislation for automobiles sold in the U.S. Trade in livestock products has suffered as a result of such pressures. Dating particularly to the mid-1970s, a combination of quotas, "voluntary" quotas, tariffs, price supports, and technical regulations have been put into place in the northern hemisphere. According to 1980 estimates, liberalizing those regulations by 25 percent (e.g., a worldwide reduction in implicit tariff rates of 25 percent) could increase total welfare in the world by $1.6 billion in 1980. The U.S., however, would be a net loser by about 11 million (U.N., FAO 1980, Table 6.) Since that date, tariffs have fallen elsewhere in the world while U.S. exports have indeed risen (USDA, FAS, FDLP, various issues), but as is often the case in these instances, the causality between the two will take some time to determine.

The preceding, however, relates to only a *partial* liberalization of trade, the lifting of tariffs and quotas. As Roningen and Dixit (1989) correctly identify, a true liberalization would also require the removal (or at least substantial reduction) of national agricultural support and supply control programs which are the underlying causes of much agricultural trade instability. In this sense, the U.S. and EC are major players, spending around $50 million on these programs in 1986 alone. In the U.S., livestock (exclusive of milk) receives little direct subsidy, but the indirect effects through grain markets means that livestock too would be influenced by a major trade-policy-induced change in the U.S. agricultural support system.

FIGURE 5-6. Example of Export Documentation Requirements (Source: Brandon 1978, pp. 84-87)

PHILIPPINES

CONSULATE GENERAL—556 Fifth Avenue, New York, N. Y. 10036. (Telephone: 764-1330). Hon. Ernesto C. Pineda, Consul General. Chicago: Hon. Thomas R. Padila, Consul General, 6 No. Michigan Ave., 60602. (Telephone: CS 2-6458). San Francisco: 445 Sutter St., 94108. 433-6666. Los Angeles: 3250 Wilshire Blvd., 90010. 387-5321.

HOURS (New York)—Documents received weekdays—9 A.M. to 12 Noon and 1:30 to 4 P.M. Returned three business days later. Saturdays Closed.

FREE ZONES—Marivales and Bataan

ARRANGEMENTS OF DOCUMENTS BEFORE PRESENTATION TO CONSULATE—In order to give better service, the Consulate would appreciate the co-operation of exporters when presenting documents for legalization. It is suggested that shippers present their consular documents as follows: (a) original consular invoice followed by a commercial invoice and packing list, stapled together; (b) blue consular invoice followed by a non-negotiable bill of lading, commercial invoice and packing list, stapled together; (c) pink and yellow same as blue set; (d) Green consular invoice followed by a commercial invoice and packing list, stapled together. All sets should then be clipped together in above sequence.

DOCUMENTS REQUIRED

According to the New York Consulate, no documentation is required for shipments with FOB-NY value of less than $400.00 except when the nature of the merchandise requires a special invoice and/or document such as: (a) Declaration of Shipper or Food and Drug Products; (b) Certificate of Fumigation or Disinfection; (c) Phytosanitary certificate, etc. In such cases, only the special invoice and/or documents should be presented to the Consulate General. Certificate of origin is no longer required.

BILL OF LADING—Three (3) non-negotiable copies of the negotiable Bill of Lading, containing freight and other charges duly signed by the carrier or its agent stamped 'FOR CONSULAR PURPOSES ONLY". (First copy of the Department of Foreign Affairs, Manila; the second for the AntiSmuggling Action Center, Quezon City; and the third, for the Consulate General).

A certified copy of the Shipper's/Exporter's Export Declaration.

Five (5) copies of detailed packing list.

Two (2) copies of the manufacturer or supplier's Export Price List, including all discounts, subsidies, bounties, direct or indirect subventions, drawbacks, rebates and remission of taxes applicable, except when shipment consists of nonstandard or specially manufactured items. (One copy for the Bureau of Customs, Manila, and the second for the files of the Consulate General. If Price List has been initially furnished, reference to that effect on the consular invoice is sufficient.

Time of submission of documents—On ocean-going vessels: The consular invoice together with supporting papers should be submitted to the consular office for processing even before the departure of the carrier or within a reasonable time thereafter.

The following periods are considered reasonable time:

1. Continental U. S. and Canada	
(East Coast	—25 days
West Coast	—15 days
2. Hawaii	— 1 week
3. Guam	— 3 days

AIR-FREIGHT SHIPMENTS: The consular invoice covering air-freight shipment shall be submitted before departure of carrier.

PARCEL POST AND AIR PARCEL POST SHIPMENTS—Three copies of the parcel post receipt should be presented to the Consulate with other consular documents for all shipments with an export value of $400 FOB or over.

102

CONSULAR INVOICE—The new Consular invoice became effective January 1 and the old form will not be accepted. The new form will be issued in triplicate and the sets must be prepared as follows: Original Consular invoice followed by original commercial invoice and non-negotiable Bill of Lading, 2nd and 3rd sets should be prepared the same way. The original and duplicate sets will be returned to shipper.

COMMERCIAL INVOICE PROPERLY accomplished in quintuplicate and duly certified under oath (notarized) by the seller, manufacturer, exporter or his authorized representative, who must be a responsible official of the exporting company. Must show FOB and CIF or C & F separately.

Specimens of signatures of responsible officers (e.g. export manager, sales manager, shipping manager, etc.) authorized to sign for the company should be filed with this office in duplicate indicating his position and signed by the head of the firm. Revised list must be submitted each time there are changes. The foregoing requirement also applies to freight forwarders and/or brokers handling shipments to the Philippines.

All copies must contain the following statements:

"I hereby certify that all information contained herein is correct; that the value declared is the same value stated in all other declarations made before or filed for official purposes in any agency in or of the exporting country; and that the amount per unit FOB declared is the same value stated in all other documents filed in connection with this exportation for official purposes of the United States."

"I further certify that this invoice is in all respects correct and true and was made at the place from whence the merchandise was exported to the Philippines. The invoice contains a true and full statement of the date when, the place where, the person from which the same was purchased, and the actual cost thereon. That no discounts, bounties or drawbacks are contained in the invoice except such as have been actually stated thereon."

PLACE OF CERTIFICATION—Consular invoices may be presented for certification at Philippine Consulates having jurisdiction over the place (1) where the merchandise was manufactured (2) where the merchandise was purchased; or (3) where the merchandise was shipped.

However, when articles have been purchased in different consular districts, there shall be attached to the consular invoice the original bills or invoice. The consular office to whom the invoice is produced for certification may require that any such original bill or invoice be certified by the consular officer for the district in which the articles were purchased.

Whenever any two of the places mentioned in the first paragraph are located within the same consular district, the consular invoice must be presented for certification at the Philippine Consulate having jurisdiction over the consular district.

SHIPMENTS TO GOVERNMENT AGENCIES—(1) All shipments to Philippine Government agencies or government-owner corporations, as well as fifty per cent (50%) of the gross tonnage of all shipments to the Philippines consisting of either U. S. surplus goods, AID commodities, or equipment financed under the credit line authorized by the Export-Import Bank, should be loaded on Philippine flag vessels, when available. (2) All the foregoing shipments which cannot be loaded on Philippine flag vessel should be accompanied by "Certificates of Non-Availability of Philippine Flag Vessels" which should be attached, stamped or printed on the consular invoices, such certificates to be duly signed by the local agent of the Philippine flag vessel and certified by the Philippine Consulate at the port of shipment. (3) A period of ten days prior to the date of sailing of a Philippine flag vessel may be allowed as the period which may be considered in determining whether or not a Philippine flag vessel is available at the port of exportation.

88a

FIGURE 5-6. (continued)

ADDITIONAL DOCUMENTARY REQUIREMENTS IN SPECIFIC CASES—

(1) Food and Drug Products—Five (5) copies, Form FA No. 53, Declaration of Shipper of Food and Drug products.

(2) Insecticides. Paris Greens, Lead Arsenates, or fungicides—Five (5) copies, Form FA No. 54.

(3) Meat and meat products—Five (5) copies, Form FA No. 55, Certificate of Ante Mortem and Post Mortem Inspection.

(4) Viruses, serums, toxins, and analogous products intended for treatment of domestic animals—Permit or veterinary license number (biological products) from the U. S. Department of Agriculture.

(5) Viruses, therapeutic serum, toxin, antitoxin or analogous products or arsphenamine or its derivatives (or any other trivalent organic arsenic compound—Permit or license to manufacture from the Federal Security Agency or its successor.

(6) Domestic animals & livestock—permit from the Director of Animal Industry of the Philippines, as well as, a health certificate issued shortly before shipment by a veterinary official of the country of origin, duly certified by the nearest Philippine consular official, stating that each of the animals is free from, and has not recently been exposed to, any dangerous and/or communicable animal diseases. A certificate of anti-rabies shot is required is the case of dogs and cats.

(7) Fruits, vegetables, seed and other plant materials —Certificate of Disinfection from the Department of Agriculture of the United States or of the corresponding department of any state of the Union.

(8) Firearms and ammunition—Permit/License from the Chief, of Constabulary, Quezon City, Philippines. (Importation now prohibited.)

(b) Sent to any foreign diplomatic and/or consular mission in the Philippines PROVIDED: that the consular invoice is accompanied by a statement or certificate of the foreign diplomatic and/or consular mission concerned showing that the merchandise is for its exclusive use and not for sale, barter or hire.

(c) Shipment of goods consigned or addressed to departments, bureaus, offices and other instrumentalities of the Government

(d) Free service does not apply to certification of quintuplicate and extra copies.

(9) Remnants—Certification from the U. S. Collector of Customs as to the quantity and prices of remnants shipments.

(10) U. S. & Canadian Wheat Flour—Information as to the amount of subsidy received by each shipment upon exportation to the Philippines.

(11) Essences, flavoring extracts and other preparations containing distilled spirits otherwise known as ethyl alcohol—Certificate of the manufacturer stating the source of the alcohol used.

(12) Resins—Specific nomenclature of the resins.

AMENDMENT AND/OR CORRECTION—After an invoice has been duly certified, any amendment and/or correction can be made thereon must be done on the shipper's/seller's letterhead in quintuplicate, duly notarized, stating the amendment and/or correction and the consular invoice number.

CONSULAR FEES

Consular invoice blanks, FA Form 48/49, per set of five .30 per set; pad of 25 sets	$ 5.00
FA form 53/54, .30 per set; pad of 25 sets	5.00
Certification of consular invoice, FA Form 48-49	30.00
Certification of Certificate of Origin	5.00
Certification of invoice with FA Form No. 53 or 54 or 55	5.00
Certification of either FA Form 53 or 54	5.00
Certification of disinfection of goods	5.00
Certification of invoice or returned Philippine goods, FA Form No. 52	5.00
Certification of amendment and/or correction	2.50
Quintuplicate copy signed manually	2.50
Extra copies rubber stamped	2.50
Extra copies signed manually	2.50

Free Certification on consular invoice of merchandise:
(a) Sent to religious or charitable organizations as donation for free distribution to the needy.

105

Using a 1986/1987 base, the authors estimate that U.S. ruminant and non-ruminant meat prices would increase by seven and two percent, respectively, while output of ruminant meat would increase four percent. The increase in total revenue to the sector would translate into even greater net benefits due to the projected sharp decline in feed input prices (7-45 percent) (Roningen and Dixit 1989, Table 7). In general, livestock producers in the U.S. and Australia/New Zealand would be the beneficiaries, while those in Europe and Japan would be the losers.

Like any study of this type, the results are sensitive to the assumptions, especially the choice of the base year. One of the major unknowns for the meat sector is how consumers would respond to the proportionally much greater price of meat compared to other foods. Trade analysis presents the anomaly that, while worldwide welfare benefits can be predicted with great surety, it is very difficult to be certain about how any single sector in a particular country would fare. This applies even to U.S. livestock producers who stand to gain the most, due to the policy-inflated costs of their major input (feed) and lack of support for their own product (meat).

EXPORT INCENTIVES

At the same time that many nations attempt to curtail their imports of meats and meat products, they and others endeavor to expand exports. Exports, of course, increase domestic employment opportunities and, as a result, efforts are made to avoid disincentives to exporting. One recent move for U.S. manufactured goods is the emphasis on "free trade" zones where products for re-exporting are allowed free of most tariffs and quotas. However, there are more particular reasons many nations stimulate exports, particularly of agricultural commodities. These are (a) the need to acquire foreign exchange (so-called hard currencies that are widely acceptable because they are readily convertible to other currencies), and (b) a desire to move stocks acquired as part of domestic price support programs. The need for foreign exchange is particularly acute at this time in developing countries and in the former Eastern Block nations. Disposing of government-controlled surpluses, on the other hand, is limited to the few nations with strong price support programs. As noted, this includes the European Community where an extensive subsidy schedule is maintained. For example, the April 1989 EC export subsidy to most other countries for frozen pork bellies was 35 ECU/100kg (20 for U.S. and Canada) (U.N., FAS, FDLP 1989, p. 16).

Export subsidies are a controversial area because they raise the demands for countervailing tariffs on imports and compensatory subsidies to third party nations (Farnsworth 1983). Complicating the process is the tendency to use favorable credit terms (or guarantees) as a form of subsidy so that the actual amount of the subsidy is difficult to compute.

For countries like Australia where meat exports are particularly critical, promotion is taken beyond the wholesale level up to retail. Under the direction of the Australian Meat Board, established in 1964 (expanded and renamed the Australian Meat and Livestock Corporation in 1977), the state takes on three responsibilities:

1. Promotion and control of meat for export and subsequent sale and distribution,
2. Promotion of internal trade,
3. Encouragement of production and consumption. (Australian Meat Board 1976)

Supported by a levy on livestock productions, the Board has the overall mandate of preserving the long-term interests of the industry. Outside Australia, this mandate takes the form of promoting the product directly to consumers. The same is true for lamb in New Zealand (Figure 5-7). Internally, the quality of stock and plant sanitation quality is carefully monitored to maintain a good image overseas.

The United States provides a variety of export assistance to agricultural producers. Much of the general assistance is directed through the Foreign Agricultural Service (FAS) of the USDA. Its four sections perform the following trade-development functions:

1. *International Trade Section:* helps formulate trade programs and agreements with the goal of expanding U.S. trade.
2. *Agricultural Attaché:* agricultural specialists operating out of some 65 U.S. embassies with the responsibilities of collecting marketing data and representing U.S. trade interests.
3. *Foreign Marketing Section:* accounting for about 90 percent of the market development budget, this group operates a number of trade fairs and joins with private trade associations and other cooperators in promoting U.S. agricultural products.
4. *Foreign Commodity Analysis Section:* analyzes worldwide supply and demand balances for major agricultural products. (McKinna 1978)

Financial support of agricultural exports is channeled through the Commodity Credit Corporation (CCC) of the USDA. The CCC offers longer terms (up to 36 months) than is considered prudent by private institutions

FIGURE 5-7. Example of an Ad Supported by the New Zealand Meat Producers Board

Order Primal Cuts As You Need Them

PRODUCT	INDIVIDUAL WEIGHT	PIECES PER BOX
LEG bone in boned & rolled	5 to 5½ lbs. 3¾ to 4 lbs.	10 12
LOIN flank off & excess kidney fat removed	1½ to 1¾ lbs.	30 to 32

for perishable consumable products. Under some conditions, the CCC will accept local (non-convertible) currency payments while reimbursing the exporter in dollars (Venadikian and Warfield 1986). The U.S. government subsequently uses the local currency to pay for embassy and other in-country expenses. As we have noted, the ability to make payments in non-convertible currencies substantially expands the ability of many nations to import.

The Food Security Act of 1985 contained a new program, the Export Enhancement Program (EEP), intended to counter unfair trade practices of competing countries in third country markets. Programs which are targeted to particular countries for specified products based on documented unfair trade practices as allowed by section 301 of the Trade Act of 1984 use a variety of procedures to lower export prices. These include an export subsidy (cash or in-kind), tax rebate, or assumption of some costs. For FY1990 $9 million were allocated for red meats, down to almost half the FY1989 level. Programs which are operated in cooperation with the U.S.

Meat Export Federation, an industry group, use funds or the equivalent in CCC commodities (Glaser 1986, pp. 41-43). To date, little detailed evaluations of these programs are available for livestock products (USDA, FAS 1989, pp. 10-11).[2]

SYNOPSIS

World trade in red meat products is a small portion of total production — less than 10 percent by volume. Live animal exchange is yet a smaller percentage of that amount. Major importers are the U.S., EC, Japan, and the Commonwealth of Independent States while the chief exporters are Australia and New Zealand, U.S. and Canada, EC, Argentina, and Germany. In the U.S., imports and exports are about four percent of the value of red meat products with trimming and processed products imported, as well as pork from Canada, and exports constituting principally lower value products and hides.

Trade theory indicates that there are public benefits to trade whenever one country is a relatively low-cost producer of a product. It is not necessary to have an absolute, but only a comparative, cost advantage. Despite this well-known situation, there are substantial pressures on national governments to limit imports. These pressures lead to the limitation of trade in several ways including tariffs and quotas (as with U.S. beef imports and in exports into Japan and the EC) as well as other so-called non-tariff barriers such as health and sanitation regulations. The GATT (General Agreement on Tariffs and Trade) is an international effort to minimize trade restrictions. The welfare implications of truly free trade, which would require the cessation of agricultural subsidies, are difficult to compute, but the indications are that U.S. livestock producers would benefit.

Study Questions

1. What are the major eras of U.S. meat imports and exports, and what caused the shifts?
2. What does the theory of comparative advantage say about the relative position of red meat production in the U.S.?
3. U.S. beef import quotas are designed to be "counter cyclical." What does that mean, and how well has it been working?

---

2. An analysis of the programs effects on wheat exports, however, indicated little impact for 1985-86 and certainly much less than a lowering of the wheat support price (Seitzinger and Pearlberg 1990).

4. The current or Uruguay round of GATT negotiations has been hung up over agricultural trade. What are the issues, and how would livestock production be affected by a reduction in barriers?

REFERENCES

Alston, J. M., C. A. Carter, and L. S. Jarvis. "Japanese Beef Trade Liberalization: Our Beef with Government Beef Trade Exports," *Choices* (1990): 34-35.

Ag. Policy Research Committee, Inc. "Perspectives: Japan's Beef Business." Tokyo, October 1982.

Australian Meat Board. "The Role of the Australian Meat Board." June 1976.

Brandon's *Shipper and Forwarder*. "Export Documentation: Requirements for Sea and Air Shipments." Section 3, November 13, 1978.

Bureau of Agricultural Economics, Australia. "Japanese Agricultural Policies: Their Origins, Nature and Effects on Production and Trade." Canberra, Policy Monograph No. 1, 1981.

European Communities, Office for Official Publications. *The Agricultural Policy of the European Community*. Luxembourg, Periodical 6/1982, 1983.

European Communities, Office for Official Publications. *The Common Agricultural Policy*. December 1977.

Farnsworth, C. H., "Reconciling Free Trade with Help to American Industries." *The New York Times*, Business Section, p. 6, November 13, 1983.

General Agreement on Tariffs and Trade, International Meat Council. "Arrangement Regarding Bovine Meat." IMC/W/161/ Rev. 1, Apr. 27, 1983.

Glaser, L. K. "Provisions of the Food Security Act of 1985." USDA, Economic Research Service, Ag. Info. Bull. 498, April 1986.

Gorman, W. D. and H. Mori. "Diaphragm Beef Imports by Japan." USDA, Economic Research Service, "Livestock and Poultry: Situation and Outlook Report." LPS-40, February 1990.

Holder, D. L. "A Conflict over Hormones." Mimeo, February 9, 1989.

International Labor Office. *Bulletin of Labor Statistics*. Geneva, 1982-2.

McKinna, D. A. "U.S. Government Sponsored Agricultural Export Market Development Programs." Cornell Univ. Dept. Agr. Econ., A.E. Ext. 78-29, September 1978.

Petry, T. "The Market Advisor." North Dakota State Univ., 1982.

Roningen, V. O. and P. M. Dixit. "Economic Implications of Agricultural Policy Reforms in Industrial Market Economies." USDA, Economic Research Service, Ag. and Trade Analysis Div., Staff Rpt. No. AGES 89-36, August 1989.

Simpson, J. R. and D. E. Farris. *The World's Beef Business*. Ames, IA: Iowa State University Press, 1982.

Seitzinger, A. H. and P. L. Pearlberg. "A Simulation Model of the U.S. Export Enhancement Program for Wheat," *Am. J. of Agr. Econ.* 72(1990):95-103.

Terpstra, V. *International Marketing*. Chicago: The Dryden Press, Third Edition, 1983.

United Nations, Food and Agriculture Organization, Committee on Commodity

Problems, Intergovernmental Group on Meat. "Production in the Livestock Sector." CCP:ME 80/.4, October 1980.

United Nations. *Statistical Yearbook*. New York, various years.

United Nations. *Yearbook of International Trade Statistics* also known as *International Trade Statistics Yearbook*. Volumes I and II, various years.

USDA. *Agricultural Statistics 1988*. Washington, DC, 1989.

USDA, Economic Research Service. "Livestock and Meat Statistics." Various years.

USDA, Economic Research Service. "Livestock and Poultry: Situation and Outlook Report." LPS-32, November 1988.

USDA, Foreign Agricultural Service. "World Livestock and Poultry Situation." FL&P, various issues.

USDA, Foreign Agricultural Service. "U.S. Trade and Prospects." FDLP, various issues.

USDA, Foreign Agricultural Service. "Livestock and Poultry Situation." FL&P, various issues.

USDA, Foreign Agricultural Service, Circular Series. "Dairy Livestock and Poultry: U.S. Trade and Prospects." FDLP, various issues.

USDA, Office of Governmental and Public Affairs. "Major News Releases and Speeches." September 23-30, 1983.

U.S. Department of Commerce, Bureau Census. *Historical Statistics of the United States: Colonial Times to 1957*. Washington, DC, 1980.

U.S. Department of Commerce, Bureau Census. *Statistical Abstract of the United States*. Washington, DC, various years.

Venadikian, H. M. and G. A. Warfield. *Export-Import Financing*. New York: John Wiley and Sons, Second Edition, 1986.

Chapter 6

World Consumption of Red Meats

After listing the major production areas and adjusting for trade, it seems unnecessary to designate per capita consumption figures by country. In a broad sense, that is true; major producing nations such as the U.S., Argentina, and Australia, all have high per capita meat consumption (Table 6-1). However, total domestic consumption data omits such major factors as population size differences among countries. Adjusting for this, it is evident that the Commonwealth of Independent States has only modest supplies available despite its large total production. Many other nations like India and Zaire, which have both limited production and large populations (relative to the productive land base), have only minuscule supplies (less than 5 pounds per capita). Some countries are so limited that kwashiorkor (a protein deficiency disease) is prevalent.[1] Indeed, all the major consuming nations are relatively wealthy, confirming an expectation that animal protein consumption is closely associated with income. Statistical analysis (see below) further confirms these expectations.

Over time, red meat production has expanded rapidly, but so has population. As a result, total per capita food production has increased by four percent from 1970 to 1980 and six percent from 1980 to 1986. Advances made by the developed market economies were slightly greater than those achieved by the developing nations (*United Nations Statistical Yearbook*, Tables 6 and 7). Meat production fared about the same. Supplies grew by 77 percent from 1960 to 1979, but world population also advanced by a healthy 47 percent. As a result, average per capita consumption grew a moderate 5.5 pounds, or just over 10 percent (U.S. Bureau Census Statistical Abstract 1962, Nos. 1243, 1250; 1982-1983, Nos. 1518, 1544.) From 1980 to 85, production worldwide rose 10.4 percent, about 20 percent over the 8.6 percent population increase (*United Nations Statistical Yearbook* Tables 1 and 99).

---

1. For a discussion of nutrition and development, see Alan Berg, *The Nutrition Factor*, Washington, DC, The Brookings Institution, 1973.

TABLE 6-1. Total per Capita Consumption of Red Meat in Specified Countries, Selected Years (Source: USDA, reprinted in American Meat Institute 1989)

Kilos per Person

REGION AND COUNTRY	1965	1975	1980	1983	1988
NORTH AMERICA					
Canada	66.3	73.2	73.6	71.8a/	68.0a/
Mexico	16.8	21.7	22.5	30.4a/	27.9
United States	81.0	83.1	83.9	80.2	78.3
SOUTH AMERICA					
Argentina	82.2	99.0	103.3	71.4b/	72.8b/
Brazil	24.4	27.2	25.3	22.4b/	21.2
Colombia	23.4	23.2	27.2	26.5a/	24.2a/
Uruguay	109.6	99.9	94.0	84.7b/	61.4a,b/
Venezuela	21.6	27.6	29.8	26.1a/	26.8a/
WESTERN EUROPE					
EC Countries					
Belgium/Luxembourg	53.3	74.9	75.0	74.8	76.8
Denmark	57.7	53.6	71.0	59.1	84.1
France	59.6	66.2	75.1	69.5	68.1
GFR	55.4	67.2	75.7	72.6	79.4
Ireland	54.6	66.5	65.9	61.8	60.7
Italy	28.3	42.2	49.5	50.9	57.0
Netherlands	45.5	55.1	61.0	54.8	61.8
Portugal	19.9	31.5	28.1	31.2	40.4
United Kingdom	63.9	58.3	58.2	56.9	54.7
Greece	30.4	44.2	51.4	51.9	61.4

Non-EC Countries					
Austria	55.6	67.8	74.3	72.4a/	73.8a/
Finland	37.0	52.6	54.9	53.3a/	53.8a/
Spain	21.8	35.9	41.5	44.8	56.9
Sweden	45.4	51.8	52.8	49.4a/	48.9a/
Switzerland	54.3	64.4	76.8	73.2a/	72.8a/
AFRICA					
South Africa	32.2	29.5	30.1	29.7b/	24.9b/
ASIA					
China (Taiwan)	17.6	22.6	38.4	29.7a/	40.3a/
Japan	8.0	17.1	21.4	19.4a/	26.0
Philippines	12.0	10.7	11.4	13.0a/	NA
MIDDLE EAST					
Israel	17.8	17.8	17.5	17.7b/	18.8a,b/
Turkey	14.0	15.0	10.2	10.7b/	NA
OCEANIA					
Australia	93.0	108.5	83.8	77.7	80.0
New Zealand	116.8	103.0	98.1	100.1b/	80.3

*Carcass weight basis, includes horsemeat
a/excludes lamb   b/excludes pork

115

REGIONAL CONSUMPTION IN THE U.S.

Per capita consumption of animal protein as a group – including red meats, poultry, and fish – increased by one-third from 1950 to 1980 (or from 172 pounds to 230 pounds). Viewed in more detail as annual changes for individual products, the trend is not as smooth since it is influenced by cyclical supply variability and relative price movements (Figure 6-1). Further variability may be seen by examining consumption patterns even closer to the individual consumer level. Due to the way data are collected, this can best be done regionally.

The most comprehensive source of meat consumption information is the USDA's Nationwide Food Consumption Survey (NFCS). The most current survey (as of this writing) dates to 1977-1978. (The subsequent survey has yet to be reported in detail.) From representative households in the 48 contiguous states, two groups of data were compiled: (1) a one-week recall of the kinds, quantities, costs, and sources of food consumed at home; and (2) individual intake records for each member of the household. Of particular use here is the household recall data. Additionally,

FIGURE 6-1. Per Capita Consumption of Meat, Poultry, and Fish

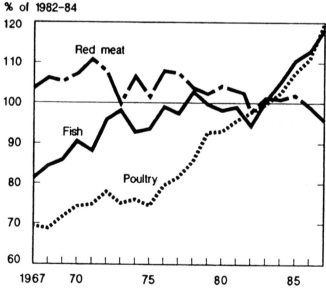

% of 1982–84

These indexes are based on pounds consumed (retail-weight equivalent) and measure quantity changes only.

information on household demographics such as age and number of members, income, and race, were collected and merged with the consumption data.[2] Overall, the file provides a good basis for both descriptive and statistical analysis of at-home meat (and other foods) consumption behavior. The survey does not, however, quantify away-from-home meat consumption, which leaves a gap in our knowledge of the total movement of meat products.

Regionally, total red meat consumption per capita is highly uniform, with a range of only three-tenths of one pound a week (Table 6-2). The composition of that consumption, however, varies substantially. The most pronounced differences are for veal and lamb, which are virtually not consumed in the South and North Central regions, and for veal, in the West also. The Northeast is the principal consuming region for these products. Pork is notably a Southern product, as expected, but the West is not the major consumer of beef, as might be anticipated. Rather, the inhabitants of the North Central area appear to be the principal "meat and potato" consumers of the nation.

Tradition is a major determinant of perpetuated regional dietary differences. The Northeasterner introduced to lamb or veal as a child is far more likely to consume both as an adult than is the Southerner brought up on a heavy pork diet. Behind these traditions lies expediency based upon regional production in earlier times of limited transport. However, preferences are no longer aligned with production, and even tradition may encompass a number of factors which themselves are more important than historical patterns. In marketing jargon, these factors are typically described as demographics.

FACTORS RELATED TO RED MEAT CONSUMPTION IN THE U.S.

Descriptive Analysis

Income

Red meat consumption varied only slightly from the lowest income household (av. $3,379) to the highest (av. $23,807), or only three percent (Table 6-3). Here again, composition masked greater differences. The lower income groups relied heavily on pork while the more well-to-do consumed beef. These trends are easier to see by looking at expenditure

---

2. For a description of the NFCS, see Rizek 1978.

TABLE 6-2. Weekly per Capita Consumption of Red Meats, 1977-1978 (Source: Haidacher et al. 1982, p. 34)

Item	REGION			
	Northeast	North central	South	West
		Pounds per capita		
Total meat	4.45	4.27	4.54	3.86
Red meats	2.62	2.70	2.66	2.41
Beef	1.67	1.77	1.63	1.67
Loin and rib				
Steaks	.30	.26	.26	.28
Roasts	.03	.02	.02	.03
Round and chuck				
Steaks	.30	.24	.24	.27
Roasts	.36	.45	.40	.34
Gound	.56	.69	.60	.62
Other	.12	.11	.11	.12
Pork	.82	.89	.99	.67
Fresh1/	.37	.37	.34	.26
Processed1/	.23	.23	.24	.16
Bacon and sausage	.22	.29	.41	.25
Veal	.07	.02	.02	.02
Lamb, mutton, goat	.06	.01	.01	.05

1/Excluding bacon and sausage.

rather than consumption data (Table 6-3). While the highest income group consumed 14 percent more beef than the lowest, the difference in per capita expenditure was nearly two and a half times the 14 percent (or 33 percent). Clearly, the wealthier shopper is buying more expensive cuts. Income nevertheless remains an imperfect explanation of food consumption. Witness the near-even levels for ground beef consumption and expenditures.

Age

One of the reasons that income inadequately describes consumption is its close association with age. Households with older heads typically have higher incomes (at least up to retirement age). Similarly, older people tend to eat less. Thus, we see the highest levels of chuck and round steak consumption among the 40-64 age group, and the highest levels of ground beef consumption in the growing years of 13-19 (Table 6-4). Consumption of these products declines for the oldest group (65 and over) while roast, veal, and lamb consumption reaches its highest levels.

Household

A second factor related to income is household size. Not only are larger households typically younger (and hence have lower incomes), they generally have more younger household members for whom meat requirements are reduced. These effects can be seen easily by examining per capita consumption and expenditures by household size (Table 6-5). Total red meat consumption declines by 25 percent moving from the two- to the six-person household, with the greatest decline (51 percent) seen with the more costly loin and ribs. The less expensive pork and ground beef decline by far less. More pronounced is the decline of one-third in expenditures over the same range of household sizes. Thus, in these cases we are seeing both a need response and an income response.

Residence

A typical stereotype of a heavy red meat consumer might be of the active rural resident with an appetite for meat to match that activity. After all, how many places outside Montana offer pork chops, mashed potatoes, and pie on the breakfast menu? On average, however, this expectation does not hold: urban consumers eat more red meat in total and (marginally) more beef and pork than their rural counterparts (Table 6-6). Again, there is probably a mix of explanations including income and age.

TABLE 6-3. Weekly At-Home Meat Consumption by Income Level, Quantity, and Expenditures, 1977-1978 (Source: Haidacher et al., 1982, pp. 27-28)

Group Average Household Income

Item	$3,379	$6,990	$10,444	$14,489	$23,807
			Pounds per capita		
Total meat	4.65	4.37	4.21	4.14	4.21
Red meats	2.55	2.58	2.57	2.59	2.64
Beef	1.55	1.61	1.65	1.71	1.77
Loin and rib					
Steaks	.16	.21	.24	.29	.38
Roasts	.02	.01	.02	.03	.04
Round & chuck					
Steaks	.25	.27	.26	.28	.24
Roasts	.34	.37	.39	.39	.41
Gound	.62	.62	.63	.62	.60
Other	.15	.13	.10	.11	.10
Pork	.94	.92	.85	.83	.79
Fresh1/	.36	.39	.34	.34	.29
Processed1/	.21	.21	.22	.22	.24
Bacon & sausage	.37	.32	.30	.27	.26
Veal	.03	.03	.03	.02	.04
Lamb, mutton, goat	.03	.03	.03	.02	.04

	Dollars per capita				
Total meat	5.05	4.95	4.93	5.08	5.51
Red meats	3.12	3.21	3.27	3.42	3.70
Beef	1.79	1.91	2.00	2.16	2.38
Loin and rib					
Steaks	.27	.36	.41	.52	.69
Roasts	.03	.02	.04	.04	.06
Round & chuck					
Steaks	.34	.36	.35	.36	.34
Roasts	.39	.43	.48	.49	.53
Gound	.60	.59	.60	.61	.62
Other	.17	.16	.13	.14	.14
Pork	1.23	1.22	1.15	1.17	1.17
Fresh1/	.48	.50	.44	.46	.41
Processed1/	.29	.31	.33	.36	.41
Bacon & sausage	.46	.40	.38	.35	.34
Veal	.05	.04	.06	.05	.09
Lamb, mutton, goat	.04	.04	.06	.04	.07

1/Excludes bacon and sausage.

TABLE 6-4. Relationship of Age to At-Home Red Meat Consumption, 1977-1978 (Source: Haidacher et al. 1982, p. 46)

Item	Average for 20-39 (Pounds)	Proportion of Average by Age Group — Standard Consumer Equivalent					
		0-2	3-12	13-19	20-39	40-64	65 & over
Total meat	4.49	0.54	0.81	0.95	1.00	1.23	1.01
Red meats	2.75	.44	.74	.90	1.00	1.22	.99
Beef	1.79	.47	.73	.93	1.00	1.20	.93
Loin and rib							
Steaks	.31	.54	.68	.81	1.00	1.16	.71
Roasts	.02	.04	.11	.88	1.00	1.76	1.78
Round & chuck							
Steaks	.29	.52	.73	1.04	1.00	1.13	.76
Roasts	.34	.27	.66	.81	1.00	1.80	1.47
Gound	.72	.51	.80	1.02	1.00	.91	.75
Other	.12	.58	.71	.85	1.00	1.45	1.35
Pork	.90	.38	.73	.84	1.00	1.22	1.02
Fresh1/	.35	.42	.89	.81	1.00	1.19	.78
Processed1/	.23	.37	.61	.74	1.00	1.24	1.24
Bacon & sausage	.32	.34	.65	.94	1.00	1.23	1.11
Veal	.03	.02	.73	.83	1.00	1.66	2.27
Lamb, mutton, goat	.02	.52	1.51	.97	1.00	1.69	3.01

1Excludes sausage and bacon.

Race and Ethnicity

One's food preferences are largely determined early in life, and ethnic preferences are passed from generation to generation. Thus, Italian-Americans and German-Americans are major consumers of veal, while those of English extraction favor lamb. However, such preferences are difficult to document because ethnic origin is not easy to observe or even to define in a "melting pot" like the United States. Thus, measured preferences are typically defined on a racial basis, particularly among whites and blacks. Hispanics are a group receiving additional attention by markets because of their relatively rapid increase in numbers (due to both immigration and births) and some common preferences and habits. Hispanics, however, come from a mixed racial background and even more varied national origins, so that as an entity, they are somewhat difficult to identify and categorize.

Blacks are relatively heavy meat eaters, whether measured in terms of volume or expenditure per capita. There is no lead only for beef, while pork consumption is ahead by nearly 30 percent (Table 6-7).

Summary

While red meat consumption shows considerable stability in the aggregate, a more detailed analysis shows considerable variability. Chief sources of that variability are income, region, race, age, and household size. A number of these factors are related to income, which is probably the single major determinant of consumption level and product selection. Separating the interdependent effects of, say, income and age requires statistical analysis, which is reviewed next.

Away-From-Home Consumption

Any current analysis of meat consumption would be remiss if away-from-home usage were left out. Red meats have long been the main fare at all three meals, and the trend has expanded following the ground-beef-based growth of the fast food sector in the 1960s. Fast food outlets constituted 31 percent of away-from-home sales in 1987, essentially double the level of 1980. But fast food constitutes only a fraction of the so-called HRI (hotel, restaurant, and institutional food) business, which had 1987 sales of $184 billion (USDA *Food Marketing Review* 1988, Fig. 29). By 1988, nearly 85 percent as much was spent on food service meals as on away-from-premise consumption, almost double the percentage from 25 years earlier (USDA, *Food Marketing Review* 1988, App. Tables 1 and 4).

TABLE 6-5. Weekly At-Home Meat Consumption by Household Size, Quanity, and Expenditures, 1977-1978 (Source: Haidacher et al. 1982, pp. 31-32)

Item	1 member	2 members	3 members	4 members	5 members	6 or more members
			Household size			
			Pounds per capita			
Total meat	4.31	4.02	4.53	4.00	3.91	3.83
Red meats	2.94	3.09	2.77	2.44	2.35	2.30
Beef	1.85	1.99	1.80	1.59	1.51	1.47
Loin and rib						
Steaks	.30	.37	.31	.26	.23	.18
Roasts	.02	.03	.03	.03	.02	.01
Round & chuck						
Steaks	.27	.29	.29	.25	.24	.24
Roasts	.42	.52	.40	.35	.34	.34
Ground	.67	.64	.65	.60	.59	.61
Other	.16	.14	.12	.11	.10	.09
Pork	.98	1.00	.91	.79	.79	.79
Fresh1/	.34	.36	.35	.32	.35	.34
Processed1/	.25	.28	.25	.20	.19	.18
Bacon & sausage	.38	.36	.32	.27	.26	.27
Veal	.05	.05	.03	.03	.03	.02
Lamb, mutton, goat	.06	.05	.02	.03	.02	.02

	Dollars per capita					
Total meat	6.51	6.22	5.50	4.80	4.54	4.23
Red meats	4.05	4.14	3.67	3.20	2.97	2.76
Beef	2.44	2.57	2.28	1.98	1.82	1.67
Loin and rib						
Steaks	.61	.67	.55	.46	.38	.29
Roasts	.04	.05	.04	.04	.03	.02
Round and chuck						
Steaks	.38	.39	.39	.34	.33	.30
Roasts	.53	.64	.51	.43	.40	.38
Gound	.69	.65	.64	.58	.56	.57
Other	.19	.18	.15	.14	.12	.11
Pork	1.41	1.40	1.29	1.11	1.06	1.02
Fresh1/	.49	.49	.48	.44	.45	.43
Processed1/	.41	.43	.39	.33	.29	.26
Bacon and sausage	.50	.48	.42	.35	.32	.33
Veal	.09	.09	.06	.06	.05	.04
Lamb, mutton, goat	.12	.09	.04	.05	.03	.03

1/Excluding bacon and sausage.

TABLE 6-6. Weekly At-Home Meat Consumption by Residence Location, 1977-1978 (Source: Haidacher et al. 1982, p. 37)

Item	Central city	Suburban	Nonmetropolitan
		Pounds per capita	
Total meat	4.76	4.14	4.15
Red meats	2.71	2.56	2.59
Beef	1.72	1.68	1.68
Loin and rib			
Steaks	.28	.29	.25
Roasts	.03	.03	.02
Round & chuck			
Steaks	.29	.25	.25
Roasts	.38	.39	.40
Ground	.59	.61	.65
Other	.14	.11	.11
Pork	.91	.82	.88
Fresh1/	.39	.33	.32
Processed1/	.21	.23	.22
Bacon & sausage	.30	.26	.35
Veal	.05	.04	.02
Lamb, mutton, goat	.04	.03	.02

1/Excludes bacon and sausage.

TABLE 6-7. Weekly Red Meat Consumption by Race, Quantity, and Expenditures, 1977-1978 (Source: Haidacher et al. 1982, p. 33)

Item	Quantities		Money value	
	Black	Nonblack	Black	Nonblack
	Pounds per capita		Dollars per capita	
Total meat	5.61	4.14	5.88	5.05
Red meats	2.89	2.58	3.53	3.38
Beef	1.67	1.67	1.95	2.12
Loin and rib				
Steaks	.23	.28	.37	.50
Roasts	.02	.03	.02	.04
Round & chuck				
Steaks	.28	.26	.36	.35
Roasts	.39	.39	.46	.48
Gound	.56	.63	.54	.62
Other	.19	.11	.21	.14
Pork	1.15	.82	1.47	1.15
Fresh1/	.50	.32	.64	.43
Processed1/	.22	.22	.29	.36
Bacon & sausage	.43	.28	.54	.37
Veal	.04	.03	.06	.06
Lamb, mutton, goat	.03	.03	.05	.05

1Excludes bacon and sausage.

For beef, an estimated 45 percent is consumed away from home. Much of this is ground beef, which comprises about 50 to 60 percent of total beef consumption (quoted in Petritz, Erickson and Armstrong, pp. 344-345). But even these figures understate the growing importance of away-from-home food consumption, for some 60 percent of supermarkets in 1989 had in-store delicatessens (up by 50 percent in a decade). According to a 1985 survey, in the deli, meat was the leading item—41 percent of sales—with boiled ham and roast beef the first and third selling products. Delicatessen shoppers tend to be younger and higher income than supermarket shoppers in general (McLaughlin, German, and Uetz 1986, Tables 2.7, 2.8, 3.1, and 3.2).

The trend to food away from home is alternatively explained as the result of rising incomes, expansion of fast food outlets, changes in household size, and differences in work outside the home, especially for wives (Kinsey 1983, p. 10). Expenditures on away-from-home meals are found to have positive income elasticities, but typically less than one. Using trend analysis, expenditures on non-home-prepared meals has been rising at about half the pace of income increase. Since income has been rising faster than population, which drives growth in food store sales, restaurants and their ilk have been taking an increasingly large share of the food dollar (USDA *Food Marketing Review* 1989, p. 2). Indeed, higher incomes beyond an "entry level" seem to contribute more to the amount spent per meal than to the frequency of dining out (Kinsey 1983, p. 11, 18). A similar inverse relationship exists with education (McCracken and Brandt 1987, Table 2). Surprisingly, full-time working wives have not been found to increase the propensity to eat away from home, possibly because of the time involved (Kinsey 1983, pp. 17-18). The most frequent consumers away from home are teenagers (15-20) and men, although age has a general negative impact on away-from-home expenditures (McCracken and Brandt 1987, Table 2).

Where the increase has been most pronounced is with the lowest income groups (Table 6-8). Breakfasts taken out of the home, for example, increased by over three times for the lowest quartile ($3,379 average household income in 1977-1978). Much of this growth is probably attributable to expanded school meat offerings and their subsidized meal programs. Budget cuts in those areas during the Reagan presidency (and subsequently) have probably reduced the numbers from the late 1970s. An aging population, several recessions, and higher transportation costs have also taken the bloom off the fast food empire. Sales in those and other HRI outlets will undoubtedly continue to expand, but not at the 10 percent annual growth rate seen in the preceding ten years.

Statistical Analysis

The preceding descriptive analysis strongly suggests what major factors affect red meat consumption. Documenting those judgments and, moreover, quantifying the degree of the effect, nonetheless requires statistical analysis. Much of that analysis utilizes a technique called regression analysis. (For an introduction, see Tomek and Robinson 1990, Chapter 15.) A typical regression analysis will be of the form:

$$Q_p = \alpha_0 + \alpha_1 P_p + \alpha_2 P_s + \alpha_3 Y + \alpha_4 D$$

where

Q_p = quantity sold of a product

P_p = its price

P_s = price of competing or substitute products (there may be more than one)

Y = disposable (or after tax) income

D = demographics (if available) such as race, age, etc.

The α's are the fitted coefficients that can be manipulated to give the estimated elasticities of own-price, cross-price, and income (see Chapter 2).

Reasoning and previous statistical results are expected to lead to the following direction of the relationships:

- Own-price: negative, as an increase in price usually reduces consumption.
- Substitute prices: positive, as more expensive substitutes usually cause a shift in demand to the product under study.
- Income: positive, since we all want more of most things and a higher income will allow us to acquire them.
- Demographics: no fixed expectation; results depend on the particulars.

Models will use either time-series data (e.g., quarterly or annual) or cross-sectional (e.g., individual buyer data across a large number of shoppers). Economists generally consider the results of time-series analysis to be short-term values because the analysis is seen as occurring while ad-

TABLE 6-8. Percent of Meals Eaten Away from Home, by Type of Meal and Selected Household Characteristics, Spring 1965 and 1977[1] (Source: Haidacher et al. 1982, p. 55)

Household Characteristic	Meal Type									
	Breakfast		Lunch		Supper		All Meals			
	1965	1977	1965	1977	1965	1977	1965	1977		
				Percent						
All households	3.7	7.2	18.9	26.7	6.9	12.0	9.8	15.4		
Income Quintile:										
1 (lowest)	2.1	7.3	15.5	26.4	3.2	7.3	7.0	13.7		
2	3.2	6.4	17.1	23.2	5.7	11.0	8.7	13.7		
3	4.1	6.2	18.5	24.3	7.1	12.9	9.9	14.7		
4	5.0	7.8	20.6	30.5	8.5	14.3	11.4	17.8		
5 (highest)	5.2	8.6	26.0	31.0	12.0	18.3	14.4	19.4		
Household size:										
1 member	3.4	5.8	17.1	19.8	11.6	14.8	10.7	13.5		
2 members	3.6	6.0	15.9	19.4	8.4	13.1	9.3	13.0		
3 members	4.6	9.4	20.1	29.5	8.0	15.2	10.9	18.3		
4 members	5.0	8.2	22.2	28.4	8.0	12.2	11.7	16.4		
5 members	3.6	5.5	20.3	29.1	6.7	9.4	10.2	14.8		
6 or more	2.6	6.8	17.0	29.7	4.1	8.5	7.9	15.1		

Race:								
Black	3.0	9.0	17.3	29.5	2.6	7.9	7.6	15.4
Nonblack	3.8	6.9	19.2	26.3	7.5	12.5	10.2	15.4
Region:								
Northeast	4.3	8.5	19.1	28.0	6.5	12.7	10.0	16.4
Northcentral	3.5	6.3	17.6	26.8	7.1	13.6	9.4	15.8
South	3.5	7.1	20.5	26.7	6.1	9.9	10.0	14.7
West	4.0	6.6	17.6	24.6	8.9	11.8	10.2	14.5
Urbanization 2/								
Urban	4.3	7.7	20.1	27.3	7.8	12.8	10.7	16.1
Rural nonfarm	3.6	6.2	18.2	24.7	7.1	10.2	9.6	13.8
Rural farm	3.1	4.6	18.0	29.7	5.3	8.6	8.8	14.6

1/ Meals prepared from home food supplies that are eaten away from home are considered at-home meals.
2/ Since only the 1965 urbanizational categories were available in both surveys, these categories are used in the comparisons.

justments are being made to price and income changes. Alternatively, cross-sectional analysis is viewed as giving long-term results. Long-term elasticities are expected to be greater than short-term ones because all the adjustments have been made.

Meat is one of the most studied commodities, and beef receives particular attention among economists. That leads to a rich array of statistical estimates over time. Regrettably, that does not always clarify the issue since many results, as will be seen, give varying (if not out and out contradictory) estimates. Making sense of these results involves evaluating statistical and non-statistical factors.

Income Elasticities at Retail

The descriptive analysis of meat consumption patterns suggested the importance of income as a determinant of retail demand. Recent statistical analysis, however, does not support that expectation for *all* meat. The group that includes poultry and fish, along with the red meats, has an estimated retail elasticity of 0. That means increases (and decreases) in consumers' after-tax incomes do not affect total quantity meat sales at all (Haidacher et al. 1982, p. 41).

Such an aggregate figure, however, camouflages many important differences among products. Estimates of retail beef income elasticities, for example, range from .47 to .07 (Table 6-9). In all cases, that means demand is income inelastic; a one-percent increase (decrease) in income leads to a less-than-proportional increase (decrease) in beef consumption. This is a double-edged sword for the industry because it means rising incomes will not increase demand to a proportional degree. Conversely, a decline in disposable income, such as was experienced during the several recessions of the 1970s and 1980s, will not decrease consumption dramatically. Thus, there is inherent stability in the market, but no basis for continued growth as seen during the period 1950-1980.

The range of income elasticities shown in Table 6-9 is partially a result of different statistical procedures. The general trend toward smaller income elasticities over time is nonetheless revealing as it suggests a population reaching the saturation point of red meat consumption. Meat consumption is typically viewed worldwide as a luxury good whose demand increases with rising incomes. However, the U.S. appears to have reached a point where that rule of thumb no longer applies. Indeed, income elasticities are estimated to be higher for most other parts of the world (Table 6-10). As a result, *volume-demand* growth (excluding that for population increase) for beef is likely to come from outside the U.S.

For pork in the U.S., the situation is more mixed, with total consump-

TABLE 6-9. Income Elasticities of Retail Demand for Red Meats, Beef, Pork, Veal, and Lamb, U.S., Various Periods, 1946-1987 (Sources: (a) Brandow 1961; (b) George and King 1971, Tables 5 and 11; (c) Haidacher et al. 1982; (d) Dahlgran 1988, Table 3; (e) Bewley and Young 1987, Table 4; (f) Eales and Unnevehr 1988, Table 4; (g) Moschini and Meilke 1989, Table 4)

Item	Income Elasticity						
	1955-57a/	1946-67b/	1977-78c/	1950-85d/	1969-83e/2/	1965-85f/	1967-87g/2/
Red Meats	NA	NA	.04	NA	NA	NA	NA
Beef	.47	.29	.07	.12	1.73	.344	1.34
Pork	.32	.133	-.06	.03	1.14	.28	.85
Veal	.58	.59	.41	NA	NA	NA	NA
Lamb and Mutton1/	.65	.571	.62	NA	1.30	NA	NA

NA - not available
1/Includes goat meat in 1977-78 estimate.
2/Elasticities of food expenditures only

133

TABLE 6-10. Income Elasticities of Retail/Red Meat Demand, Selected Countries and Years (Sources: Canada: Tryfos and Tryphanopoulos 1973, Table 1; Saudi Arabia: Coyle et al. 1983; Remaining Countries: Regier 1978)

	Canada 1954-70	EC-9	Japan 1963-70	Argentina	Saudi Arabia 1980[1]
Beef	4.09	.62	.89-.97	.35	2.0
Pork	-.01	.52	1.24-1.79	NA	NA
Veal	.21	NA	NA	NA	NA
Lamb	-.26	NA	NA	NA	1.9

[1] Elasticity of import demand
N A - Not Available

134

tion actually *declining* with income increases up until about 1980, and increasing since. Under definitions used by economists, this means pork was an inferior good. However, more current associations of pork as a "white" meat may have contributed to its increase in popularity (see Chapter 19).

By comparison, both veal and lamb show much larger domestic income elasticities, although more recent estimates are missing. In part, this is due to the price situation: these products are typically more costly than, say, pork. Hence, assuming the taste for those products is present, higher incomes will generate greater sales. As a result, movement of these products can be expected to advance in the future provided, of course, consumers show continued increase in spending power.

In Table 6-9, the two columns flagged with a "2" are results for expenditure rather than elasticity estimates. Expenditure elasticities refer to purchases out of the portion of the budget set aside for food. Economists are coming to believe that purchase decisions are multi-fold: a decision on how much to spend for food and then what to buy with that allowance, or a choice to buy pork followed by a decision on how much to spend on that pork (see, e.g., Popkin, Guilkey, and Haines 1989). The evidence suggests that shoppers are willing to spend more of an increased food budget on beef than on pork.

So measured, the "income" elasticity is — in effect — a quantity elasticity, measuring how quantities purchased vary with income or expenditures. In such measures, a recognition is missing that the shopper may move, in a manner of speaking, higher on the hog as higher incomes permit greater selectivity at the meat counter. The shopper may choose the pork loin over the belly or the beef rib steak over the chuck. These quality-related changes can be computed as *quality elasticities*. Empirical estimates confirm expectations: choicer cuts are associated with higher incomes (Table 6-11). As anticipated, beef loins and ribs show the greatest quality response with ground beef about half that. Figures for pork are difficult to interpret because the product listing is so broad as to combine both lower and higher value items within each category.

Another way of reflecting the same phenomena is to examine demand for particular products (e.g., hamburger) as opposed to types of meat (beef), so-called desegregated demand. Eales and Unnevehr (1988) found an expenditure elasticity for hamburger of − 1.57 (an inferior good) compared to + .34 for all beef. The more select "table cuts" have an expenditure elasticity of 1.56, again compared to the all-beef level of .34 (Tables 4 and 5).

The quality elasticity figures indicate that the product mix will change

TABLE 6-11. Income and Quality Elasticities of Retail Demand for Red Meats, Beef, Pork, Veal, and Lamb, U.S. NFCS Estimates for 1977-1978 (Source: Haidacher et al. 1982, Table 22, p. 41)

Item	Quantity (Income) Elasticity	Quality Elasticity[1]
Red Meats	.04	.10
Beef	.07	.11
Loin and Rib		
Steaks	.55	.10
Roasts	.43	.10
Chuck and round		
Steaks	-.07	.01
Roasts	.12	.08
Ground	-.09	.05
Other beef	-.13	.13
Pork	-.06	.08
Fresh[2]	-.16	.04
Processed[2]	.14	.14
Bacon and sausage	-.11	.05
Veal	.41	.13
Lamb, mutton, goat	.62	.17

[1]Difference between the expenditure and income elasticities.
[2]Excludes bacon and sausage.

as incomes increase (or decrease). Packers and retailers have some flexibility in how the carcass is cut so that the different needs over time can, in part, be accommodated through the production process. Nevertheless, the primals are established by nature, not human tastes and preferences, so that other balancing mechanisms will be needed. These can be exports/imports of particular items or relative price adjustments which limit the demand for one cut while stimulating that for another. How different and more complex this system is than that used in many developing areas where a pound of beef is the same price whether it is a T-bone or neck trimmings! That unitary price system is, however, not without its justification; the two cuts from an older, grass-fed animal can indeed taste remarkably similar.

Price Elasticities at Retail

While income may provide a final cap on what can and cannot be afforded, shoppers show additional selectivity among species and cuts in the way they respond to price changes. This is measurable as the price (sometimes called the own-price) elasticity of demand. Recalling that the price of a product and sales are negatively related, a value greater than one (in absolute value) means that a price increase (decrease) will lead to a more than proportional decrease (increase) in purchases. For these cases, in which products are known as price elastic commodities, a price decrease will cause total sales revenue to increase. For example, suppose a product has a price elasticity demand of 1.2. A retail price reduction from $1.25 to $1.19 (almost a five-percent decline) increases sales by $.048 \times 1.2 = 5.8$ percent. Total revenue also increases as $$1.19 \times 105.8 - $1.25 \times 100 = $125.9 - $125 = +$.90$. Of course, it is not necessarily more profitable to reduce prices for price elastic items, but it will always lead to higher total sales revenue.

Among food products (as we shall see), the price elasticity of demand is typically less than one, and is quite often close to zero. That simply means that price has a limited effect on how much of the product is consumed. More formally, demand for a price inelastic item falls less than proportionally to the magnitude of a price increase. A price increase of ten percent for an item with a price elasticity demand of .5 causes the quantity purchased to fall by only five percent ($.5 \times 10\% = 5\%$). What of total sales revenue? It, too, will increase with a price increase as sales fall less than the price is increased. The situation is ready-made to the benefit of agricultural producers if quantity can be restricted (and thereby price increased). Yet, quantity restriction is one thing farmers have been unable to accomplish.[3]

Estimated domestic red meat retail price elasticities range from .41 to 2.35 (in absolute value) depending on the product and time period (Table 6-12). However, for the volume products — beef and pork — the values are typically below one, indicating demand is relatively insensitive to price. These estimates have generally remained constant within a wide range over the last score of years.

Showing far greater price sensitivity, at least by this estimate, are lamb and veal. A possible explanation is that these products are used as special treats when the price is low. Certainly that kind of use would apply to veal with its substantial supply variability during years when low prices cause diversion of meat-type cattle calves to veal slaughter. An elastic demand is beneficial in such situations since a lower price will cause demand to expand more than proportionally.

Due to the statistical procedures used to estimate them, the elasticities in Table 6-12 are known as short-run elasticities. Elasticities, as noted previously, are generally expected to be smaller in the short-run when tastes and limited familiarity with other products restrict the choices of a shopper. Long-run price elasticities for beef and pork, have been estimated at 1.0 and .75 respectively (Tomek and Cochrane 1962). Over the long-run, the amount of beef purchased is possibly somewhat more sensitive to the price level than is the case in the short term. In that instance, the long term is the period it takes for the shopper to adjust fully to a price change (nine to twelve months for the Tomek and Cochrane estimate). Retail pricing of meat should be done with a sensitivity toward the long-run effects. Pork shows less difference between the short- and long-run effects, probably because shoppers adapt themselves much more rapidly to pork price changes, taking only about three months. This may be a response to the instability of pork prices in recent years.

Like income elasticities, demand elasticities are frequently larger outside the United States. Available estimates (Table 6-13) only partially support this expectation. Japan, for example, shows far larger elasticities than the U.S. For an explanation, one can cite the sharply higher meat

---

3. This comment is made to point out the situation faced by farmers compared to individual firms that dominate production in an industry. For independent farmers to collude on supply reduction is a "restraint of trade" prohibited by the Sherman Act of 1890. The 1922 Capper-Volstead Act gives farmers organized into cooperatives some specific exemptions from the Sherman Act (and other anti-trust law). For a discussion of these limited exceptions, see Jesse et al. 1982.

TABLE 6-12. Retail Price Elasticities for Red Meats, Beef, Pork, Veal, and Lamb, Various Periods, 1947-1987 (Source: Refer to Table 6-9)

Item	Retail Price Elasticities						
	1955-57[a]	1946-67[b]	1953-77[c]	1950-85[d]	1969-83[e]	1965-85[f]	1967-87[g]
Red Meats	NA	NA	.68[1]	NA	NA	NA	NA
Beef	.95	.64	.67[2]	.13	1.78	.57	1.50
Pork	.75	.41	.73	.12	1.42	.76	.84
Veal	1.60	1.78	NA	NA	NA	NA	NA
Lamb & Mutton	2.35	2.23	.69[3]	NA	1.53	NA	NA

Note: The negative sign is omitted in all cases.
NA - Not available.
[1]Estimate for 1950-77.   [2]Includes veal.   [3]Includes goat meat.

139

TABLE 6-13. Retail Price Elasticities for Red Meats, Selected Countries and Years (Source: Harris 1988, p. 35)

Retail Price Elasticities

Country and Period	Beef	Pork	Veal	Lamb & Mutton
Germany 1960-69	.60	.55	NA	NA
Japan 1963-70	1.76-1.78	1.27-1.95	NA	NA
1965-79	1.27	1.08	NA	NA
Canada 1954-70	.52	1.05	1.40	1.80
Italy	.45	1.66	NA	NA
Spain	.59	.20	NA	NA
Australia	.79	NA	NA	1.40[1]
Argentina	.56	NA	NA	NA

NA - Not available
[1]Lamb only.

prices (especially beef) in Japan and the heavier reliance on seafood as a source of animal protein. The remaining countries all show beef price elasticities close to those in the U.S., suggesting that beef is a commodity, a frequently purchased product there also. Poorer countries and those with smaller domestic meat resources would no doubt show different price responses, but due to data limitations, estimates from those areas are scant.

The preceding are all *aggregate* elasticities for composites of beef, pork, and lamb cuts. Aggregation tends to smooth the extremes and give a single value that may not be reflective of the situation for individual items. There are disaggregate studies available, as shown in Table 6-14. The first and third of these studies are based on data collected at the store level, the first through a special tabulation, the second using scanning records. This data focus means these results must be interpreted slightly differently from others using the data of entire markets. Basically, one would expect that store-level demand is more elastic than that for a market-wide study. In addition to the other substitution options, shoppers at individual stores have the choice of selecting among stores. Data are also collected on a weekly basis, making these very short-run elasticities.

Store specials further complicate the picture because a favorable price can cause significant additional sales that may be put in the freezer and used in subsequent weeks, depressing later sales. But despite these difficulties, only store-level analysis gives a real indication of how shoppers respond to prices and other factors. The store-level data also make it possible to collect information on how pay weeks and other factors influence sales.

The first two studies indicate relatively large elasticities, as would be expected. The third is quite out of that pattern, but because it is the first large-scale study to use scanning data the results are difficult to interpret in that there are some replications. The Capps study (1989) does indicate only a limited "pay-period" effect, a distinct response to newspaper advertising and seasonality in sales, but it is not in agreement with the Marion and Walker results (1978). Considering the small geographic scope of these studies, it seems prudent not to generalize the results to a great degree.

Elasticities of Substitution at Retail

At the meat counter, it is common to see a shopper picking up a sirloin steak only to put it down and examine a porterhouse. Or he/she may be torn between buying beef short ribs or country-style spare ribs for a family barbecue. Each of us has a preference among the several meats and the various available cuts, a preference that is tempered by the relative prices

TABLE 6-14. Demand Elasticities for Individual Meat Cuts, Various Periods, 1977-1989 (Source: [1]Marion and Walker, Table 2. Elasticity range reflects different weeks and stores, year approximate; [2]Eales and Unnevehr, Table 5; [3]Capps, Table 2)

Cut	1977[1]	Retail Price Elasticities	
		1965-85[2]	1986-87[3]
Beef round	4.19-7.66		
Beef loin	2.98-7.39		
Pork loin	.56-2.68		.83
Hamburger		2.59	.15
Beef "table cuts"/Steak		.68	.72
Pork/chops		.56	.70
Ham			.36

of the competing products. Are the beef ribs worth the extra 25, 40, or 75 cents a pound over the pork? Or should we cut our expenditures significantly by taking home chicken rather than either of the red meats?

Economists refer to the aggregate of these numerous individual choices as elasticities of substitution. Elasticities of substitution are positive as a higher price for, say, pork will shift demand to beef so that more beef is consumed.[4] Substitution elasticities range from zero to one with a zero value indicating no substitutability between two items (e.g., an orthodox Muslim would show no substitutability from beef to pork). A number close to one means the shopper shifts readily between two products based on price differences. A negative sign signifies that a rise in the price of one item will reduce the consumption of the other.

Conceptually, most products are substitutes for each other. A price check at a Cadillac showroom on the way to the supermarket may encourage us to buy that crown roast of lamb if, indeed, the price increases have put the deVille out of reach. But, as a practical matter, many items are so dissimilar as to make an evaluation between them questionable and statistically problematic. From another perspective, some economists have described the consumer budget allocation process as a two-step procedure of increasing particularization. For the first round, gross allocations could be made for transportation, food, shelter, and other major groups. Substitution, the second round of the process, is then limited within the major groups: Cadillacs for Chryslers; lamb for veal.[5] Both approaches lead to the same result; substitution elasticities are commonly estimated only for relatively similar products like the meat group.[6]

A .10 substitution elasticity of pork for beef means that a one-percent increase in the price of pork will raise the demand for beef by .10 percent. From Table 6-15, it is evident that there is a distinct interaction among the prices for all the major red meat products. Nominally, the smaller volume products (lamb and veal) show the strongest responses to changes in the prices of substitutes, but the statistical precision of these estimates is the weakest. Significantly, for even the major meat items, the substitution is

---

4. If a product has a negative sign then it is called a complement. As an example, a higher price for applesauce would depress the demand for pork among those who like to eat the two together.

5. For a theoretical treatment of the consumer allocation process, see Lancaster 1966, application to food in Haines, Guilkey, and Popkin 1988.

6. Sometimes non-meat protein like cheese is included in the analysis. Cheese used in macaroni and cheese or similar dishes would fit into the same food group as meat, at least in the minds of many consumers.

TABLE 6-15. Domestic Elasticities of Substitution for Meat at Retail, Various Periods, 1947-1987 (Source: Refer to Table 6-9)

Item	Period	Pork	Veal/	Lamb & Mutton	Chicken
Beef	1955-57a/	.10	.06	.04	.07
	1946-67b/	.08	.03	.04	.07
	1953-77c/	.12	-.1	NA	.04
	1950-85d/	.03	NA	NA	.01
	1969-83e/	.17	NA	.27	.08
	1965-85f/	.17	NA	NA	.05
	1967-87g/	-.08	NA	NA	-.13

Item	Period	Beef	Veal	Mutton & Lamb	Chicken
Pork	1955-57a/	.13	.04	.03	.07
	1946-67b/	.08	.01	.06	.03
	1953-77c/	.16	-.1	NA	.10
	1969-83e/	.53	NA	.30	-.26
	1965-85f/	.31	NA	NA	.01
	1967-87g/	.13	NA	NA	-.09

Veal		Beef	Pork	Mutton & Lamb	Chicken
	1955-57a/	.38	.19	.07	.14
	1946-67b/	.36	.20	.07	.17

Lamb & Mutton		Beef	Veal	Pork	Chicken
	1955-57a/	.62	.41	.17	.22
	1946-67b/	.59	.89	.07	.23
	1953-77c/	NA	NA	NA	NA
	1969-83d/	.64	NA	.23	-.23

Chicken		Beef	Veal	Pork	Mutton & Lamb
	1955-57a/	.23	.16	.06	.04
	1946-67b/	.20	.12	.06	.04
	1953-77c/	.16	.28	-1	NA
	1969-83e/	.26	NA	-.21	-.23
	1965-85f/	.052	NA	.02	NA
	1967-87g/	-.02	NA	-.07	NA

NA - Not available.   1Beef and veal combined estimate.

substantial enough that no one product can be priced independently of the others. Stated differently, a price change for one product will necessitate a round of price changes if the market is to be cleared for all items.

It is important to note that effects are not necessarily symmetrical. A one-percent price increase for pork, by one estimate, raises the demand for chicken by nearly three times the amount (in percentage terms) that a similar price increase for chicken increases demand for pork. More recently, chicken has entered with a negative sign, indicating its demand is becoming quite distinct from that of the red meats. Thus, red meat producers can see a sort of "ratcheting away" of markets as successive price increases and decreases lead to ever greater poultry consumption.

With some individual country differences, substitution elasticities for other countries are roughly twice those of the U.S. (Regier 1978, pp. 39-40). That indicates an even stronger price interrelationship than seen here and probably relates to the smaller per capita consumption throughout much of the rest of the world. However, a smaller base means a larger percentage change, even if the absolute shift is constant.

Changes in Taste

Statistical analyses of price responses assume tastes remain constant; that is, the only causal factor is change in price. Although assuming that taste is constant is reasonable — tastes, after all, are generally quite stable — it is not always valid. Even a slight change of tastes can have a substantial effect on demand, prices, and estimated elasticities. Nevertheless, taste variations are difficult to identify, largely because they are so gradual.

After examining the matter at length, Tomek is convinced that some form of structural change, possibly tastes, occurred for beef in 1958-59 (1977). Goodwin, Andorn, and Martin (1968) argue that price itself can cause preferences to shift. Following their logic, low beef prices during the liquidation phase of the cycle created a taste for beef. When prices rose again during rebuilding of the herd, consumers were willing to spend proportionally more for beef than previously.

Whether the Goodwin, Andorn, and Martin (1968) supposition is correct or not, the measure of changed preferences used in that study is the proper one. As Ikerd (1982) notes, the test is not whether the consumer is willing to pay higher prices — that may only describe a movement up the demand curve. A shift in product preferences is observable only if the consumer is willing to pay higher prices *and* buy more of that item. Based on this criteria, Ikerd detects major changes in market share among beef, pork, and chicken over the past two decades. However, these adjustments could be cyclical and other supply changes, and not underlying changes in preferences. Stated differently, the decline in beef consumption from the mid-1970s into the early 1980s is a result of beef supply limitations. The shortfall was filled by increased pork and chicken consumption made possible by substantially lower prices for those products. Comparing pork and chicken, he found:

> . . . about 60 percent as much chicken as pork has been sold at chicken prices about 60 percent as high as pork prices. But to move 80 percent as much chicken as pork, it has required a drop in broiler prices to about 45 percent as high as pork prices. (1982, p. 17)

Thus, there *was* little statistical evidence of substantial changes in tastes among the major meat products up until the early 1980s.

Recent advances in analytical techniques, combined with a larger observation period, have made it easier to identify structural shifts in demand. That term can be interpreted to mean *movement of* the demand curve, generally to the right or left and/or pivoting so the curve is more or less steep. Eales and Unnevehr conclude their analysis shows a "preference shift away from beef and toward chicken after 1974" (1988, p. 530). Moschini and Meilke place the change just a few years later but find much the same effect (1989, p. 256-59). "Structural change is significantly biased against beef, in favor of chicken and fish, and it is neutral for pork." "Biased against beef" explains part of the 3.5-percent decline in beef consumption since the mid-1970s. According to Moschini and Meilke's estimates, the demand for beef became more elastic in the current period than the former (1989, Table 4). (Figures from the post-change period are presented in the preceding tables.) However, most studies indicate that beef demand became less elastic, and chicken, a stronger substitute (Eales and Unnevehr 1988; Chavis 1983; Hudson and Vertin 1985).

The *cause* of these changes is more difficult to ascertain. Moschini and Meilke find these results consistent with dietary concerns (1989, p. 260). That is also in line with the change in eating habits of women 1977-85 who "shifted from higher to medium and lower fat beef, pork and poultry products" (Popkin, Guilkey, and Haines 1989, p. 958).

SYNOPSIS

Considered on the most aggregate level, red meat consumption is quite stable within countries across time. Production increases are largely offset by population growth to give constant or slowly increasing per capita figures. Aggregation, however, tends to obliterate the important differences among species and by region and income level. More detailed analyses, based on a 1977-78 at-home consumption USDA survey, indicates the following notable differences:

- Income — rising incomes have little effect on the quantity consumed, but do have an impact on the quality (cut);
- Age — the largest volume consumers are the 40-64 age group, except for ground beef;
- Race — blacks tend to eat slightly more red meat and spend more per capita than non-blacks.

Away-from-home consumption is significant for red meats (comprising about 45 percent of beef sales), the bulk of which is ground. Income is related more to what is spent on food outside the home than the amount consumed.

Statistical (regression) analysis substantiates the descriptive approach by indicating, in general, that red meats — as a group and individually — are income and price inelastic in the U.S. Elasticities in other countries tend to be greater, although estimates are limited.

Statistical techniques have helped settle the controversy over changes in meat demand. Recent tests generally concur that preferences began moving from beef to chicken in the mid- to latter 1970s. In economic terms, that means the demand curve for beef has been moving to the left as consumers are willing to buy somewhat less beef at the same price than in the past.

Study Questions

1. How do the income, price, and cross-product elasticities affect red meat consumption and prices in the U.S. and other countries?
2. Distinguish between a quality and quantity income elasticity. What are the practical effects of these two measures?
3. There is much talk about a "structural change" in meat demand. **What is it, what is the evidence, and what are the ramifications?**
4. Why is demand typically more elastic in the long than the short run, and how might that relationship be significant to the red meat sector?

REFERENCES

American Meat Institute. *Meatfacts*. Various editions.

Berg, A. *The Nutrition Factor*. Washington, DC: The Brookings Institution, 1973.

Bewley, R. and T. Young. "Applying Thelil Multinomial Extension of the Linear Logit Model to Meat Expenditure Data," *Am. J. Agr. Econ.*, 69(1987):151-57.

Brandow, G. E. "Interrelations Among Demands for Farm Products and Implications for Control of Market Supply." Penn. Agr. Exp. Sta. Bull. 680, 1961.

Capps, O. "Utilizing Scanner Data to Estimate Retail Demand Functions for Meat Products," *Am. J. Agr. Econ.* 71(1989):750-60.

Chavis, J. P. "Structural Change in the Demand for Meat," *Am. J. Agr. Econ.* 65(1983):148-53.

Coyle, J. R., M. E. Burfisher, J. B. Parker, H. H. Steiner, A. Abou-Bakr, A. J. Dommen, and M. E. Kurtzig. *Food Import Demand of Eight OPEC Countries*. USDA, Economic Research Service, Foreign Ag. Econ. Rpt. No. 182, June 1983.

Dahlgran, R. A. "Changing Meat Demand Structure in the United States: Evidence from a Price Flexibility Analysis," *Northcentral J. Agr. Econ.* 19(1988):165-76.

Eales, J. S. and L. J. Unnevehr, "Demand for Beef and Chicken Products: Separability and Structural Change," *Am. J. Agr. Econ.* 70(1988):521-32.

George, P. S. and G. A. King. "Consumer Demand for Food Commodities in the U.S. with Projections for 1980." Giannini Foundation Monograph 26, Univ. Calif., Davis., Div. Ag. Sciences, March 1971.

Goodwin, J. W., R. Andorn, and J. E. Martin. "The Irreversible Demand Function for Beef." Oklahoma Agricultural Exp. Sta. Tech. Bull., 1968.

Haidacher, R. C., J. S. Craven, K. S. Huang, D. M. Smallwoodland, and J. R. Blaylock. "Consumer Demand for Red Meats, Poultry, and Fish." USDA, ERS, NED, AGES 820818, September 1982.

Haines, P. S., D. K. Guilkey, and B. M. Popkin. "Modelling Food Consumption Decisions as a Two-Step Process," *Am J. Agr. Econ.* 10(1988):543-52.

Hudson, M. A. and J. P. Vertin. "Income Elasticities for Beef, Pork and Poultry: Changes and Implications," *J. Food Distrib. Res.*, 1(1985):25-31.

Ikerd, J. E. "The Battle Among Beef, Pork and Poultry for the Consumer's Meat Dollar." Oklahoma State University, Dept. of Agr. Econ., A. E. Paper 8233, March 1982, p. 17.

Jesse, E. V., A. C. Johnson, Jr., B. W. Marion, and A. C. Manchester. "Interpreting and Enforcing Section 2 of the Capper-Volstead Act," *Am. J. Agr. Econ.* 64(1982):431-443.

Kinsey, J. "Working Wives and the Marginal Propensity to Consume Food Away from Home," *Am. J. Agr. Econ.* 65(1983):10-19.

Lancaster, K. "A New Approach to Consumer Theory," *J. Pol. Econ.* 74(April 1966):132-57.

Marion, B. W. and F. F. Walker. "Short-Run Predictive Models for Retail Meat Sales." *Am. J. Agr. Econ.* 60(1978):667-77.

McCracken, V. A. and J. A. Brandt. "Household Consumption of Food-Away-From-Home: Total Expenditure by Type of Food Facility," *Am. J. Agr. Econ.* 69(1987):274-84.

McLaughlin, E. W., G. A. German, and M. P. Uetz. "The Economics of the Supermarket Delicatessen." Cornell U., Dept. Agr. Econ., A. E. Res. 86-23, September 1986.

Moschini, G. and K. D. Meilke. "Modeling the Pattern of Structural Change in U.S. Meat Demand." *Am. J. Agr. Econ.* 71(1989):253-61.

Petritz, D. C., J. P. Erickson, and J. H. Armstrong. *The Cattle and Beef Industry in the United States: Buying, Selling, Pricing.* Purdue University Cooperative Ext., CES Paper 93, undated.

Popkin, B. M., D. K. Guilkey, and P. S. Haines. "Food Consumption Changes of Adult Women Between 1977 and 1985." *Am. J. Agr. Econ.* 71(1989):949-59.

Regier, D. W. "Livestock and Derived Feed Demand in the World GOL Model." USDA, ESCS, Foreign Ag. Econ. Rpt. No. 152, September 1978.

Rizek, R. "The 1977-78 Nationwide Food Consumption Survey," *Family Econ. Rev.* USDA, Sci. Ed. Adm., Fall 1978, pp. 3-7.

Tomek, W. G. "Empirical Analyses of the Demand for Food: A Review." Cornell University, Dept. of Agricultural Economics, Staff Paper, April 1977.

Tomek, W. G. and W. W. Cochrane. "Long-run Demand: A Concept and Elasticity Estimates for Meats," *J. Farm Econ.* 44(1962):717-730.

Tomek, W. G. and K. L. Robinson. *Agricultural Product Prices.* Ithaca, NY: Cornell University Press, Third Edition, 1990.

Tryfos, P. and N. Tryphonopoulos. "Consumer Demand for Meat in Canada." *Am. J. Agr. Econ.* 55(1973):647-52.

United Nations. *Statistical Yearbook 1985/1986*, New York, 1988.

U. S. Bureau Census, "Statistical Abstract of the U.S." Washington, DC, 1962, 1982, 1983.

USDA, ESCS. "The Future Role of Cooperatives in the Red Meats Industry." Mkt. Res. Rpt. 1089, April 1978.

USDA. *1989 Agricultural Chartbook.* Ag. Handbook No. 684, March 1989.

USDA, Economic Research Service. "Food Marketing Review 1988." Ag. Econ. Rpt. 614, August 1989.

Chapter 7

Prices

In a world of instant communication, with farm prices and Wall Street quotes beamed virtually instantaneously on teletypes and over the radio, it is tempting to think in terms of one world price for cattle, hogs, and sheep. That may be true for such homogeneous and storable products as unrefined sugar, but for livestock the reality is quite different. Live animals are sold in local markets due to the obvious difficulty of long-distance transport as well as national restrictions on imports/exports and the limited flow of information across markets. Further complicating the price compilation process is the variability of animal types and grades across and within markets. The net effect is a series of prices representing a multitude of individual transactions. Compiling those prices into a meaningful aggregate figure is a significant statistical problem made especially uncertain by the many prices that are unreported. Such a task is clearly impractical on a worldwide level. And even at the national level of aggregation, the problems are daunting. Thus, reported national prices are usually "representative" prices rather than the (possibly) more correct average or median price.

"Representative" is a vague term that is used differently in each nation. Typically, one or several markets are taken to represent the national cattle, hog, or sheep market. The United States, for example, uses one of the major terminal markets (often Omaha) to describe the national market. How well prices from one market describe overall livestock prices — especially in countries like the U.S. where only a small percentage of animals move through those facilities — is discussed later on in this chapter. At this point, we shall use the available data, recognizing that while prices are generally consistent intra-country, inter-country prices are only roughly comparable.

This chapter discusses several aspects of livestock prices beginning with reported prices for the world and the U.S. Then, prices are broken down along several lines including seasonality, factors affecting prices, and pricing efficiency. The chapter ends with a discussion of one major aspect that determines the effective price for producers: shrink and its components.

WORLD PRICES

Wholesale cattle prices in major producing countries show several distinct patterns over the past 25 years (Table 7-1). All prices show the same general price patterns, with sharp increases in the mid-1970s as weather conditions and the Russian/U.S. grain deal tightened feed supplies and lead to higher costs and prices. Beyond these mega-trends, individual countries show substantial variations. Until recently, Argentine prices were universally lower than other countries, a reflection of limited fresh export markets (due to the prevalence of hoof-and-mouth disease), the lower finish of the steers, and internal agricultural policies that keep prices low. The EC countries represent the opposite extreme. Agricultural policy was involved by maintaining prices through linking them to the cost of production rather than to world prices.

The U.S. provides a good benchmark for open market value since the sector operates with few market restrictions. Nevertheless, it should be recognized that domestic agricultural policy partially restricts feed grain production, thereby raising feed prices and production costs compared to what they would be in a totally open market. Having a similar, if much smaller, effect are import quotas that maintain domestic prices at a higher level.

Hog prices, too, accelerated in the mid-1970s. The EC price support policy is evident in the latter period as prices remained notably higher and followed a different pattern than in the more open U.S. market (Table 7-1). The twenty-cent-per kilogram price decline during the late 1970s and early 1980s, and the subsequent farm depression placed a financial strain on the sector which is still evident. Many traditional rules-of-thumb that governed producers' decisions proved inaccurate, leaving them uncertain and cautious. Heavy accrued losses also slowed the expansionist urge.

Sheep prices are difficult to compare with the existing data series. Seemingly, much of the inter-country price differentials are due to quality differences. The relative stability of New Zealand prices compared to those received by producers in Australia is also evident. The prime lamb market is probably more predictable, particularly as a result of New Zealand's efforts to develop overseas markets.

More generally, the inter-country comparisons highlight three points about livestock prices. First (at least for the market economies), it is an

TABLE 7-1. Wholesale Prices for Cattle, Sheep and Hogs, Selected Countries and Years (Source: U.N., FAO, Production Yearbook, 1960, 1970, 1980, 1989)

- ¢/kg. -

	1955	1960	1965	1970	1974	1976	1978	1980	1982	1985	1988
CATTLE											
U.S. (Omaha)	51.1	57.8	57.5	66.8	92.2	86.2	115.3	147.5	142.4	129.0	152.8
United Kingdom	42.7	43.1	49.5	55.0							
Germany	49.1	55.5	74.5	75.0							
Argentina	N A	17.9	30.3	27.6							
Argentina (young bulls)					45.1	39.4	46.6	101.4	58.6	39.5	152.1
EEC					101.7	113.6	137.6	182.3	157.3	119.8	174.9
Australia					75.9	62.1	77.5	169.7	121.1	119.1	159.2
SHEEP											
Argentina ($/head)	N A	7.32	9.83	7.09							
U.S.	47.1	44.1	55.0	63.7							
Australia					50.7	22.4	52.9	79.5	37.6	29.6	52.5
New Zealand (prime)					63.9	61.5	76.3	100.1	112.0	85.7	N A
										280.0	403.1
HOGS											
Argentina	N A	19.6	34.9	34.9							
France	54.6	50.4	52.9	N A							
U.S. (Omaha)	33.4	35.4	47.5	48.0	81.3	98.3	108.2	89.9	122.1	99.9	95.7
EEC					113.7	134.0	139.3	184.3	158.1	120.6	143.0

NA - Not Available

153

interconnected marketplace, so that major changes affect virtually all country prices. Second, protectionist trade practices have a substantial impact on domestic prices. (See also Table 6-9.) And finally, the product characteristics vary from country to country. As a result, there are price differentials reflecting not only the different value of the items but also the slight variation in the product markets they compete for.

U.S. PRICES

Livestock prices have, and continue to show, considerable variability (Figure 7-1). Several factors are at work that will only be touched on here. Predominant is the farm sector, near depression in the mid-1980s, which caused red meat producers to hold down cattle numbers to lows not seen in 20 years. Cattle prices were further depressed by the dairy Whole Herd Buyout program of mid-decade. Despite good returns since 1984, lamb producers kept flock sizes down (abetted by dry weather in the North-

FIGURE 7-1. Prices Received by Farmers for Cattle, Hogs, and Lambs, 1980-88 (Source: USDA, *Agricultural Chartbook 1989*, Chart 196)

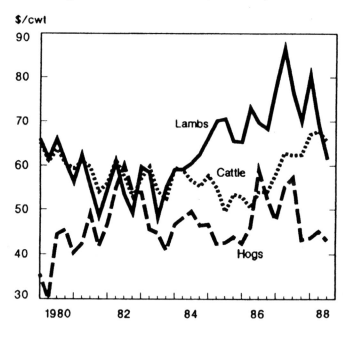

west), and prices responded strongly to sharply higher grain prices. The rebuilding phase contributed to higher prices, but as supplies reached toward record levels, the combined effects of abundant total meat supplies and reduced consumer buying power caused prices to drop sharply. By mid-1980, prices moved upward again, following a sharply lower production (declining in the neighborhood of ten percent for the first half of 1982). Profitability had been restored, but heavy losses made many feeders unable or—at best—wary of increasing placements (USDA, ERS, "Outlook and Situation," various issues).

The above discussion identifies how short-term price changes are particular to the conditions of the time. The outcome of these several effects in combination is difficult to decipher; it is apparent why accurate price projections are so elusive. Nonetheless, there are components of the pricing mix, including marketing costs and seasonality, that are more regular. Recognizing them can help in understanding overall price movements.

COMPONENTS OF PRICES

While actual prices may be difficult to predict, there are a number of components of these prices that are more regular and, thus, predictable. The principal factors to be considered are price spreads between several market levels and seasonality. Each will be explored in depth.

Price Spreads

Livestock producers and other farmers are regularly galled by the high and rising marketing bill. This farm-to-retail price spread increased by one-third from 1972 to 1987, or from a 38 percent to a 25 percent share of the retail food dollar going to the farmer (Figure 7-2). For livestock producers, the situation is modified a bit, with the farm share comprising 52 percent of the consumer's choice beef meat dollar in 1987 (44% for pork) (USDA *Agricultural Chartbook 1989*, Chart 125). The limited processing typical for meat products allows a greater portion of the retail value to pass through to the producer.

The major component of the marketing bill, and the fastest rising one over time, is labor. Other significant costs are packaging and transportation (Figure 7-3). During the 1970s, stagnant and declining productivity contributed to the wage-cost spiral. General food manufacturing posted gains, but food retailing showed a decline of over ten percentage points during the decade and has remained largely fixed through much of the 1980s (Figure 7-4). Much of the decline is attributable to expanded ser-

FIGURE 7-2. Farm-to-Retail Price Spread, 1976-1987 (Source: USDA, *Agricultural Chartbook 1989*, Chart 127)

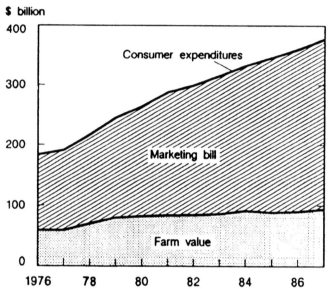

Data for domestically produced farm foods purchased by civilian consumers for consumption both at home and away from home.

vices in the form of extended store hours and additions of labor-intensive deli and bakery departments. In fact, productivity would have declined more but for the fact that wages fell since about 1984. Meat processing in total had a positive advance of over 40 percentage points from 1970 to 1985, largely the result of substantial technological advances and some adjustments in wage structures. (See Chapter 17.)

Producers often view an increase in the price spread as money out of their pocket. That notion assumes consumers' expenditures are fixed and the producer is the residual claimant of any amount not absorbed by retailers, wholesalers, processors, and anyone else serving as a middleman. While compelling, that view is not entirely correct and, indeed, depends on the source of the cost increase. If costs rise because of successful new product development, then sales expand and the producer can sell more product at a higher price. The quality, microwave-prepared dinners such as Le Menu are an example of recently developed meat dishes.

Cost increases due to external factors (such as utilities) or inefficiencies (such as declining productivity) have an entirely different effect. The im-

FIGURE 7-3. Components of the Farm-Retail Market Bill, 1987 (Source: USDA, *Agricultural Chartbook 1989*, Chart 128)

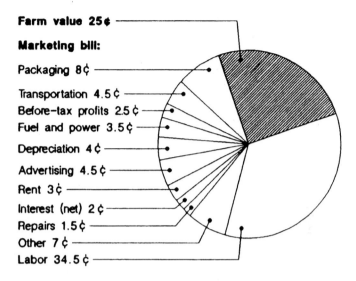

Farm value 25¢

Marketing bill:

Packaging 8¢

Transportation 4.5¢
Before-tax profits 2.5¢
Fuel and power 3.5¢

Depreciation 4¢

Advertising 4.5¢

Rent 3¢

Interest (net) 2¢
Repairs 1.5¢
Other 7¢
Labor 34.5¢

Other costs include property taxes and insurance, accounting
and professional services, promotion, bad debts, and miscellaneous items.

pact depends on two factors: demand conditions (e.g., elasticities) and the relative bargaining power between producers, and processors and retailers. The demand elasticities reported in Chapter 6 show that an increase in the price of the major red meat products will typically raise total revenues in the short run. Longer term higher prices will cause some erosion of market in favor of chicken (Table 6-15). How these increased revenues are shared among the players in the market chain depends, in part, on their relative bargaining position, with the more powerful operator receiving a proportionally greater share. Analysis of this issue is an involved matter that is treated in detail below and in Chapter 15. For now, suffice it to say there is no strong evidence that producers receive a disproportionately small share of any price increase, although there is no unanimity of opinion regarding this matter. Thus, there is no reason to believe that livestock producers are losers from rising farm-to-retail price spreads and may even benefit, at least over a moderate range. Viewed from a social perspective, however, a cost increase or decline in productivity causes social loss and inflationary pressure which, over enough products and a sustained period of time, is decidedly damaging to the farm sector.

FIGURE 7-4. Labor Productivity in Meat Processing and Retailing, 1967-1986 (Source: Dunham 1988, Tables 14 and 15)

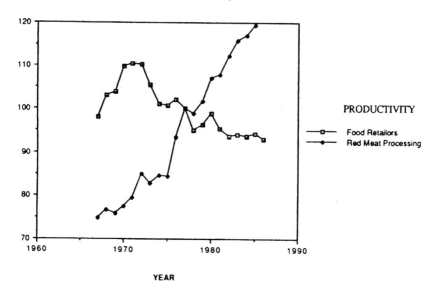

YEAR

Seasonality

Consumers have distinct product preferences for meat: steaks and chops in the summer for grilling, roasts and joints during the cooler winter months. Production also falls into cycles that follow crop production patterns. Lambs and feeder calves tend to be born in the spring, while range-fed stock are heavily culled in the late summer or fall when natural grasslands dry up or are snow covered. The interaction of these demand and supply shifts imparts a distinct seasonality to most livestock marketing and prices.

Cattle

The effect of seasonality on the production side can be seen in its purest form by examining slaughter numbers on a monthly basis. For cattle (Figure 7-5), there is a slight seasonal pattern, with the highest marketings during the fall (especially October) and a slight decline in early summer. The same pattern appears when considering the maximum monthly sales during the 1970-1980 period. The early summer lull and fall increase thus

FIGURE 7-5. Monthly Commercial Cattle Slaughter, 1989, 1985, and 1980 (Source: USDA, Ag. Stat. Board, various issues)

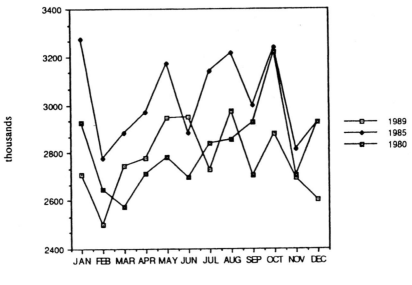

Column 1

appears rather robust. However, for any one year, the pattern can show considerable variability, thereby making a seasonable marketing strategy a chancy matter.

Monthly price patterns are clearly influenced by the supply trends evaluated above. Prices, however, reflect a number of additional factors. Some of these are demand-related, while others refer to production characteristics such as monthly variations in market weights. The latter matter is largely standardized by examining prices within a narrowly defined animal class (e.g., medium frame #1 steer calves), but small variations over the season can nevertheless occur which have a substantial impact on average prices. Prices, therefore, should not be interpreted as an indication of the varying value of the *same* animal over the year.

Farmgate prices reflect the seasonal pattern of combined slaughter with a pattern of high summer prices (Figure 7-6). This is probably enhanced by the popularity of cookouts during the summer months. This is especially true in August when price levels hold, despite the increased slaughter, compared to the previous several months. Winter weather problems

FIGURE 7-6. Index of Monthly Variations in Choice Steer Prices, Omaha, 1987, 1985, and 1980 (Source: USDA, ERS, "Livestock and Meat Statistics 1984-1988," Table 114)

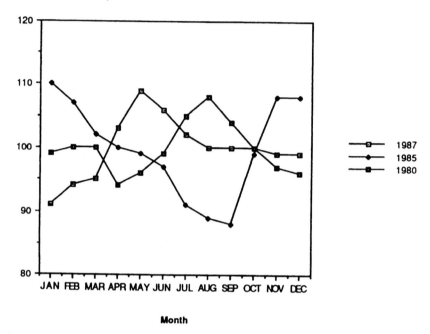

Month

can disrupt shipments causing temporary price increases that show up as maximums.

Feeder Calves

Feeder sales data are more difficult to capture than those for slaughter stock. Feed placements are available and show very high late fall purchases (Figure 7-7). Monthly price data, however, show a mid-year low, and otherwise considerable annual variability (Figure 7-8).

The grazing season used to be important to determining calf availability, but is no longer as significant now that there is a year-round feedlot industry. Winter weather, however, probably diminishes interest in placing calves during these months. Feeder calf prices are heavily influenced by shifts in the profitability of feeding and by seasonal shortages, while commercial realities dictate that a feeder fill the lot. Thus, while the cow-

FIGURE 7-7. Feedlot Placements, Monthly Percentages, Seven-State Total, 1984 and 1988 (Source: USDA, Ag. Stat. Service, Cattle, Jan. 1990, pp. 56-62)

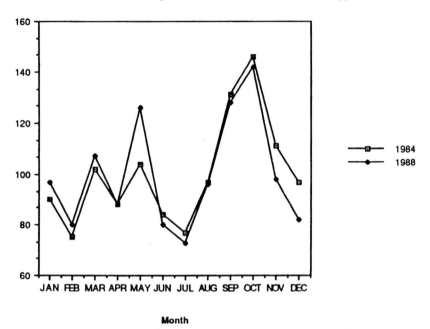

calf operator can maximize average prices by following traditional seasonal price patterns, the real rewards are from astute projections of demand.

Hogs

Hog slaughter is highly uniform throughout the year, with a slight "up tick" in the final quarter (Figure 7-9). Some increases in marketings do appear in the fall, a likely result of the heavier spring farrowing (traditionally March-April) compared to the fall (especially September). The variation between the two farrowing periods has, however, been declining to a point where it is close to five percent (USDA, ERS "Livestock and Meat Statistics," various issues, Table 30).

Slaughter prices show considerably more variability than does slaughter

FIGURE 7-8. Index of Monthly Changes in Feeder Calf Prices All Weights and Grades, Kansas City, 1980, 1985, 1987 (Source: USDA, ERS, "Livestock and Meat Statistics 1984-88," Table 111)

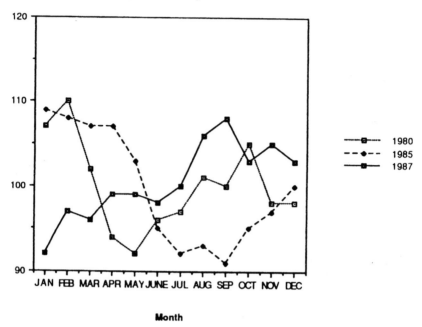

Month

(Figure 7-10). Total variation in the average index is over 20 percentage points for the years shown. The peaks generally correspond to the low points of slaughter in August, but the volatility means the seasonality is not a good predictive device.

Sheep and Lambs

Commercial slaughter of these animal classes is highly uniform over the 11-year average (Figure 7-11). Maximum variation is only 15 percent over the three years shown. To the extent there is an increase, it is in the fall when the spring lamb crop is marketed while some fed or shorn lambs may also be available.

The segmentation of the market into spring lambs and sheep suggests distinct price patterns for the two products. This is similar to veal and beef sales except spring lambs account for a greater percent of the total volume. Documentation is difficult because separate slaughter statistics are

FIGURE 7-9. Quarterly Pork Production, 1989 and 1990 (Source: USDA, ERS, Outlook '90 Charts, 1990, p. 47)

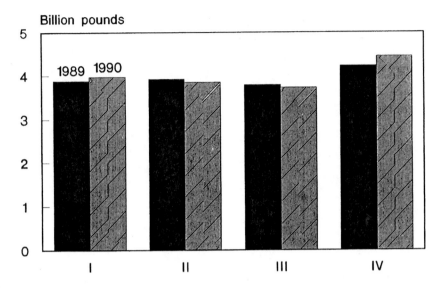

not kept for lambs and sheep. The effect of the two-product markets is, however, evident in the seasonal price patterns. Lambs show a notable increase through May or July (Figure 7-12). Although they are not shown here, choice feeder lamb prices display much the same seasonal price pattern. This is not surprising, as the two uses compete very directly with each other.

Slaughter sheep prices show a highly seasonal pattern with variations of over 60 percent in the index (Figure 7-13). Spring and summer prices are steady but lower, probably because of competition from spring lambs. The patterns are, however, highly erratic from year to year and are not a ready guide for production scheduling.

Managing Price Seasonality

The caveats placed on the description of seasonal price patterns makes the situation perhaps more confusing for the producer who wishes to time sales to get the best price. However, beginning with a focus solely on price is inappropriate since the objective is increased margins net of production costs. A seasonal production strategy thus needs to consider both cost variation by month along with expected price movements. Such an

FIGURE 7-10. Index of Monthly Slaughter Barrow and Gilt Prices, Seven Market Average, 1980, 1985, and 1987 (Source: USDA, ERS, "Livestock and Meat Statistics 1989," Table 122)

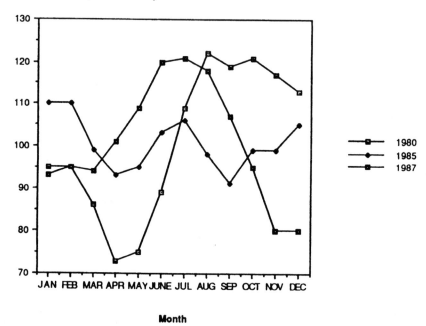

Month

analysis may indeed reveal that the most profitable time to sell is when the price is lowest. This would happen especially for grazing stock if feed had to be purchased when the natural pasture became inadequate due to weather changes or grazing pressure. The operator is strongly advised to consider production costs because they are typically easier to forecast than the more ephemeral future price.

Yager, Greer, and Burt (1980) have prepared an example of how counter-seasonal production can increase returns. In their example, which considers carrying beef cull cows from the normal fall/early winter selling point to the spring, net returns per head increased from $20 to $55 depending on weight, prices, and feeding regime. Producers will recognize this as a very significant increase over long-term average returns. This figure should, however, be viewed as maximum value since the traditional late autumn and spring culling periods impart significant seasonal shifts in unfed beef supplies and prices. Deferring marketing to a seasonal low supply point, such as early summer, then increases the expected price by a

FIGURE 7-11. Monthly Commercial Sheep and Lamb Slaughter, 1980, 1985, and 1987 (Source: USDA, ERS, "Livestock and Meat Statistics 1989," Table 68)

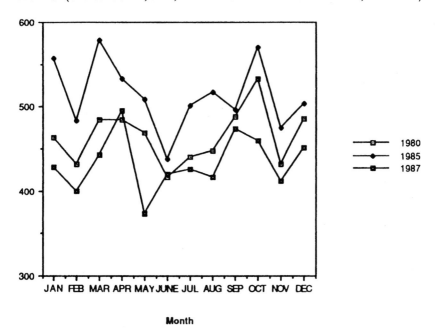

Month

substantial amount. For other species and classes without the pronounced seasonal change, the potential for benefitting from staggered sale dates is inherently less. Some potential nevertheless exists and should be considered.

Yager, Greer, and Burt (1980) use a sophisticated analytical technique called "dynamic programming with a stochastic price factor" to come up with their estimates. This procedure allows the simultaneous consideration of (a) feeding for maintenance or gain, (b) minimizing feed costs, (c) expected seasonal price patterns, (d) grade-related price relationships, (e) discount rates, and (f) capital and overhead costs, so that the optimal strategy may be determined. Such analytical capabilities are beyond the reach of most livestock producers. A simplified analytical procedure, while technically suboptimal, may nevertheless indicate the profit potential of adjusting the market-ready timing of the stock. The process is a simple one of evaluating the difference between production costs and expected price for some non-customary sale date, being careful to consider all capital and facility costs as well as units of labor and management

FIGURE 7-12. Indexes of Monthly Changes in Slaughter Lamb Prices, Omaha, 1980, 1985, and 1987 (Source: USDA, ERS, "Livestock and Meat Statistics 1989," Table 125)

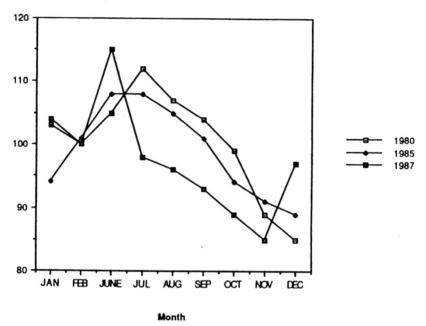

time. Procedurally, it is similar to the kind of analysis which is (or should be) done when any animal is put on feed or included in the brood herd. Price uncertainty can be moderated by using the futures market, when suitable contracts exist, or by means of a future contract with a fixed price. (See Chapter 14.) The timing of sales is one factor the producer has considerable control over, and it provides one possible means of increasing profits in a sector with a large number of constraints on production and marketing options.

FACTORS AFFECTING PRICES

Conceptually speaking, anything that causes a movement along or shift in the supply or demand curves affects prices. Price, after all, is determined by the intersection of these curves, so that influencing them changes the equilibrium price. This is to say that most anything, or everything, can affect prices to a degree. Our attention here is directed to a

FIGURE 7-13. Indexes of Monthly Changes in Slaughter Sheep Prices, 1980, 1985, 1987 (Source: USDA, ERS, "Livestock and Meat Statistics 1989," Table 126)

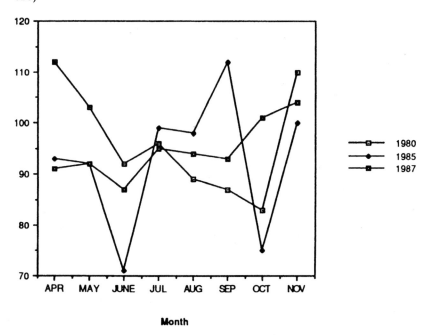

Month

much more limited and specific range of topics. In turn, the following topics are addressed: (1) where prices are determined, (2) estimated price elasticities, and (3) effects of imports on prices.

Where Prices Are Determined

To the theoretical economist sitting at his or her desk, this is a non-issue: prices are set by the consumer-determined final demand curve as it intersects the supply curve. Product prices leading up to the final demand level (which, for livestock, include live animal sales, carcass prices, etc.) are derived from the final demand and hence are known as derived demand. In the field, the issue quickly becomes clouded. Producers could control shipments and thus affect prices. (While this is unlikely for livestock, it is possible and observable in many industries. Note, for example, the maximum target share for BMWs in the U.S.) Or processors acting as monopolists could raise prices, again adjusting the release of product to the market.

It should be emphasized that a monopolist (or monopsonist) somewhere in the supply system does not affect what will be called basic demand (the way a consumer values pork versus beef versus chicken), but the monopolist certainly affects price. Figure 7-14 is a simple example of how this works. For the same shifts in supply (costs) (S_1 to S_2), price rises more in the competitive market (P_{C1} to P_{C2}) than in the monopoly market (P_{M1} to P_{M2}),[1] although, of course, the customer always pays more in a monopolized market than in a competitive one. Clearly, the seller has acted in a way that affects price.

FIGURE 7-14. Effect of Monopoly on the Price Determination Process

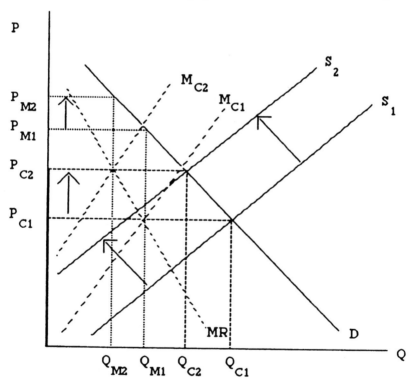

---

1. This happens because the monopolist is operating on the more elastic portion of the demand curve. The monopolist does not wish to raise prices "too much" because total revenues are falling.

Another way that price determination can occur at a market level other than retail is if standard markups are used. Consider, for example, a case in which a retailer attaches a fixed 30-percent margin to the wholesale price to determine his retail price level.[2] Real price determination is then done at the wholesale level, with the retail price derived from the wholesale rather than vice versa as "the textbook" would have it. Again, the wholesaler is not entirely divorced from the retail demand relationships that will ultimately dictate price. But under this scheme, small price changes can and are initiated by wholesalers rather than as a reaction to a change at the retail level. This example indicates that a supply shift can lead to a greater price change in a competitive market than in a monopolized one even though the monopoly prices will be higher.

The monopoly and fixed retail markup arguments are plausible, just as the textbook explanation is. The resolution of these multiple depictions is then an "empirical matter," to use a favored term among economists. In all likelihood, several factors are present and the empirical analysis helps to determine which is the most significant among them.

In a detailed evaluation of this issue, Barksdale, Hilliard, and Ahlund (1975) used cross-spectral analysis on monthly prices from 1949 to 1972. Spectral analysis is a technique that evaluates cycles, or pairs of cycles, over time. If one series regularly leads another, then it may be inferred that the earlier event is indeed the causal one. Other more formalized tests of causality also exist. From their analysis, Barksdale, Hilliard, and Ahlund state emphatically, "The obvious conclusion, therefore, is that beef prices are not established at the retail level" (1975, p. 311). Rather, it is wholesale prices that are the determining influence, with retail level adjustments lagging by three weeks. The remaining prices in the system — those for wholesale, slaughter and feeder animals — were found to be so closely matched in time that no causality could be inferred. Using a more recent (1983-1985) data set and different statistical procedures, Schroeder and Hayenga (1987) reached much the same conclusions for beef as well as pork, and Hahn (1989) concurs. Wholesale price changes led to retail adjustments for the two species, while retail and wholesale beef prices appeared codetermined. For pork, wholesale prices did appear to lag behind live annual prices but, as the authors note, this could be a function of the greater degree of processing of pork (1987, p. 175). Futures prices are

---

2. This is indeed a common practice among food retailers and the only practical one when prices for 10,000 + items must be set. With meat, the retailer must adjust for retail trim loss, if any.

sometimes considered to be a further price-determining factor, especially because they are used in some future sales contracts.

What does this finding mean? Simply that particular attention must be given to the price formation process at the wholesale level since it also prevails in retail price setting. No similar studies exist for the other red meat products, but it seems likely that they follow the same process as beef.

Wholesale and Farm-Level Elasticities

The retail elasticities examined earlier have a direct relationship to wholesale and farm-level elasticities (those which occur at the earlier stages in the marketing "chain") These elasticities, as a rule, are smaller than retail ones. One reason for this is the existence of the marketing margin between the several levels, thereby making the wholesale elasticity smaller even if the demand curves are parallel. Other, more theoretical, explanations also exist that show farm-level elasticities will never exceed retail ones and are generally smaller (in absolute value).[3]

Small (in absolute value) wholesale and farm level elasticities are of significance for the way they describe the sector. Small price elasticities mean that a large price reduction is required to move proportionally more product. Conversely, small income elasticities (Table 6-9) suggest that demand will remain quite stable if incomes rise (or fall). Smaller cross elasticities mean that demand for one product is not highly dependent on another's prices. In total, small elasticities describe a stable market in terms of volume but one with significant price swings. Such a situation is a very apt description of many livestock product markets.

Estimated farm-level red meat price elasticities are in agreement with our observations of the industry (Table 7-2). (No comparable figures are available for the wholesale level.) In all cases, the farm-level elasticities

---

3. In simplified and modified terms,

$$E^W = E^R + E^T$$

where E^W represents wholesale price elasticity (negative)

    E^R represents retail price elasticity (negative)

    E^T represents elasticity of price transmission of a change in wholesale prices to retail (positive)

Then, since $E^T = E^W - E^R$, the wholesale elasticity can equal but not exceed the retail value. In most cases (e.g., when some proportion of the price change is transmitted each period), where E^T is a positive fraction, the retail elasticity will be smaller than the wholesale estimate.

TABLE 7-2. Farm-Level Own Price Elasticities for Beef, Pork, Veal, and Lamb, Selected Years, 1946-1967 (Source: See Table 6-9)

Item	1955-1957	1946-1967
Beef	.68	.42
Pork	.46	.24
Veal	1.08	NA
Lamb and Mutton	1.78	1.67

Note: The negative signs have been omitted.
NA - Not available.

are smaller than those estimated for the retail level. The cross elasticities (not reported) show the same relationships.

Imports

Meat imports are very upsetting to the domestic industry that sees them as lowering prices and profits. In a purely conceptual sense, imports do lower domestic prices because they are a (nearly) perfect substitute for domestically produced products. Thus, the questions are: (1) how much, in fact, do exports lower domestic prices, and (2) are there any other aspects of imports that offset, or at least mitigate, the price effect? Imports are likely to be species specific in their effects, depending on volume, composition, and other factors.

For pork, 1987 imports of 1.2 billion pounds were eight percent of domestic production, not an inconsequential number (USDA, ERS "Livestock & Meat Statistics" 1989, Table 107). Canada, which supplied nearly half that amount, was found to subsidize its pork exports and a countervailing duty was imposed on live hogs beginning in 1985. While causality is difficult to document, the imposition of the duty did coincide with a stemming of the growth of these imports (Figure 7-15). In May 1989, a countervailing duty of three cents per pound was imposed on U.S. pork imports from Canada. Tariffs of this type will be phased out over a ten-year period in accordance with the U.S.-Canada free trade agreement.

With lamb, imports are typically identified as being from New Zealand or Australia and, in many markets, are frozen. These factors, combined with the smaller frame sizes predominating "down under," mean domestic and imported lamb are only partial substitutes for each other. The obverse production seasonality that causes the two sources not to compete contemporaneously for the consumers' dollar further limits substitutabil-

FIGURE 7-15. U.S. Hog Imports from Canada, 1980-1990 (Source: USDA, ERS, Outlook '90 Charts, 1990, p. 46)

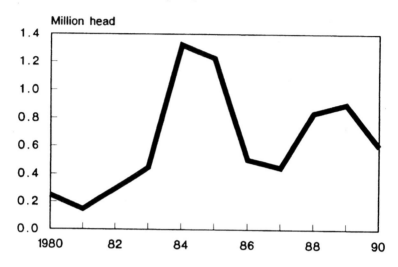

ity. For these reasons, there is no strong reason to believe that lamb imports, at their historical and present levels, depress domestic prices. And imports may strengthen the domestic industry in the long run by keeping prices lower, especially if the imports are countercyclic.

Imports are largely a matter for beef, and several studies have been undertaken to measure the price impact of imports. Beef is not a homogeneous product and the import effect is most sensitive for non-fed beef. Within this category, a 700-million-pound import increase in 1972 (35 percent) was estimated to reduce domestic hamburger retail prices by about ten percent (Freebairn and Rausser 1975, p. 687).[4] One reason for this rather small response is Ehrich and Usman's (1974) finding that imports are not perfect substitutes for domestic, low-grade beef. No explanation was given for why that would be the case. Estimated effects on retail prices by substitute prices of other meats were even smaller than for hamburger.

Livestock producers are, however, typically not retailers, so it is instructive to examine the effects of imports on farmgate prices. Fed beef

---

4. The actual long-term impact of the hypothetical 700-million-pound import increase is an eight-cent decline when national average prices were about 75 cents a pound (1975, p. 687).

prices were projected to decline by about six percent while cull prices had an estimated slide of 15 percent. As feeder calf prices were also depressed by imports (by five percent), calf producers bore a disproportionate effect of imports since both their products lose value as a result, giving a weighted value of ten-percent (Freebairn and Rausser 1975, p. 687). While this is not negligible, the evidence does not substantiate domestic producers' claims of ruin as a result of imports. Indeed, imports may even help domestic producers. Since imports rose when domestic supplies were low (and prices high), the imports helped to maintain the supply at stable prices. This can help the growth of the sector, or at least limit the loss of market share to competing products.

PRICING EFFICIENCY

Pricing efficiency has such an esoteric sound that one can easily doubt its significance. Yet, in a market economy where prices both allocate income and — to a large degree — determine production levels, prices have an overwhelming effect. How closely the exchange prices used in the industry reflect true supply and demand conditions, then, is of interest to producers and consumers alike. While there is no general agreement on the terminology, pricing efficiency can be described as having two components: (1) whether the price discovery procedure used is adequate to reflect true demand conditions free from errors and manipulation, and (2) whether the bargaining position of buyers and sellers is sufficiently balanced for the price to be a competitive one, not providing an unduly large share to either group. In this section, the first issue, that of the institutional framework of price discovery and its possible ramifications, will be analyzed. The question of the market's competitiveness will be left to Chapters 15 through 18. In those chapters, the related issue of the technical efficiency of the sector and its price effects will also be considered.

Price formation takes place at each market level, and complete analysis would include every one of them. That, however, is a rather overwhelming task which may be unnecessary. Instead of analyzing every level, the evaluation here is limited to possibly the most significant one, the wholesale level. Wholesale price determination is chosen for two reasons: (1) evidence indicates it has a strong causal effect on retail prices, at least for beef (see above); and (2) on face value, this pricing system is the most problematic in terms of standards used by economists. For clarity, wholesale prices shall refer to carcass or subcarcass prices (e.g., boxed products) received by a packer.

As a basis for comparison, it is helpful to sketch out what an economist would consider to be an "ideal" price formation process. This would conform as closely as practically possible to the perfect competition requirements of good market information, homogeneous products, and many buyers and sellers (each of whom handles an insignificant portion of the total volume). For livestock, this kind of market is approximated by the large terminal market where price information is public knowledge and widely distributed.[5] It is no wonder the economics profession mourned the greatest of these markets, the Chicago yards, during their decline and final demise in the 1970s.

Thin Markets

The current reality of pricing carcass meats is quite distinct from the ideal outlined above. Most trades are made directly between packers and users, or brokers, in what is known as "private treaty" sales. Clearly, these sales are not in the public record, and indeed exchange prices must be collected directly from the participants in a transaction. This market news service is provided through two major publications: the "Yellow Sheet" (formally titled the *National Provisioner Daily Market and News Service*) (the "Green Sheet" reports weekly prices) and the *Meat Sheet* (from the Meat Market Research Reporting Service, the "Pink Sheet"). A third, the USDA's *Market News*, was discontinued in mid-1990 after more than 60 years. Of these, the "Yellow Sheet" is by far the most heavily used, particularly for carcass beef.

Compiling information on all of the thousands of daily trades is clearly impossible. By any standard, the three reporting services fall far short of the ideal: for beef carcass prices both the "Yellow Sheet" and the *Meat Sheet* quotes are based on less than two percent of total federally inspected steer and heifer slaughter. The *Market News*, which had a larger sample in proportional terms, still polled less than five percent of all carcass beef movement (USDA, AMS, P&SA 1978, p. 6).

The use of "formula pricing" by the trade further compounds the price discovery process. A price formula is usually an agreed-upon relationship to a reported price for some time in the future, such as two cents under the "Yellow Sheet" quote for Thursday. Formula prices are heavily used in the industry, accounting for an estimated 70 percent of carcass beef sales,

---

5. The concept of the public market is ingrained in the rules of our major marketing institutions, the stock exchanges, and futures markets. In these exchanges, prices are announced by "public outcry" so they become part of the public record. It is a minor irony that the outcry is often so vociferous that real price information is relayed among the participants by hand signals.

10 to 20 percent of boxed beef sales, and 40 percent of fresh pork sales. Formula trading does have internal benefits to firms, especially smaller ones, including lower transaction costs and protection from paying sharply higher prices than a competitor (USDA, Meat Pricing Task Force 1979, p. 19). However, since such trades do not involve any independent price discovery, the amount of available price information is effectively reduced by the volume traded under price formulas. A circular situation develops, one which economists refer to as thinly traded markets, or thin markets for short.

The situation for carcass beef is likely to deteriorate (e.g., become "thinner") as more and more product is handled in boxed form [up to about 80 percent in 1987 (Duewer 1989, Figure 2)], reducing the amount of carcass beef to report. Boxed beef, the common term for cryovac-wrapped sub-primals packed in cardboard boxes, has not been an issue in the thin market debate as some 80 to 90 percent of trades are negotiated (Hayenga 1979, p. 1). One observer feels these prices are negotiated because the value of boxed products like beef loins is very high, so a few percent difference in price is meaningful to the buyer (Hayenga, Iowa State University, personal communication). The proportion of negotiated boxed trades may, nevertheless, decline in the future as the industry becomes more familiar with how these products are priced. The greater reliance on negotiations in carcass pork trade is not readily explainable, although it has remained quite stable over time.

Concerns with thin markets can be categorized in three areas: (1) the accuracy with which the price reporting services reflect true supply and demand conditions, (2) the potential for manipulation of quoted prices, and (3) the perpetuation of geographic pricing patterns that may no longer represent regional comparative advantages (National Commission on Food Marketing 1960, p. 58). Based on rising concerns during the late 1970s, concerns which might in part be attributed to sharply rising prices earlier in that decade, several investigations of the "Yellow Sheet" were undertaken by the USDA. The major conclusions of these analyses were as follows (USDA, AMS, P&SA 1978, p. 6; USDA, Meat Pricing Task Force 1979, p. 18, 25):

1. No substantive evidence of price manipulation was documented;
2. No statistical difference was found between the "Yellow Sheet" quote and the sample of prices used to determine that quote;
3. No price difference in packer-to-packer and packer-to-retailer trades was observed.[6]

---

6. See also similar conclusions in Marsh and Braster (1985).

Do these conclusions exonerate current pricing practices for carcass red meat, especially those of the "Yellow Sheet?" Not entirely, and certainly not for all market observers. As long as market information is based on such a small sample, the potential for errors or manipulation remain and suspicions will continue. Particularly troublesome are packer-to-packer trades that allow the same party to serve alternatively as both seller and buyer. Packer interchanges serve a valid function as firms strive to balance the kill and boxing line demands, for example, but the potential for manipulation is heightened. No group is likely to skew prices "much" or "for long," but even a minor move of one cent per hundredweight for one week would have caused the reallocation of $48,000 in the beef industry in 1987. As an additional protective measure, it seems prudent to adopt the Secretary's Meat Pricing Task Force recommendation of having the "Yellow Sheet" identify the degree to which packer-to-packer prices are used in developing a price quote. The other price reports currently contain that information. Overall, however, there is no clear evidence that reported prices are inaccurate or manipulated. Indeed, that is essentially the conclusion reached by Judge Kazen in December 1988 when he dismissed 12-year-old charges of price fixing brought by groups of cattle feeders against a combination of packers, retailers, and the publisher of the "Yellow Sheet."

Live Animal Price Formation

There are two characteristics of price formation at auction that affect pricing efficiency and warrant comment. The first of these is the theoretically based expectation that buyers at an auction will, in effect, bid away their consumer surplus. The consumer surplus, roughly speaking, is the area below the demand curve but above the price line (the shaded area in Figure 7-16). The consumer surplus is (approximately) the area under the demand curve above the market price P.

In simplified terms, the figure relays the following information: one buyer is willing to pay price P', a second price, P'', and so on. However, market equilibrium is at price P, so that any amount above that which (some) consumers were willing to pay but did not need to accrues to consumers — as their surplus. The astute observer will recognize that the only way for auction participants to "bid away" that surplus is for prices to change (probably decline). Buyers who valued the cattle, hogs, or sheep particularly highly (perhaps because of an acute need for raw product and an unwillingness to go home empty-handed) would bid first and drop out. Successive buyers with lower valuations in the same quality animals would then become effective bidders and so on until all animals were sold.

Professor Buccola (1982) investigated this hypothesized behavior at 42

FIGURE 7-16. Approximate Consumer Surplus

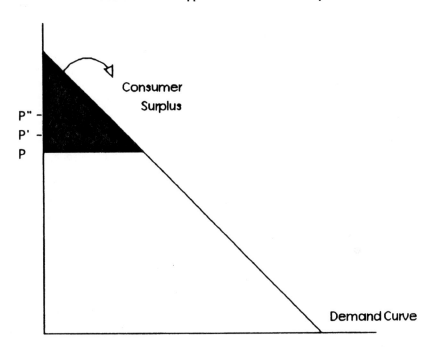

successive yearling steer auctions in Virginia. Sales averaged about 600 head with a range of weights and breeds. Over 1958 to 1979, prices declined approximately 16 cents per hundredweight as the lot number was increased by one within each major grade category. That meant a price spread of $3.20 per hundredweight between the first and last lot sold within each category, with the amount increasing toward the end of the sample period. These results are difficult to generalize since Virginia is a relatively minor feeder producer (about 22nd in 1980) and one could expect that auction behavior could be quite different at very large markets selling to a possibly more homogeneous group of buyers. However, Schroeder, Jones, and Nichols confirmed that finding in a 1988 study of a Kansas feeder pig auction where the prices declined 3.01 and 5.37 percent during the third and fourth quarters of the sale (1989, p. 261). In any case, the implication for sellers is clear: to be prudent, the seller should get his or her animals sold early in the marketing day. Being first may present some risks as buyers sense the "mood" of the market, but soon thereafter gives the best chance of the highest price.

A second characteristic of public markets is their increasing thinness, or smaller volumes handled, over the past two decades. (For details, see Chapter 13.) Has this reached a point where prices at some markets are no longer representative? Tomek (1980) evaluated this issue by focusing on cattle sales at the old Denver terminal market where volume dropped from 860,000 head in 1955 to 10,000 in 1968, the last year it was included in the Agricultural Marketing Service price reports. (The market closed in 1972.) Viewing prices from a statistical perspective, Tomek describes sales as supplying information to the market so that prices are set with less precision as volume declines (1980, p. 443). At some point, the precision of those results may become unacceptably small, meaning the seller is advised to seek another market with a better likelihood of paying the "going price." For Denver during the 1960s, only about 50,000 steer sales per year were found necessary to establish prices with good precision. Market volume dropped below this level in Denver only after 1966, so that it was apparently a viable market at least through that year.

For the producer selecting possible markets, these results suggest a relatively moderate volume is significant to offer a "fair" price. These cautionary notes should be added: first, it is important that the volume be concentrated within one or a few types or grades. It does not particularly matter which class or grade predominates, only that it is, say, choice cattle as opposed to a smattering of grades from canner and cutter through prime. Yet, since public markets are well suited to odd-lot sales, the tendency is to attract a range of grades. Second, as Tomek carefully points out (1980, p. 442), his estimate was for the 1960s while commodity prices have become more volatile in succeeding decades. Volatility is significant because the more variable the price, the more information is required to peg that price in a narrow range. Indeed, in the formulation Tomek used, variance (a measure of variability) is directly related to the volume requirements. The seller must then use discretion in selecting market outlets and, above all, carefully scrutinize those with small volumes of the class and species to be sold. A simple way to carry out analysis is to plot local prices and those of a major national market (e.g., Omaha). The two may be separated (for example, if there is a transportation differential), but that separation should remain rather constant across time.

EFFECTIVE PRICES:
SHRINK AND OTHER COMPONENTS

Which is preferable, $50.50 per cwt for slaughter hogs or $48.00? A trivial question, certainly, if all other sale terms are equal but, in practice, other conditions are generally not the same. Rather, the seller must choose between alternative sales where the shipping distance, payment for transportation, feeding arrangements before shipping, weigh-in point, and other factors are all different. Under such conditions, the seller must calculate the net value at the farm before an accurate comparison can be made among the competing outlets. Some costs, such as trucking, can easily be compared. To compare a sale including trucking costs with one not including trucking costs, the cost of transport is simply deducted to place prices on an even farmgate basis. Other costs are neither as obvious nor as easy to calculate, yet ignoring them could lead to significant losses for the producer. This subsection treats several such costs, beginning with shrink.

Live Animal Shrink

Shrink is the weight loss caused by a reduction in body fluids through excretion or respiration. Under normal conditions, moisture is lost and restored periodically. Only under unusual circumstances, such as extreme stress or the absence of water, is the weight loss substantial enough to affect the value of an animal (value is the product of price per pound and weight). Marketing often involves such stress and absence of water. Principal causes of stress are sorting, loading, and transport, although unloading and waiting at a market can cause additional weight loss. The amount of weight lost prior to the final weighing must be considered when computing the total costs of marketing.

For a typical sale, weight loss will be in the range of two to six percent. Using averages is, however, unsatisfactory because weight loss is a highly nonlinear function of time and distance (Figure 7-17). Non-linearity is a factor because the greatest loss for fed cattle occurs during the first seven hours. Loss increases at a fairly steady rate for the next ten hours when it takes another jump before leveling off again. For slaughter hogs, 1.09 percent of liveweight is lost for transit from 10 to 30 miles, but only .46 percent for the next 20 miles (30 to 50 miles) (Brandt 1986, p. 300).

Actual loss is a function of several other factors including temperature, handling, fill, weight, and sex, as well as species and class. For example, feeder animals often lose more weight during marketing than does mature stock. This is probably due to recent weaning, which is itself a stress. Heavily filled animals lose weight faster than those off feed and water for a period, and heifers are often thought to be more susceptible to shrink than steers, although the evidence for this is unclear. Measuring all the factors affecting shrink and the interactions among them is a daunting task. However, a number of factors have been measured and can be summarized as follows:

FIGURE 7-17. Shrinkage Increases as Hours in Transit Increase (Fat Cattle) (Source: Brownson)

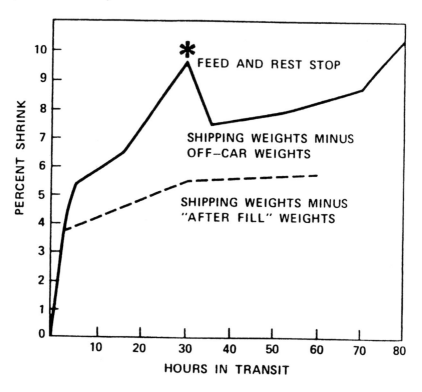

Cattle (summarized in Raikes and Tilley 1975; Brownson; St. Clair 1976):

1. With feed and water freely available, morning weights are 2 percent less than evening weights;
2. Withholding from feed and water for 12 hours causes a 2.13 percent shrink. However, cattle on grass and silage lose closer to 4 percent and range cattle closer to 5 percent;
3. The greatest loss occurs in loading, unloading, and the first few miles of transport;
4. There is little loss difference among the breeds;
5. Unusually high and low temperatures increase shrink;
6. Actual percent shrink increases with weight;

7. Fill back restores some 35% of the lost weight;
8. Calves are heavy shrinkers according to the following table (Brownson, Table 3):

Conditions	Percent Shrink
8-hour drylot stand	3.3
16-hour drylot stand	6.2
24-hour drylot stand	6.6
8 hours in moving truck	5.5
16 hours in moving truck	7.9
24 hours in moving truck	8.9

Hogs (Brandt 1986):

1. Shrink increases with distance, but at a decreasing rate;
2. Unusually high and low temperatures both increase shrink;
3. Animals sorted one to two days prior to shipping lose more weight than those sorted just before loading;
4. Heavier animals lose more weight than light ones.

Based on this information, it is evident that the major means of minimizing shrink is to reduce transport distance. When that is not possible, the shrink-related distance difference should be considered as a cost. Shrink can be reduced during extreme weather periods through adequate protection of the animals by avoiding overcrowding. On long hauls, the rest stop/fillback is economically beneficial to the producer. It is best not to mix unfamiliar animals, but when that is done it should occur just prior to shipping, not a day or more earlier.

One situation to avoid is the acceptance of multiple discounts. Consider, for example, a cattle feeder with no scales at the lot. An overnight stand will cause approximately a two-percent loss while a five-mile haul to the scales, with loading and unloading, will double that at least. If, on top of that, a three-percent *pencil shrink* is imposed (see below) the total is seven percent as above, an excessive amount. On the other hand, an excessively filled animal will be apparent to buyers and will be heavily discounted.

Tissue Shrink

With much transport-related weight loss readily restored through fillback, one might reasonably ask why buyers implicitly discount so heavily for shrink. Slaughter buyers care little for loss of fluids since they have modest value for the packer. The packer is more concerned with tissue shrink, the loss of carcass weight. Studies on slaughter cattle have shown that hot carcass weights are unaffected when feed is withheld for up to 48 hours. However, when feed and water are *both* withheld for 12 hours prior to slaughter, hot carcass weights drop about one percent (Raikes and Tilley 1975, pp. 83, 87). This would have a great effect on the packer. In simple terms, a live steer selling at $67.50 per cwt has a carcass value of $110.65 at a yield rate of 61 percent. A one-percent loss in liveweight can then be priced at $.01 \times \$67.50 = \$.67$ per cwt while the carcass loss would be $.01 \times \$110.65 = \$1.11$, or 65 percent more. This example is not entirely appropriate since a packer would not attempt to recover the entire live animal value through the carcass. The hide and by-products would be important contributions. Yet, the significance of carcass weight loss to the packer is evident.

As Raikes and Tilley (1975) show, the withholding of feed and water prior to sale is beneficial to the packer for it reduces his/her costs more than the loss in profits. Knowing that, producers should negotiate the withholding agreement by demanding a premium. However, the knowledgeable seller will realize there is no premium the packer can afford that would make the agreement worthwhile for the feeder. Using the numbers above, a two-percent live weight loss for the producer costs $13.50 in lost value. For the packer, however, the benefit is $13.50 - $11.06 (the loss of 1 percent of hot carcass weight), or $2.44, but the producer would not accept that for the stand. Thus, no mutually agreeable negotiated premium is possible, but some price accommodation should be made.

Pencil Shrink

Animals sold off the farm or out of the feedlot suffer no shrink. Buyers obviously recognize this situation and attempt to price the unshrunk cattle, hogs, or sheep relative to those shipped for a distance. This makes perfect sense, but the way the discounts are actually handled is a little confusing. Rather than quoting a lower net price, buyers typically offer a price minus a percentage known as "pencil shrink." "Pencil shrink" is simply a reduction in the quoted price. The value of the discount can be determined simply as follows: a four-percent pencil shrink (4%) means the seller retains 96 percent of the sale value. The pencil shrink is intended to serve in the place of an overnight stand without feed and water and should not be accepted *in addition* to a stand.

Other Discounts

The other commonly used form of discounting is cutbacks. Cutbacks refer to accepting a portion of a lot at a lower price. For example, a buyer may offer $65.00 per cwt for lambs with a ten-percent cutback at $62.50. A common reason for a buyer to offer a cutback is a concern that not all animals will grade choice. Whether justified or not, the net effect is a lower price for the producer. The seller must know the net price before a reasonable selection among buyers can be made. In any case, the seller should not accept a percentage cutback with the dollar amount unspecified at the time of sale.

In an idealized market, the packer buyer would not penalize the seller for non-tissue shrink. That is, if a buyer knew that an animal had dropped two percent of its weight through fluid loss, then he or she would be willing to pay about two-percent more for that animal than for one with no fluid loss. After all, most of that fluid is valueless anyway. The buyer, however, does not know what has occurred and must believe information provided by the seller or identify the loss through visual identification. A skilled buyer can recognize some of the effects of shrink and will usually be able to judge carcass yields within one percent. Risks in estimating carcass yield increase with instances of excessive loss or fillback, and even when a hide carries mud or manure. Buyers react to risk by being cautious, so that any uncertainty is likely to be passed back to the seller in the form of a lower bid. The seller's strategy should be to avoid selling slaughter stock when shrink exceeds the norm and never to overfill stock. Thus, shrink discounts for slaughter stock, at least at moderate levels of shrink, are the results of imperfect information that the seller cannot overcome entirely.

Shrink is less important in the feeder stock market since fluid can be regained rapidly. However, there are extra costs associated with substantial shrink. The feeder must allow time for the feeder stock to recover from this significant stress before entering a cycle of regular rapid gain. Even modest shrink has its costs because some time is required for the stock to return to original weight. Time costs include not only the costs of occupying facilities such as feeding structures, but also costs related to the charge for borrowed funds. These are nominal. Another reason for discounting feeder stock is the inexperience of some feeder buyers who are only in the market periodically and who do not become as experienced as the professional buyer. As a result, the seller should view shrink as a real and major cost of marketing and should plan marketing strategy accordingly. Major components of that plan are the location of stock weighing in relation to the farm, the length of time that feed and water are withheld, and the amount of fillback permitted.

SYNOPSIS

Prices have a major role in market economies where they both allocate income and weigh heavily in production decisions. Despite that importance, the "true" price is difficult to identify even though world prices tend to move together in markets that are not heavily controlled and/or protected.

Prices are affected by a range of factors including production and processing efficiency (rising), retail efficiency (which has been declining as service levels increase), seasonality (in general not a reliable basis for gauging production timing), the number of firms involved in the market (the structure), and imports (measured effects are minor, with some producer benefits possible).

Studies of the price formation process have identified the wholesale level as the critical point rather than the contact with the consumer at the retail level as would be predicted by theory. One aspect of pricing that has long been controversial is the wholesale price reporting services, and the "Yellow Sheet" in particular. However, several studies have found no improprieties or inaccuracies in the operation of that report, and a long-standing law suit against it (and others) was dismissed in 1988. Concerns nonetheless remain, as its significance declines along with the share of carcass trade in the industry.

The "Yellow Sheet" is one example of what economists refer to as "thin markets," the use of a small proportion of observable sales to represent a market. Small public markets can have similar non-representativeness problems. At public markets prices appear to decline throughout the sale, thereby giving a premium to being early in the sales ring.

The effective (net) price to the producer is also affected by weight loss or shrink. There are three major forms: live weight shrink, tissue shrink, and "pencil shrink." While the first two are related to actual losses, the third is a marketing agreement for direct sales that is intended to substitute for an overnight stand without food and water. Shrink is traceable to a number of factors, of which loading and unloading and transport distance (although increasing at a decreasing rate) are paramount. Many factors can be minimized by careful planning. Producers should be especially careful not to accept multiple source/forms of shrink.

Study Questions

1. Identify the major factors affecting livestock prices over time. Which of these are long-term, short-term or irregular effects?
2. Why do some species and classes show much greater seasonal variability in prices and marketings than other animals?
3. Should livestock producers be concerned about where prices are determined?
4. What can be said definitively about "thin markets" for livestock prices?
5. How do shrink, "pencil shrink," cutback, and holding off feed and water all interrelate?
6. What is the net cwt price for a $60.50 slaughter lamb weighed 50 miles from the farm when the feeder accepts a three-percent "pencil shrink" and a ten-percent cutback of $1.50? Assume a 2.15-percent weight loss during the first 50 miles of transit.

REFERENCES

Barksdale, H. C., J. E. Hilliard, and M. C. Ahlund. "A Cross-Spectral Analysis of Beef Prices," *Am. J. Agr. Econ.* 57(1975):309-315, p. 311.

Brandt, J. A. "Economic Consequences of Factors Affecting Liveweight Shrink on Hogs," *North Central J. Agr. Econ.* 8 (1986):296-304.

Brownson, R. "Shrinkage in Beef Cattle." Cornell Beef Production Reference Manual. Cornell University, Department of Animal Science, Fact Sheet 4310, undated.

Buccola, S. T. "Price Trends at Livestock Auctions," *Am. J. Agr. Econ.* 64(1982):63-69.

Duewer, L. A. "Changes on the Beef and Pork Industries," *National Food Review* 12 (January-March 1989):5-8.

Dunham, D. "Food Cost Review, 1987" USDA, Economic Research Service, Ag. Econ. Rpt. No. 596, September 1988.

Ehrich, R. L. and M. Usman. "Demand and Supply Functions for Beef Imports." University of Wyoming, Division of Agricultural Economics, B 604, January 1974.

Freebairn, J. W. and G. C. Rausser. "Effects of Changes in the Level of U.S. Beef Imports," *Am. J. Agr. Econ.* 57(1975):676-688.

Hahn, W. F. "Assymetric Price Interactions in Pork and Beef Markets." USDA, Economic Research Service, Tech. Bull. No. 1769, December 1989.

Hayenga, M. L. "Pork Pricing Systems: The Importance and Economic Impact of Formula Pricing." U. Wisconsin, NC-117, WP-37, August 1979.

Marsh, J. M. and G. W. Braster. "Short-Term Adjustments in Yellow Sheet Carcass Prices for Red Meats," *Am. J. Agr. Econ.* 67(1985):591-99.

National Commission on Food Marketing. *Organization and Competition in the Livestock and Meat Industry*. Tech-Study No. 1, June 1960.

Raikes, R. and D. S. Tilley. "Market Loss of Fed Steers During Marketing," *Am. J. Agr. Econ.* 57(1975):83-89.

Schroeder, T. C., J. M. Jones and D. A. Nichols. "Analysis of Feeder Pig Auction Price Differentials." *North Central J. Agr. Econ.* 11(1989):254-63.

Schroeder, T. C. and M. L. Hayenga, "Short-Term Vertical Market Price Interrelationships for Beef and Pork." *North Central J. Agr. Econ.* 9(1987):171-80.

Tomek, W. G. "Price Behavior on a Declining Terminal Market," *Am. J. Agr. Econ.* 62(1980):434-444.

St. Clair, J. S., "Marketing Alternatives and Costs for Wyoming Cattle," U. Wyoming, Ag. Exp. Station, *Research Journal* 108, November 1976.

United Nations, FAO, Production Yearbook, various years.

USDA. *Agricultural Chartbook*. Ag. Handbook No. 684, March 1989.

USDA, AMS, P&SA. *Beef Pricing Report*, December 1978, p. 6.

USDA, Economic Research Service. "Livestock and Meat: Outlook and Situation." Various issues.

USDA, Economic Research Service. "Outlook '90 Charts." AGES 9001, February 1990.

USDA, Economic Research Service. "Livestock and Meat Statistics: 1984-88," Stat. Bull. 784, September 1989.

USDA, Nat. Ag. Stat. Service, Ag. Stat. Board, "Livestock Slaughter, Summary." Various issues.

USDA, Nat. Ag. Stat. Service, "Livestock Slaughter: 1989 Summary." Mt. An. 1-2-1, March 1990.

USDA, Nat. Ag. Stat. Service, Ag. Stat. Board. "Cattle: Final Estimates 1984-88." Stat. Bull. No. 798, January 1990.

USDA "Report of the Secretary's Meat Pricing Task Force" June 15, 1979, p. 19.

Yager W. A., R. C. Greer, and O. R. Burt. "Optimal Policies for Marketing Cull Beef Cows," *Am. J. Agr. Econ.* 62(1980):456-467.

Chapter 8

Livestock Cycles

Livestock cycles are perhaps the most significant issue in the hog and cattle industries. Cattle producers have seen the asset value of their steer, cow, and calf herds vary by 50 percent (in dollars adjusted for inflation) since World War II solely due to cycles (USDA, ESS, Crop Reporting Board, "Cattle on Feed" various issues, p. 26). Meat packers have withstood sustained periods of below-capacity operations in an industry where full utilization is the key to prosperity. And the consumer, while spending an approximately fixed share of income on meat, has had to make large adjustments in quantities and, particularly, types of meat consumed over successive cycles.

Clearly, there are advantages to packers and consumers from a stabilization of supply. Producers, though, will benefit the most by minimizing the periodic price declines and subsequent losses. Despite the obvious need for stabilizing cycles, there is not even broad agreement on what causes livestock cycles. Rather than emphasizing understanding and planning, the livestock producer has been cajoled into exercising "self-constraint," a hopeless approach because many producers do not recognize complicity in perpetuating cycles. To them, cycles are caused by outside forces like the weather, the government, or imports (Kendell and Purcell 1976). Here, I emphasize their need to understand the several causes of livestock cycles and then to develop policies for minimizing the impact on individual operations and on the entire cattle and hog sectors. Sheep are excluded from the analysis because their numbers are proportionally so much smaller than the other two species.

THEORIES OF LIVESTOCK CYCLES

To that familiar adage about the certainty of taxes and death can be added livestock cycles. The periodic cyclical movements of numbers on farms, and prices, has been documented in the United States since systematic data collection began in the time of President Lincoln. Yet, even without a century's worth of data to spread on the table, the regular ups and downs of hog numbers were observable. In 1876, Brenner described the "advance and decline" in hog prices over 20 years to be "as alternately certain as the diurnal revolutions of the earth upon its axis" (quoted in Breimyer 1959, p. 760).

Exogenous Causes

The persistence and regularity of these cycles suggest — even to the casual observer — the existence of causes neither random nor totally exogenous. For many years in the United States, the cause was thought to be fluctuations in the corn supply. Since corn is the principal hog feed, unexpected production shortfalls necessitated reducing feedings, while good crop years allowed farmers to hold more pigs for feeding. Acting in part on this perception during the 1930s, the Department of Agriculture embarked on production and price stabilization programs for corn and other grains. With market supply controlled by manipulating government corn stock levels, the hog cycle should have disappeared according to this simplistic view. It indeed did not, although some greater regularity is evident.

The reason for the general failure of crop stabilization to harness livestock cycles can be readily understood by recognizing that feed prices are but one component of the producers' profitability mix. Even with input costs constant, short-run imbalances of supply and demand can cause wide price movements. Indeed, the emergence of the government as a defacto buyer, seller, and holder of grain made it possible to expand and contract herd sizes more rapidly (Meadows 1970, pp. 38-40).

Endogenous Causes

On this battlefield of observable empirical evidence, the so-called exogenous school was wounded but not, as we shall see, mortally. In its place, as a kind of temporary victor, stood the endogenous school.[1] To a large degree, this group places responsibility for the cycles with the producers themselves. While this position may seem unduly critical, it really is not. Rather, producers are seen as small players in a large system. These players have two key characteristics: (1) production is adjusted according to changes in profitability, and (2) each individual contributes such a small

---

1. The several explanations for livestock cycles are expressed in this way to enhance readability rather than to suggest any strict chronology of ideas.

portion of the total supply that he or she does not consider the effects of changes from his or her farm or ranch on aggregate production or prices.

Both practices make sense from the perspective of the individual operator. When profits from cow-calf or sow farrowing operations turn negative, as they periodically do, it is only reasonable to cut back on numbers to control variable costs. This decision may either be made independently by the producer or induced by a lending agency that refuses to grant the full requested amount of a production loan if, say, the equity value of the breeding herd or feeder crop has declined. For feeders, it can be more profitable to sell grain directly rather than through stock so that reductions in output follow promptly. More precisely, the production decision is not based solely on current prices but rather on the expectation of future prices and profitability when the animals are market-ready. However, there is a strong correspondence between current prices and future expectations, one which Walters (1965) described as "spontaneous optimism" in response to high and rising prices. In a broader sense, there is an inertia among all of us when it comes to forecasting a future different from the present; when times are good, there seems no reason they should change, while bad times generate their own pessimism.

Agricultural producer's vision of their independent actions having no aggregate effect on supply and prices is also entirely rational. Individual decisions do not necessarily affect aggregates. It is only when each individual choice is made using the same criteria and the same information that the decisions become de facto group decisions that have an enormous impact on overall production. Thus, when most livestock producers make decisions based largely on current prices, the aggregate effects are large even when the individual ones are small.

Yet, under these conditions, cycles would not be generated if production decisions were immediately correctable and excessively high output were balanced with cutbacks, or insufficiently low production speeded up. For livestock, that is not possible since 27 months are required to breed a calf and bring it to slaughter weight (or about 5 1/2 years if the breeding herd is first expanded). For hogs, the production lag is one year. With these lags, misjudgements made in the past will persist over several years. And since prices are inversely related to current production (see the section on price elasticities in Chapter 7), prices fluctuate widely.

A convenient way to conceptualize the working of livestock cycles is to use the so-called cobweb model (Figure 2-14). The model depends on ten specific assumptions that can be abbreviated as follows:

1. Supply and demand can be summarized by conventional curves;
2. There is a lag between initiating production and realizing the commodity; and
3. Production decisions are based on current prices that are expected to continue indefinitely. (Meadows 1970, pp. 10-11)

To see the effects of the model, assume the system is investigated at a point of disequilibrium such as A in Figure 2-14. Any number of factors, such as bad weather or an import ban could leave the system at A. There, the sector is in equilibrium with consumers paying P_A. In the next (second) period, increase production to B. Now consumers exercise their sovereignty and bid only P_B for that amount. Disappointed, producers respond in the third period by cutting back to C. Price goes up to P_D and so on until a steady equilibrium is achieved (approximately) at P_M.

The system converged (the oscillations declined over time) in this case because the figure was constructed with the supply curve steeper than the demand curve (e.g., supply is less elastic than demand). If the two had equal elasticities, prices would oscillate indefinitely. Alternatively, if supply is more elastic than demand, the cycles become more violent until the system "explodes." Revisions of the model have eased the more stringent assumptions, in particular, the requirement that producers instantly and unequivocally adjust production based on the most recent price information. It has been shown that system stability (e.g., dampened oscillations approaching equilibrium) holds for a wider range of relative supply and demand elasticities (reviewed in Meadows, pp. 15-18). Even with these modifications, the cobweb model is based on rigid and implausible assumptions. In recent years, replacing the simplistic cobweb models — with greater complexity and reality — have been dynamic commodity models and their depiction of cycles as harmonic motion (Meadows 1970, Chaps. 4 and 5; Larson 1964).

A somewhat less charitable way to describe the perseverance of cycles is as "irrational" behavior on the part of producers. If everyone anticipates their existence, then why operate in a way that perpetuates them? Hayes and Schmitz (1987) investigated the issue and found that for hogs the degree of "cobweb type" behavior was high during the first half of this century but declined after about 1941. That change they surmise was due to the improved information available that assisted producers in recognizing when cycle turns had occurred. Moreover, larger operations could better profit from planned countercycle production.

Due to the quicker and better managed response to price changes later in the century, price savings moderated, and along with them, the economic incentive to produce countercyclically. If the producers contrive to operate procyclically, it is not through ignorance (". . . a significant portion of producers do notice cycles and eventually act to eliminate them." [Hayes and Schmitz 1987, p. 770]) or irrationality but simply because countercycle production is insufficiently profitable when the delays in adjusting herd size are considered.

Combined Causes

Implications of the more intricate models shall not concern us here. The cobweb explanation does, though, hold one further insight: cycles are the result of both exogenous and endogenous factors. As Figure 2-14 shows, once the system is out of equilibrium (an exogenous cause) the perceptions of producers (an endogenous matter) combined with biological factors slow the return to equilibrium. This combined explanation for cycles agrees closely with observed market behavior. Forces for a price break, say, may build over time, but no action is triggered until some exogenous event causes it. Possible causes are weather, imports, government policy, price freezes, etc. Then slaughter increases sharply and prices sag. At this point the cycle builds on itself. Lower prices reduce interest in breeding stock so that fewer heifers and gilts are held for breeding and become available for meat. In the final stages, brood herds are liquidated, further expanding the volume of meat on the market, although in this latter case, it is lower grade meat.

So in a way, every explanation of livestock cycles is correct. The producer sees exogenous events as the causal factor and, indeed, they can cause the cycle by acting as a triggering mechanism. The precipitation of the cycle would not be significant if it were not for the biological lags in rebuilding supplies. And the lags would not create sustained cycles if it were not for the systematic response of producers to current prices.

The important part of understanding the causes of cycles, though, is not determining whom to blame; it is recognizing the early symptoms of a cycle while something can still be done to minimize losses and maximize profits. This entry into the real world requires a more detailed knowledge of what happens during cycles, particularly when repeated shocks occur or when hog and cattle cycles converge. To obtain this detailed understanding, we now turn to a review of recent cycles.

THE ANATOMY OF LIVESTOCK CYCLES

Cattle

Cattle (and hog) cycles can be described as a combination of two stages: expansion and liquidation. Since 1896 (at least until quite recently) cattle cycles have averaged nine to twelve years with an expansion period of six to eight years and declines of three to ten years. Expansion is more regular due to biological factors. Once the decision to increase is made, a gestation period of nine months and growth and feeding periods of 14 to 19 or more months, depending on growth characteristics and feeding program, must be allowed. Further expansion is made possible by holding back the heifer calves for breeding. This requires an approximate doubling of the 27 month period for raising slaughter animals so that the growth period typically lasts 5 1/2 years. Liquidation, on the other hand, is entirely controllable by livestock interests. The rapidity of the kill-off is determined by profitability which in turn is strongly affected by outside factors. The record reduction (and record losses) of the mid-1970s was initiated principally by a severe drought. Other factors, among them high grain prices and the wage/price freeze, further precipitated the decline.

As a result of these factors, the eight cycles observed since record keeping began in the United States display more regularity in the growth periods than the downswings (Table 8-1). The liquidation phase has, however, shown a tendency to shorten. In a small degree this is due to the (relative and absolute) decline in the dairy herd (see Figure 4-4A), which, at one time, lent some stability to total cattle numbers. The reduction in

TABLE 8-1. Cycles in Total Cattle Numbers, 1867-1988 (Sources: Reported in Beale et al. 1983, Table 1; USDA, ERS, "Livestock and Meat Statistics, 1984-88," Table 1-6, "Livestock and Poultry: Situation and Outlook Report," Feb. 1984, Table 3)

	--1,000,000 HEAD--		
Year	Numbers at Low Point	Year	Numbers at High Point
1867	28.63	1840	60.01
1896	49.20	1904	66.44
1912	55.67	1918	73.04
1928	57.32	1934	74.36
1938	65.25	1945	85.57
1949	76.83	1955	96.59
1958	91.18	1965	109.00
1967	108.78	1975	131.83
1979	110.86	1982	115.44
1989	99.6		

the decline period, however, demonstrates greater stability of the cycles rather than increased instability, as a rapid reduction might suggest. The total variation during the 1938-1949 cycle was 20 million head, or 31 percent from the low. For 1958-1967, it was only 18 million of a larger herd (or 19 percent) and only two percent from 1967 to 1979 (Table 8-1). This last cycle has shown the greatest inter-cycle stability to date, but using the high points masks the record 23-million-head kill-off from 1975 to 1979.

Since 1979, the "cycle" has not followed traditional patterns. The "rebuilding" lasted but three years, 1979-1982, while the drawdown has been prolonged. Indeed, it is better depicted as a prolonged downturn in cattle numbers extending over approximately 15 years. This might better be described as a restructuring of the industry than a cycle-based phenomenon. The apparent shift in consumer preferences among meat products that occurred in the mid-1970s is one reason for the restructuring (see Chapter 7).

The modest January 1, 1990 increase in numbers suggests the beginning of a new cycle. This one is predicted to peak mid-decade at 108-110 million head, reflecting a one- to two-percent annual increase compared to earlier trends of around three percent (USDA, ERS, Situation and Outlook Rpt. 1990, LPS-40, p. 13).

Other observers have described cattle cycles as consisting of three segments: (1) rapid growth, (2) deceleration, and (3) turnaround (Hasbargen and Egertson 1974). The turnaround stage and the rapid growth stage incorporate two aspects of the expansion stage. Deceleration combines parts of liquidation and expansion phases. Considering the 1949-1958 cycle, rapid growth occurred during 1950-1952. Deceleration extended from 1953-1956, while the turnaround carried over to the next cycle (that is, beyond the point of minimum herd size), or 1957-1960. For the 1958-1967 cycle, the respective periods are 1961-1963, 1964-1965, and 1966-1970 (Figure 8-1). Other observers have further delineated cycles into four phases: (1) rising, (2) high-constant, (3) falling, and (4) low-constant. This characterization better describes price movements than changes in the standing herd.

Prices are, of course, closely (if inversely) correlated with production and herd size. Price cycles nevertheless display some of their own characteristics and idiosyncrasies. The principal one of these is the lead/lag relationship between production and prices. As a direct result of increased herd size and beef production, prices weaken and begin to decline. Producers who were unable to anticipate the turning point in prices — one which is typically precipitated by some exogenous event — have started

FIGURE 8-1. Cattle Inventory Cycles, 1896-1990 (Sources: USDA, Extension Service 1983, Figure 2; ERS, "Livestock and Meat Statistics," Tables 1-6 and ERS, "Livestock and Poultry: Situation and Outlook Report," Feb. 1989 and Feb. 1990, Table 3)

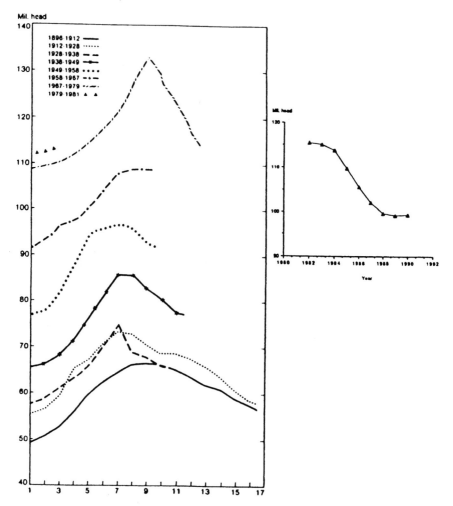

calves through the cycle. These animals are nonetheless grown to slaughter weights (albeit to somewhat lower weights) so that maximum production lags minimum prices by about two years (Figure 8-2). No parallel lag in prices is evident when production reaches a minimum and begins upward.

The second significant characteristic of price cycles is the correlation between fed and feeder prices. The simple correlation between the two is strong, as indeed it must be, with calf prices a major component of slaughter cattle costs and, hence, prices. Moreover, calf prices are really predicting cattle values — the expected price when slaughter weight is reached — and it is apparent that *on average* predictions are quite accurate. However, the simple correlation shows only that the two move together: when one increases, so does the other. Correlation says nothing about how far the movement goes, and it is there that the two quite literally diverge (Figure 8-3). Feeder calf prices simply move more than fed cattle prices. (Feeder steers occupy a position between the two.) The reason for this is quite simple: calves are the residual claimant to the consumers' beef dollar.

The concept of a residual claimant can be explained through a simple example. These numbers are picked for simplicity rather than realism. Suppose the market price for a 1,000-pound steer is 50 cents per pound, and the feeder estimates a pound of gain (including overhead and profits) costs 25 cents. What, then, is the value of a 500-pound feeder calf? The steer is worth $1000 \times \$.50 = \$500$. Subtracting the cost of gain: $\$500 - 500 \times \$.25 = \$500 - \$125 = \$375$. The calf is worth $375, or $.75 per pound. The cow-calf operator would not get the full value because, among other things, the feeder faces a risk in his/her investment and seeks a "risk premium" in the projected profit, but the procedure does give a close approximation of the calf's value.

Suppose the market price increases to $.60 a pound, what is the calf worth? Assuming nothing else changes, the steer value is $1000 \times \$.60 = \$600$, and the calf $= \$600 - \$125 = \$475/500 = \$.95$ per pound. Similarly, if the slaughter price falls by a dime (or feeding costs rise by the same amount), the value of the calf drops 20 cents per pound. The point is that at the end of the market chain the value of feeder calves is whiplashed by market forces. Nor is it necessary for prices to actually change to get this effect. If feeders anticipate a price decline or rise in feeding costs, they bid calf prices down, and vice versa.

It is because of these two facets of feeder calf prices, the exaggerated movements, and anticipatory component that feeder prices behave as they do. First, calf prices are more volatile (Figure 8-3). Second, calf prices rise in anticipation of the building phase of the cycle only to fall more

FIGURE 8-2. Total Production of Beef in the United States, the Steer-Corn Price Ratio at Chicago, and Deflated Prices of Choice Steers (All Weights) at Chicago, 1946-1964 (Source: Ehrich 1966, Chart 4)

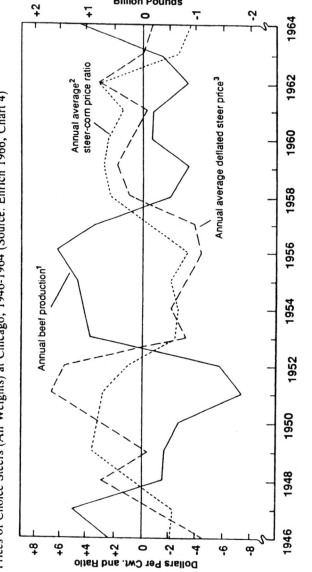

Data are expressed as deviations from trend. Source: U.S. Dept. of Agr., Econ. Res. Serv., Livestock and Meat Statistics (Stat. Bul. 333), July 1963, and supplements.

[1] Carcass weight of total United States' beef production, excluding veal.

[2] Price per hundredweight of beef steers divided by the price per bushel of No. 3 yellow corn.

[3] Deflated by the all-commodity index of wholesale prices (1957-59: =100)

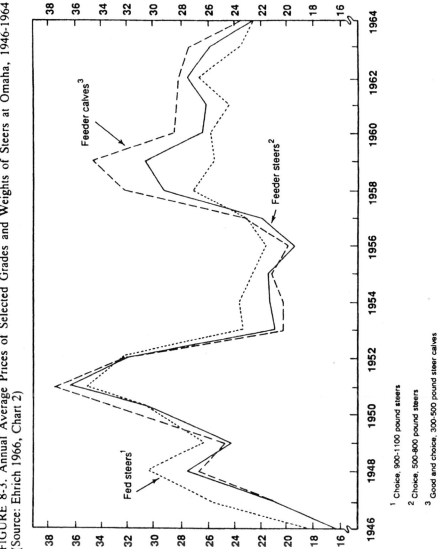

FIGURE 8-3. Annual Average Prices of Selected Grades and Weights of Steers at Omaha, 1946-1964 (Source: Ehrich 1966, Chart 2)

Feeder calves[3]

Feeder steers[2]

Fed steers[1]

[1] Choice, 900-1100 pound steers
[2] Choice, 500-800 pound steers
[3] Good and choice, 300-500 pound steer calves

rapidly during liquidation. Indeed, the record kill-off during the 1967-1979 cycle led to record losses for the cow-calf sector, particularly in 1976. That year, producers in the northern corn belt lost an estimated $120 per calf (Beale et al. 1983, Table 6).

Hogs

Hogs, too, follow cyclical trends, historically averaging four years, and composed of two years of expansion and two of liquidation (Figure 8-4). Perhaps because the cycles are shorter than for cattle, and hence provide a larger data base, economists have shown great interest in measuring the periodicity of these cycles. They have been described alternatively according to the simple cobweb model, the multi-frequency cobweb model, and as displaying harmonic motion and rational expectations (Jelavich 1974; Larson 1964; Jameson 1982). Indeed, it was for hogs that the first regular cycling of supplies was recognized way back in 1876 (Brenner 1907).

Whatever the best statistical description of the cycles may be, analysis and observation suggest two characteristics of importance: (1) the cycles are attributable to producers' responses to price while production response lags for biological reasons, and (2) the cycles are less regular than for cattle, ranging from three to seven years.

Point one says hog and cattle cycles have similar origins. Hog cycles are less pronounced, however, because the gestation period is shorter (four months vs. nine months for cattle), and principally because litter size ranges upwards of 20. (The average is about 7.7 pigs per litter.) Thus, a sow can produce 15 to 40 pigs a year compared to one calf for a cow. As a result, the buildup phase is much more rapid, allowing hog feeders to be more responsive to changes in profitability. Thus, hog cycles, at the extreme, fluctuate by only some +/−15 percent from low to high. Cycles do affect sector profitability, but not to the degree seen for cattle.

The variation in cycle deviation means, however, it is difficult to use them as a planning tool. If the farrower plans for a two-year expansion period only to have liquidation begin after one year (as in 1982-1983, Figure 8-4), he or she is off the mark by 100 percent. In the past, the hog-corn price ratio was a good indicator of the profitability of feeding and, hence, the numbers of animals produced. Traditionally, a fall ratio of around 12.7 signaled expanded spring farrowing (Figure 8-5).

Over the years, several factors have changed, thereby minimizing the predictive value of the hog-corn ratio. Cost increases for labor, fuel, and capital have reduced the feed energy cost of the enterprise to only about 30 percent (USDA, ERS, Situation and Outlook 1990, Table 22). Hog far-

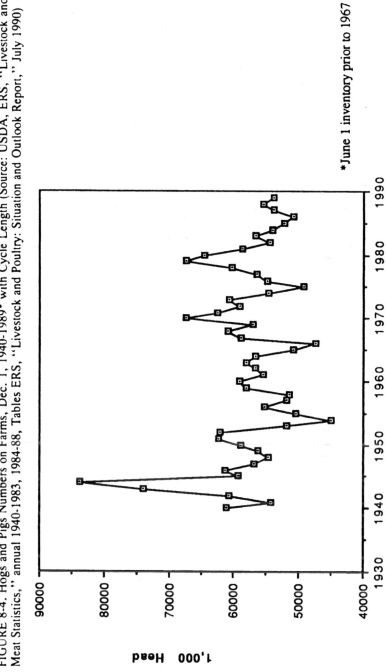

FIGURE 8-4. Hogs and Pigs Numbers on Farms, Dec. 1, 1940-1989* with Cycle Length (Source: USDA, ERS, "Livestock and Meat Statistics," annual 1940-1983, 1984-88, Tables ERS, "Livestock and Poultry: Situation and Outlook Report," July 1990)

*June 1 inventory prior to 1967

FIGURE 8-5. Relationship of Spring Farrowings to Fall Hog-Corn Ratio Previous Year, 1925-1961 (Source: Harlow 1962, p. 23)

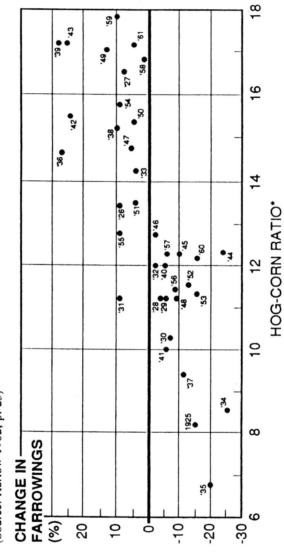

* RATIO OF PRICE OF HOGS PER HUNDREDWEIGHT TO PRICE OF CORN PER BUSHEL, SEPT.-DEC. PREVIOUS YEAR, U.S. FARM BASIS.

rowing and finishing have become more specialized and capital-intensive so that producers are not as flexible in varying production levels. But, perhaps most importantly, federal price support programs have done much to stabilize corn prices since the 1930s. This means that the denominator of the ratio is relatively constant year to year, so that variation is due largely to the change in hog prices themselves. This is not to say that corn prices do not vary: changes in exports are one source of change but large domestic stocks dampen the swings most years. The principal matter in determining future profitability is, then, projecting pork prices (Harlow 1962, p.23). Studies of producer price expectations have shown them to be more complex than simple extrapolations of current prices (West 1969, p. 34; Hayes and Schmitz 1987). Most likely, producers consider the supplies of competing meat, particularly beef, when making price projections.

PROFITING FROM LIVESTOCK CYCLES

The combination of cyclical animal numbers with the resultant cycle in prices leads to a distinct profitability cycle for cattle and hogs. Profit cycles are more difficult to graph because profits and means for measuring those profits vary from region to region and individual to individual. The results are nonetheless very real — and very catastrophic — regular periods of net losses. Not only are the losses in themselves costly, but they also require the livestock operator to maintain capital reserves (equity) to survive those periods by securing loans. The average return on that equity is often low, making livestock enterprises less profitable overall.

The trick to managing cycles is planning for them. The principal offense is anticipating cycle changes so that, ideally, numbers are low when prices break but high when they strengthen. Techniques for anticipating changes in the cycle are discussed here, with emphasis given to cattle because of the relative severity of the cycle for that species. However, no anticipation can be perfect because turns in a cycle are generally precipitated by external events: drought, export grain sales, etc. Thus, even the best production planning leaves a degree of uncovered price risk. Procedures for dealing with price risk are described in Chapter 14.

Indicators

Eight indicators are generally recognized as signalling turning points in the cattle cycle (USDA, ESS, Crop Reporting Board 1981, pp. 31, 35 and 40; Beale et al. 1983, p. 11). While any one indicator may be inaccurate, when taken together, most effectively indicate when cattle prices are at their highs or had reached them the previous year. Concurrently, they signal when prices are strengthening and the buildup can be anticipated. The eight indicators, in brief, are:

1. *Year of the Cattle Cycle* — Historically, the expansion phase has taken six to eight years while prices break one to two years before numbers begin their decline. Thus, on average, a price break can be anticipated five to seven years following the beginning of the cycle. This is perhaps the least precise measure.

2. *Annual Expansion in All Cattle Numbers* — Using the past as a guide again, total beef consumption has been averaging a two-percent gain per capita in the long term. Thus, whenever supply grows by more than two percent for two or three years in succession, a price break is imminent.

3. *Annual Expansion in All Cow Numbers* — In an analogous way to all cattle numbers, growth in cow numbers greater than two percent a year will lead to an oversupply after several years. Alternatively, steady-state cow numbers have been about 24 per 100 persons. That number rose to 26.8 in 1975 when prices reached their low, but was down to half that by 1988.

4. *Ratio of Annual Cattle and Calf Slaughter to January 1 Inventory* — When the ratio is less than 36 percent, the herd is growing at a rate above two percent per year. When this happens for two to five years in succession, a price break is nearly certain. With consumption expanding at a somewhat lower rate than two percent annually, a ratio of 37 percent is a better target figure than 36 percent.

5. *Ratio of Annual Cattle and Calf Slaughter to Previous Year's Calf Crop* — A ratio of 85-88 percent is an equilibrium value. Below that, herd growth is excessive; above it, liquidation is underway. With the 1980s trend to lower consumption, a growth rate nearer the 88 percent figure is the better measure.

6. *Ratio of Annual Cow Slaughter to January 1 Inventory of All Cows* — As a rule of thumb, slaughter of more than 15 percent of the cow herd signals liquidation, while slaughter of 12-13 percent indicates rebuilding. Rebuilding over several successive years signals pending supply and price problems.

7. *Ratio of Cow and Heifer Slaughter to Steer Slaughter* — Traditionally, female slaughter must be 82 percent of steer slaughter or the herd is growing excessively, but with the growth rate of demand declining, the danger level is now about 90 percent.

8. *Steer-Corn Price Ratio* — The steer-corn ratio, the live steer price

divided by the price of a bushel of corn, is a simple indicator of sector profits. Historically, it has signaled an equilibrium at around 25, with a decline preceding a price break and onset of the liquidation phase.

Historically, these indicators have effectively signaled price breaks. For example, in 1972 six of the eight were in the danger zone, and, by the following year, seven were (Table 8-2). Prices declined sharply from 1973 to 1974 and again from 1974 to 1975. The indicators, then, effectively projected when prices will decline and, alternatively, when they will increase. During the last, most unusual cycle, these classical ratios performed poorly in predicting either turning point in the cycle.

Short-Term Projections

Short-term projections are possible from the USDA series, "Cattle: Cattle on Feed," released in January, April, July, and October. This series reports on the 23 major feeding states that typically represent 95-97 percent of all fed cattle marketings. Placements during any three-month period can be assumed to average 600 pounds. At a gain of three pounds per day, six months are required to bring a calf to a market weight of 1,050 pounds. Thus, average net placements indicate marketings as follows:

Average Net Placements During	Expected Slaughter During
August-October	January
September-November	February
October-December	March
November-January	April
December-February	May
January-March	June
February-April	July
March-May	August
April-June	September
May-July	October
June-August	November
July-September	December

Dividing the number by .96 (the approximate percentage supply of the 23 states in the report) gives the estimated total marketings each month. Sharp movements are indicative of price adjustments. Longer term projec-

TABLE 8-2. Cattle Cycle Indicators (Sources: Beale et al. 1983, Table 5; USDA, and ERS, "Livestock and Meat Statistics, 1984-88," "Livestock and Poultry: Situation and Outlook Report," Feb. 1989 and 1990)

Year	Year of cycle	INVENTORY INDICATORS			SLAUGHTER RATE INDICATORS				Prices received by farmers	
		Percent annual growth rate		Percent of Jan. 1 inventory slaughtered	Slaughter as percent of prev. year's calf crop	Cow slaughter as percent of cow herd	Cow & heifer slaughter as percent of steer slaughter	Beef steer/ corn ratio	Cattle	Calves
		-- total inventory	--cow herd						$/cwt	$/cwt
	(1)	(2)	(3)	(4)	(5)	(6)	(7)	(8)		
1950	2			37		15	82		23.30	26.30
1951	3	+5	+4	32	74	14	84		28.70	32.00
1952	4	+7	+5	32	78	14	77		24.30	25.80
1953	5	+7	+7	39	96	17	81		16.30	16.80
1954	6	+2	+5	41	95	18	94		16.00	16.50
1955	7	+1	0	41	93	19	100		15.60	16.80
1956	8	-1	-2	42	97	19	92		14.90	16.10
1957	9	-3	-3	42	95	19	90		17.20	18.70
1958	1	-2	-3	37	85	18	90		21.90	25.30
1959	2	+2	0	34	82	14	76		22.60	26.70
1960	3	+3	+2	36	89	12	78		20.40	22.90
1961	4	+2	+2	35	88	13	81		20.20	23.70
1962	5	+3	+2	35	87	12	77		21.30	25.10
1963	6	+4	+3	34	85	12	76		19.90	24.00
1964	7	+3	+3	36	93	11	72		18.00	20.40
1965	8	+1	+2	38	94	14	94		19.90	22.10
1966	9	0	-2	38	93	17	94		22.20	26.00
1967	1	0	-1	37	93	16	87		22.30	26.30
1968	2	+1	0	38	94	14	90		23.40	27.60
1969	3	+1	+1	37	92	14	91		26.20	31.50
1970	4	+2	+2	35	88	13	82	22.8	27.10	34.50
1971	5	+2	+2	35	87	13	81	28.9	29.00	36.40
1972	6	+3	+2	33	84	12	78	25.5	33.50	44.70

Year	(1)	(2)	(3)	(4)	(5)	(6)	(7)	(8)		
1973	7		+4							
1974	8	+3	+4	30	76	12	80	16.0	42.80	56.60
1975	9	+5	+5	32	82	14	83	13.8	35.60	35.20
1976	10	+3	-3	36	92	20	123	15.6	32.30	27.20
1977	11	-3	-5	38	97	19	121	17.9	33.70	34.20
1978	12	-4	-5	39	101	19	112	23.3	34.40	36.90
1979	1	-5	4	38	96	17	109	28.4	48.50	59.10
1980	2	0	0	31	84	12	90	27.7	66.00	88.80
1981	3	+3	+4	33	86	13	93	20.7	66.90	86.67
1982	4	-1	-1	33	84	13	91	26.1	63.84	72.43
1983	5	0	-3	34	89	17	103	23.2	64.22	72.43
1984	6	-3	0	35	91	19	105	20.0	62.52	68.16
1985	7	-4	-6	36	97	22	110	22.6	65.34	68.91
1986	8	-3	-5	39	97	20	110	25.8	58.37	68.65
1987	9	-3	0	38	100	23	109	41.4	57.74	70.67
1988	10	-2	-2	38	97	19	100	35.7	64.60	69.67
1989	1	0	0	37	93	19	98	28.8	70.15	78.02
1990	2	0			91	18	101		72.52	81.50

Warning indicators (see underlined data):

(1) Fifth to seventh year of cycle

(2) Growth rate above 2% for several years

(3) Growth rate above 2% for several years

(4) Less than 37% of inventory slaughter

(5) Slaughter less than 88%

(6) Less than 13% of cow herd slaughter (Less than 15%, 1950's; less than 14%, 1960's and early 1970's)

(7) Female slaughter less than 85% prior to 1970; 90% from 1970 forward

(8) The historical equilibrium point has been 25

205

tions are possible through examining the calf numbers, but that indicator is already captured in number five above.

SYNOPSIS

Livestock cycles have existed since data have been kept — and longer. They cause major disruptions and losses in the sector, yet there is no universal agreement on how to avoid them, or even their cause. Two sources receive the most attention: (a) exogenous factors, including weather and imports; and (b) endogenous matters that refer to producers' responses to market conditions. The best explanation appears to be a combination of the two, with internal decisions (acting within biological response delays) creating the cycles once they are triggered by an outside event.

Historically, cattle cycles have lasted nine to twelve years, including a six- to eight-year expansion and three- to ten-year decline. Despite this regularity, the magnitude of the cycles (as measured by a percentage change from high to low) has been declining over time. The most recent complete cycle, running from 1979 to 1988, has not followed the traditional pattern due to a prolonged drawdown phase which is possibly attributable to the "structural change" in demand. For biological reasons hog cycles are shorter duration, about four years, and of less magnitude than cattle. The hog-corn price ratio is an increasingly poor predictor of hog cycles due to the diminishing role of feed costs in profitability.

A number of indicators have been identified to predict cattle cycle turning points in particular. While any one may not be valid, as a group they are indicative of pending supply and price changes. In the very short run, data for animals on feed are indicative of supplies coming to market, which are a major determinant of price.

Study Questions

1. Describe livestock cycles and identify whose "fault" they are.
2. How and why have cattle/hog cycles changed over the century?
3. Using the most current data available, identify where the cattle cycle is currently, and project the nearest high (or low) point in slaughter.
4. What happened following 1979, and how can the cycle be predicted again with any assurance?

REFERENCES

Beale, T., P. R. Hasbargen, J. E. Ikerd, D. E. Murfield, and D. C. Petritz. *Cattle Cycles: How to Profit from Them.* USDA, Extension Service, Misc. Pub. No. 1430, March 1983.

Brenner, S. *Brenner's Prophecies of Future Ups and Downs in Prices.* Cincinnati: R. Clarke Company, Edition 16, 1907.

Breimyer, H. F. "Emerging Phenomenon: A Cycle in Hogs," *J. Farm Econ.* 41(1959):760-68.

Ehrich, R. L. "Economic Analysis of the United States Beef Cattle Cycle." University of Wyoming, Science Monograph 1, April 1966.

Harlow, A. A. *Factors Affecting the Price and Supply of Hogs,* USDA, Econ. and Stat. Analysis Division, Tech. Bull. 1274, 1962.

Hasbargen, P. and K. Egertson. "Cattle Cycles and Economic Adjustments in the Beef Industry," In *Cattle Feeders Days Report,* University of Minnesota, December 1974.

Hayes, D. J. and A. Schmitz, "Hog Cycle and Countercycle Production Response." *Am. J. Agr. Econ.* 69(1987):762-70.

Jameson, M. H. "Rational Expectations and the U.S. Hog Cycle: Statistical Tests in a Linear Model." In G. C. Rausser (ed.), *New Directions in Econometric Modeling and Forecasting in U.S. Agriculture.* New York: North Holland Publishing Company, 1982, pp. 257-280.

Jelavich, M. S. "Distributed Lag Estimation of Harmonic Motion in the Hog Market," *Am. J. Agr. Econ.* 56 (1974):38-49.

Kendell, K. and W. Purcell. "The Beef Cycles in the 1979s." Oklahoma State University, Agricultural Experiment Station Bulletin B-721, March 1976.

Larson, A. B. "The Hog Cycle as Harmonic Motion," *J. Farm Econ.* 46(1964):375-386.

Meadows, D. L. *Dynamics of Commodity Production Cycles.* Cambridge, MA: Wright-Allen Press, Inc. 1970, pp. 38-40.

USDA, Economic Research Service. "Situation and Outlook Report."LPS-34, Feb. 1989, LPS-40, Feb. 1990 and LPS-42, July 1990.

USDA, Economic Research Service. "Livestock And Meat Statistics." Various issues.

USDA, ESS, Crop Reporting Board. "Cattle on Feed." Various issues.

USDA, Extension Service. "Cattle Cycles: How to Profit from Them." Misc. Publication #1430, March 1983.

USDA, Statistical Reporting Service, Crop Reporting Board. "A Handbook on Surveying and Estimating Procedures." ESS-13, July 1981, p. 26.

Walters, F. "Predicting the Beef Cattle Inventory," USDA, *Agr. Econ. Res.* XVII, No. 1, 1965.

West, D. A. "Swine Producers' Price Expectations and the Hog Cycle." North Carolina State University, Dept. of Economics, Economic Res. Rpt. No. 10, October 1969.

PART III:
UNDERSTANDING LIVESTOCK MARKETING

Chapter 9

Functionaries in the System

Several years ago, *The New York Times* carried a story on the beef industry titled "On the Trail of a Steer: From Birth to Burger." The article described the two-and-a-half-year odyssey of a steer beginning in Missouri and travelling through other states, owners, and functionaries before it was finally consumed in New Jersey (Serrin 1978). This marketing system is much more complex than the system used earlier at Faneuil Hall, the first permanent livestock market in the States (see Chapter 3). It is tempting to view this modern system as excessive and, therefore, overly costly. This observation is partially correct. The payments that the marketing intermediaries and others receive come, in the long run, out of the producers' shares since the producer is the residual claimant (see Chapter 8 for a discussion of this term). However, this interpretation ignores the importance of the necessary services provided by each successive owner and functionary. The services may be extensive, such as slaughter and processing, or seemingly minor, such as market news, but all act together to advance the livestock enterprise. Without all of these middlemen, producers would have two alternatives: they could perform the service themselves by internalizing the operation, or they could dispense with the service.

Early pioneers chose the latter alternative and, until about 60 years ago, farms in the United States were largely self-sufficient. Farmers produced many of the necessities for family members and their own production needs as well as selling largely finished products to the ultimate consumer. Indeed, most farmers in the developing world still operate in this way. Recently, producers in the industrialized world have chosen to specialize, that is, to concentrate on one or a few stages in the production process. Necessary inputs such as fuel, machinery, food for the farm family, and marketing services are bought, while semifinished products are bought, sold, and resold. The producer gains as a result of the process of specialization. For example, Adam Smith applied this concept to the manufacture of pins and recognized that production could be made less costly by reducing the number of tasks performed by each worker. He called this labor specialization.

The process by which livestock producers benefit from specialization is somewhat different; they have access to equipment and inputs such as fertilizer and machinery that are impossible or impractical to home-produce. Producers also benefit by better using land resources so that, for example, calves can be weaned on grasslands and subsequently shipped to areas of abundant grain. And finally, producers gain by reducing the number of operations in which they must be proficient. Thus, producers may be expert on one species rather than three or four, or they may emphasize production in only one stage of the life cycle, or they may choose to concentrate on feeding only, rather than feed production and feeding.

This does not mean that specialization is necessarily advantageous. For example, the prevailing practice of combining feeder pig production with fattening strongly suggests there are no significant benefits to separating the two activities. However, specialization is a trend and one which has justifications beyond the size economies of a particular activity or machine. Indeed, the principal benefits from specialization for the livestock producer probably come from the more efficient use of land and management resources.

Specialization does place the producer in a different position vis-à-vis the market. Whereas the self-sufficient farmer may have sold ten head per year, the specialized producer must sell 5,000 or 10,000 or even 100,000. The value added per head decreases as purchased inputs increase. At the same time, the margin declines and the producer gets less out of each animal. Increased volume is necessary to make up the difference. The larger numbers, however, make the producer a major livestock seller. Ten head could perhaps be sold efficiently as a part-time effort, but 10,000 head demand considerable effort and attention. It is in this situation, as in the other parts of the livestock operation, that the producer uses specialized selling services rather than attempting to do everything himself or herself.

It is not feasible in today's complex and integrated national and world markets to do without any marketing services. Doing without places the producer in the situation of 100 years ago, when a dealer well informed about prices could make as much on a quick trade as a producer did from a year's hard work. Today, price information is a necessity for a properly functioning market. For example, grade standards, another marketing service, expand the market by making it possible to buy sight-unseen. Marketing services help the customer, upon whom the system ultimately depends, buy healthful and predictable products.

The purpose of this chapter is to describe the major functions and functionaries of the livestock marketing system. These include buyers, graders, and regulators, all of whom define the environment within which the producer operates. The emphasis in this chapter is on understanding how and why the system operates as it does. In Chapter 13, this information on marketing activities will be interpreted and incorporated into a marketing strategy.

PRINCIPAL FUNCTIONARIES IN THE SYSTEM

This subsection describes the functions of the livestock buyer or his/her representative, perhaps the most important participant in the system from the producer's perspective. The principal and obvious function of all buyers is that of taking title in exchange for payment. When choosing from the several types of buyers, the producer should consider three principal differences. First, buyers have some discretion over price, especially on the down side. Carefully selecting among buying groups can have a significant impact on price. Second, a related matter is costs; more efficient buying systems, as represented by particular buyers, are lower cost than others. These buyers have the potential for paying higher prices. They may not necessarily pay higher prices, because price is competitively determined, not cost based. Most important is the package of services associated with each type of buyer; buyers' services differ by time of payment, assurance of payment, location of sale, and a host of other factors that distinguish one form of transaction from another. The activities of the major buyer groups are described below.

Packer and Feeder Buyers

The principal characteristic of the packer buyer is that he/she works for the packer. Thus, virtually all animals are intended for immediate slaughter. Therefore, the packer buyer looks for stock with good carcass characteristics and freedom from any impairments that could lower the value of the meat or hide. Packer buyers are highly skilled and operate under considerable pressure. As a group, they must provide stock to keep the plant in operation while actually taking ownership only a few days ahead of planned slaughter. They are also buying for carcass characteristics which must be judged from examining live animals. Often they must operate within a margin of $+/-$ 1/2 percent on estimated yield percentage. The yield percentage is the proportion of liveweight that goes into the cooler after slaughter.

Characteristics that affect yield include breed, fill, and cleanliness, as well as the major factors of configuration and fattiness. Choice is made based on judgement and experience, with the emphasis on experience. Firms constantly improve their buyers' judgment by giving the buyers feedback on comparisons of actual yields to estimated yields. They also pay their buyers well. In fact, in small firms the owner is often the principal buyer, while in medium-sized firms the buyers are among the highest paid in the company.

Firms differ as to how much discretion is permitted to buyers. Many companies place narrow limits on bids by having the head buyer establish the limit at the beginning of the day. Improvements in communication have changed this process so that the head buyer is kept abreast of market trends throughout the day and can adjust bids accordingly. It is for this reason that major buyers have phones at ringside and CB radios in their cars (Figure 9-1). This communication equipment is also used to relay changes in buying quotas as a day's market develops. Pricing has become more centralized because of these better communications and the resulting enhanced coordination among the buyers. The individual buyer is then responsible for judging the relative worth of the stock, rather than absolute prices. It is important for the producer to recognize the limited price discretion of the individual buyer when negotiating a transaction.

The high cost of supporting a buyer in the field means a packer must use his/her time efficiently. This, in turn, determines where they will be found. Smaller numbers of head are available at many terminal and auction markets (see Chapter 13 for a description of these markets), so that they are relatively expensive places to buy. Hence, packer buyers concentrate on direct purchases from feeders. There are other advantages and

FIGURE 9-1. Packer Buyers Use CB Radios to Coordinate Field Operations (Source: IBP)

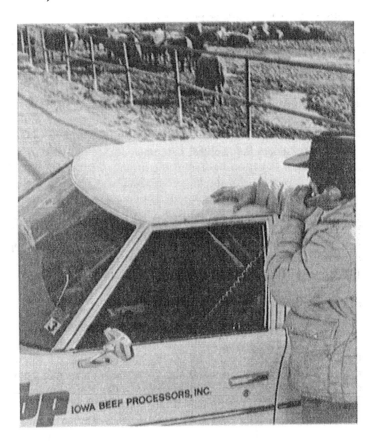

disadvantages between public and direct markets (see Chapter 13), but many of the decisions are based on costs.

Selling to a packer buyer can be an intimidating experience. A producer, a part-time seller at best, confronts an experienced, full-time operator who often has the support of a large organization. To some extent, the producer can minimize the buyer's advantage by being as knowledgeable as possible and by adopting an effective strategy. But common interest can override opposing goals. Above all, buyers are looking for a consistent quality product. The quality sought does not have to be high, but the buyer must know what quality is being bought. To a large degree, quality can be observed by an experienced person, but even the best buyer makes errors without additional information. The information preferred is previous experience with a producer's stock. A producer with consistent quality animals has an advantage when selling repeatedly to the same buyer. The buyer might hesitate to take advantage of one particular situation if that action could jeopardize an important long-term source of animals. When buyers and sellers can see beyond the current transaction to the long term, their mutual dependence becomes more evident. One of the key requirements of a good marketing program is establishing these longer term relationships.

Like packer buyers, feeder stock buyers can either work for a feeder or, indeed, may be feeders themselves. Circumstances, however, frequently make employing a buyer an impracticable arrangement. A salaried individual is too costly, even for the largest feeders who are nevertheless not in the market regularly. Maintaining market contacts, on the other hand, presents a problem for the buyer when purchases are sporadic. Geography presents another problem, especially for feeder calves. With calf production and feeding often separated by one or two thousand miles (Figures 4-8 and 4-9), the costs of buying become substantial. As an alternative, would-be buyers turn to a commission representative, the order buyer.

Buying Stations

A variant on direct buying at the farm is the use of buying stations (Figure 9-2). Buying stations are owned by packers and serve as concentration points for livestock. In many areas, stations (sometimes known as concentration points) are less expensive for a packer than buying direct. The packer does not have the time and transportation expenses of moving buyers and trucks over wide areas. Those expenses are particularly high when production is dispersed or when there is a predominance of small (less than truckload) lots. Thus, the packer shifts some costs to the producer by expecting him or her to ship to the station where lots may be concentrated. In this respect, the packing plant itself can serve as a type of buying station where the producer may ship and receive payment.

Buying stations provide a good alternative for producers in areas where it may not be economical for buyers to visit the farm. Nonetheless, farm sales remain the option of choice for two reasons. First, the station price must reflect transportation expenses incurred by the seller as well as the shrink loss during transit (see Chapter 7). Second and more important, the producer is in a weaker bargaining position once his animals have been loaded and shipped. Rejecting a bid usually means hauling back to the

FIGURE 9-2. Typical Country Buying Station

farm and finding a market that will pay a price high enough to cover the additional costs.

Despite the apparent advantage for packers who operate buying stations, the number has been decreasing over the past two decades. Numbers can only be estimated, because buying stations are not regulated as distinct entities and hence figures are unavailable. Declines may be traced to increases in farm size and specialization, making truckload sales commonplace. At the same time, the maintenance of a buying facility with regular weekly hours is costly. Squeezed from below by costs and from above by declining volume, many buying stations are closing.

Order Buyers

Order buyers are commission agents who buy on the account of the feeder or packer. Order buyers are brokers and may be so called, although that term is not often used. The commission may be on a percentage or per head basis, but the important characteristic is that the order buyer never takes actual title to the animals he or she handles. This type of buyer does not accept any of the price or death loss risk inherent with ownership. Rather, the order buyer provides a service for buyers who find it impractical to carry out that function on their own.

Order buyers engaged in interstate commerce or selling to firms engaged in such commerce (as are almost all packers) must be bonded and follow certain fair marketing practices as stipulated by the Packers and Stockyards Administration (P&SA) of the USDA (see below). In addition, many states have bonding requirements that extend to operators involved in intrastate trade.

In some situations, the order buyer can function more efficiently than the direct buyer. This frequently occurs when buying entails a protracted assembly and sorting process. An extreme example is the buying of bob dairy calves for special white veal feeding programs. Dairymen send most bull calves to the local auction barn shortly after birth. The result is a scattering of calves of different grades and conditions across a number of markets. For the feeder to assemble a lot of 50 uniform head would take a protracted effort of visiting a series of markets over several days. Purchased calves would either have to be held at the individual markets until the total number could be picked up or assembled daily during the buying period. In either case, the transport costs from market to market and then to the farm would be high. Time costs for the buyer/operator are also high, because most of a day's work would have to be charged to the 10 or 30 head purchased at each auction.

Under these circumstances, the order buyer can be more efficient in terms of both time and expense. He or she buys a range of grades, meaning more animals are available at each market and the time value of each purchase is correspondingly lower. Assembly and distribution are also more efficient with the use of a larger truck and possibly better routing and handling arrangements. In a competitive market, these cost savings accrue to the producer directly when the order buyer purchases for a feeder, or, indirectly, as a higher price when a packer's buying costs are reduced. There are trade-offs, however, so that the producer may suffer some loss in competition, especially in the smaller markets. An order buyer representing three packers is unlikely to compete as fiercely as would three representatives of separate packers. Sellers in some markets notice a price drop when the largest buyer (often an order buyer) leaves the ring for a cup of coffee. On the other hand, if the order buyers were not present, the total selling activity at many markets would be reduced. In summary, order buyers provide an important service to the livestock industry even if their activities are controversial.

The number of active order buyers is difficult to estimate accurately. According to the P&SA, which combines them with dealers (described below), there were nearly 6,000 in 1986, the last year for which these data are reported (USDA, P&SA, Stat. Rpt. 1987, Table 21). Of these, the west North Central area had the greatest share (over 25 percent). Percentages do not tell the complete story. For example, beef packers use salaried buyers to procure an estimated 85 to 95 percent of their needs (Ward 1979, p. 11). At the same time, the fringe producing states have greater numbers than their proportion of stock. This partially substantiates an expectation that order buyers support the industry in areas where direct purchases are impractical.

Further complicating the computation of the number of animals handled by these groups is the common practice of exchanging ownership several times before slaughter. Gross figures reveal that dealers and order buyers handled nearly one in two hogs sold in 1980, but they could have bought and sold some animals repeatedly at several points in the growth process (USDA, P&SA, Stat. Rpt. 1989, Tables 3 and 24). These two types of market agents handled $20 B in livestock during 1987, making them a major force in the sector.

Livestock Dealers

Like order buyers, livestock dealers serve packers and producers. The main difference between them is dealers take ownership. That is, they are entrepreneurs and risk-takers in their own right. Both state and federal regulations treat dealers and order buyers alike.

Dealers also serve important, if specialized, functions. For example, smaller cattle feeders in the Corn Belt may wish to select calves personally and rely on dealers to augment the local supplies by shipping in animals from Montana and Wyoming. Also, dealers are able to provide more personalized services such as direct cash payment and transport. Some even advance terms on sales. For this, of course, the buyer pays. Smaller scale dealers lack a full complement of equipment so that weights may be estimated rather than measured.

Dealers can also serve a salvage function. The larger feeder may find it necessary to remove sick stock promptly or risk contaminating other animals. Rather than going to low-valued slaughter (possibly for non-human consumption), the dealer may bring those animals back to health and thereby benefit the entire industry. In these ways, the dealer is operating on the fringe of the industry, adding value through services rather than numbers. These services can be invaluable, especially for smaller or isolated producers.

The number of active dealers can only be estimated. P&SA data do reveal that dealers handle about twice the value of stock compared with order buyers. Dealers usually operate on the fringes of the industry, in areas where production is limited or great variability is seen among animals. That is, dealers provide specialized services and are most in demand where marketing is not a routine function. Dealers are particularly useful where culls are sold off on an occasional basis.

Livestock Trucking

The trucker links the participants in the livestock system. This represents a change from years ago when cattle were driven to market, and then later when train transport predominated. Today, virtually all livestock moves by truck, which can typically offer more rapid and reliable transport.

Producers who sell at the farm or use their own trucks are seldom involved with itinerant truckers. The remainder, and this means the great bulk of the sector, do rely on independent trucking firms. Truckers are contracted either directly or, for shipments to a public market, through an agency. Once the rate and dates are agreed upon, the haul is made and the contract is settled, with the trucker being paid directly in case of a public market sale. This is all straightforward. Further adding to the seeming homogeneity of this activity is the highly uniform rate structure, most notably in states that regulate such rates. During the 1970s, state regulation of rates was most pronounced in the West and Southwest (Boles 1976, p. 21). Given the apparent similarity of trucking firms, producers are often very casual about selecting among them. This may be a mistake since the trucker can have a substantial impact on shrink, condition, and health. Stressing through improper loading can cause bruising and excessive shrink. In extreme cases of heavy exercise shortly before slaughter, the carcass becomes a "dark cutter" of a deep red color. While no tissue damage is done, the cosmetic value of the meat is reduced. Disease is an issue for animals going back on feed, especially younger ones with weak immune systems. A common form of infectious disease known as "shipping fever" is obviously associated with the stress and contact made during transport. Attention should be paid to the cleanliness of the equipment. These issues are more acute for partial loads when unfamiliar animals (often of mixed species) are mingled. It is especially important that the trucker give particular attention to proper sorting and penning procedures.

In each of these situations, truckers' services can differ. A producer should get information on care and reliability of an individual trucker or firm before loading up his or her animals. Neighbors and market personnel, if a public market is to be used, are obvious sources. A trucker's level of cargo liability coverage is critical in the case of loss or accident. While the chance is small — a mid-1970s survey found less than one percent of the loads had claims placed against them and most were for injury rather than death (Boles 1976, p. 10) — a heavy loss can nonetheless be catastrophic if compensation is inadequate. Producers should inquire about the cargo insurance carried by a trucker. The minimal levels of cargo coverage mandated by many states probably do not keep up with inflating livestock values. Departments of agriculture in states with mandated insurance are good sources of information.

Producers can help truckers by providing adequate loading facilities, by sorting and penning the animals properly in an accessible location, and by being present when the pick-up is made. Truckers are often called upon to do nearly impossible things on the farm, inevitably leading to conflict and misunderstandings.

Because livestock carriers come into contact with producers infrequently, and then only briefly, they are mistrusted. Among the many myths of the transport sector is that individuals go in and out of business regularly, and therefore they are not interested in offering good service or in preserving a reputation. The facts point to quite a different industry, one which is indeed quite stable. The average years in service of surveyed firms in the mid 1970s was 18 years, and nearly 90 percent of them had been in business for at least three years (Boles 1976, p. 2). My own work in New York, where the industry is quite different, nonetheless showed much the same pattern; by and large, companies are long-standing, some operating over several generations (Lesser 1980, p. 11). Livestock carriers do tend to be independently minded because access to the industry is relatively easy and inexpensive. Commercial realities, however, mandate good service if a firm is to survive in the longer term. Firms tend to be small; many operate an average of only 5.2 trailers (Boles 1976, p. 4). This is probably a reflection of the limited size economies of the sector as well as the need for personalized service.

Government regulation is one factor that lends stability to the sector. As mentioned, several states control tariff structures, others stipulate minimum insurance requirements, and some determine entry based on a needs assessment. Limiting entry and mandating rate structures have the general effect of maintaining higher costs (see Council Ag. Sci. and Tech. 1979, pp. 30-32). Shippers of unprocessed agricultural products have been exempt from federal trucking regulation since the original legislation went into effect following the Depression. Of course, livestock haulers are subject to driver safety regulations covering hours driven per day and per trip.

The relative costs of live animal delivery have declined relative to the movement of meat as packing plants have been built closer to production points. Nevertheless, the total live animal transportation input for a Plains calf fed out in the Corn Belt has been estimated at $7.50 in 1989-1990. Under the same estimates, hogs are far lower—around 50 cents (USDA, ERS, Outlook and Situation Jan. 1990, Table 17). In terms of net margins, these direct costs are substantial. Considering possible indirect costs argues for a better understanding of this sector and for more care in the selection of a carrier.

Marketing Intermediaries

Marketing intermediaries facilitate the distribution of the product. They consist of two principal groups: middlemen and "facilitators" (Kotler 1980, p. 45). Facilitators assist in the sales and distribution of the product; they do not take title and, in many cases, they may not handle or even see the product. Included in this group within the livestock sector are market news agents, inspectors, risk-takers (bankers and speculators), and regulators. As a class, they fit into the principal functionaries of the system while, as a group, they receive the sharpest criticism, probably because their contribution is more ephemeral than, say, that of a buyer. Rather than offer a round endorsement for the class—one which is generally justified but certainly not in each instance—the functions of the major facilitators are described in detail below. We begin in Chapter 10 with perhaps the most significant, if controversial: market news.

SYNOPSIS

The livestock producer's principal contact with the market will be through buyers employed by packers and feeders. These are highly skilled individuals operating within small margins when estimating yields; a regular supply of animals of constant quality eases their task and can raise the long-term price of stock. Most of these purchases are made directly at the feed lot or farm. The other place where these buyers can be contacted is *buying stations*, but these have been declining in numbers. Producers take a greater risk when shipping to a buying station because the costs of the return haul must be absorbed if a deal is not struck.

Order buyers and *dealers* serve as intermediaries between the producer and feeder or packer. They are more costly, but they provide useful services to smaller producers with limited marketing experience and those in remote areas poorly served by packer buyers. The two can be distinguished by noting that the order buyers do not take title to the animals (they receive a commission), while dealers do.

Study Questions

1. How do the several classes of buyers complement and compete with each other?
2. A principal distinction between livestock dealers and order buyers is the taking of title. Why is that such a key factor?
3. What factors are important in selecting among livestock carriers?

REFERENCES

Boles, P. P., "Operation of For-Hire Livestock Trucking Firms."USDA, Economic Research Serivce, Ag. Econ. Rpt. No. 342, July 1976.

Council for Agricultural Science and Technology. "Impact of Government Regulations on the Beef Industry."Ames, Iowa, Rpt. No. 79, October 1979.

Kotler, P. *Marketing Management: Analysis, Planning and Control.* Englewood Cliffs, NJ: Prentice-Hall, Inc., Fourth Edition, 1980.

Lesser, W. "Marketing Fat Cattle and Feeder Calves in New York." Cornell Univ., Dept. Agricultural Economics, A.E. Res. 80-34, December 1980.

Serrin, W. "On the Trail of a Steer: From Birth to Burger," *The New York Times* 12/10/78, Section 3, page 1.

USDA, Economic Research Service. "Livestock and Poultry: Outlook and Situation Report." Various issues.

USDA, Statistical Reporting Service. "Scope and Methods of the Statistical Reporting Service." Mis. Pub. 1308, July 1975.

Ward, C. E. "Slaughter-Cattle Pricing and Procurement Practices of Meatpackers." USDA ESCS Ag. Info. Bull. No. 432, December 1979.

Chapter 10

Market News — Sources and Uses

In the historical sketch in Chapter 3 situations were described in which shrewd buyers would ride out to meet incoming drovers. After buying the cattle, hogs, or sheep at a low price, they would move the livestock the few remaining miles to market and resell at a substantial profit. An abuse? Yes, certainly, but one not possible when all the market participants are well informed of the going price. Subsequently, sellers and buyers had easier access to market information because of the increased trade done at terminal and auction markets. At these markets, buyers competed among themselves and a full-time seller acted in the place of producers. Beginning in the 1960s, the volume of business done at those markets began to decline. It appeared that livestock producers were in the same position as a century before. However, during the interim a broad-based market news program had been established by the USDA and private firms. This chapter covers the conceptual views of information, an overview of what information is available, and an evaluation of the use of that information by producers.

First, however, it is important to distinguish between short- and long-run information. By economists' definition, everything is variable over the long run. For livestock producers, this time span may be understood as the period when production decisions are made: retaining a gilt for breeding, placing a steer on feed, etc.[1] Producers, at that juncture, make the important decisions of what to produce and how much to produce. The short-run decisions must be made when those animals are nearly ready to move to market. At that point, the producer must determine exactly when

---

1. The long term may indeed span a far longer period than a single feeding cycle. When National Farms invested $50 M in a hog farrowing and finishing facility, that was indeed a long-term decision (Houghton and McMahon 1984). For the purposes here, however, it is convenient to distinguish the long and short term based on feeding and marketing decisions, respectively.

to ship and what market to use. In this subsection we consider primarily the *short-run* issues.

Information is available that assists the making of long-run production decisions. Data on trade, slaughter, and animals on feed are the responsibilities of the USDA's divisions of Foreign Agricultural Service, Economic Research Service, and the National Agricultural Statistics Service, respectively. Bulletins are released regularly and critical information such as crop estimates are available through wire services (Figure 10-1). An annual outlook conference, typically in late November, is supplemented by six updates per year for livestock and poultry. Market price information for live animals is handled by the Market News Service of the Agricultural Marketing Service, USDA. Price collecting and dissemination is often done in cooperation with state agencies. A collection of public and private efforts covers the reporting of meat and futures prices.

JUSTIFICATION FOR PRICE REPORTING

Because of the agricultural recession following the Civil War, increased attention was given to means of stabilizing farm incomes and consumer prices. Today, that seems an obvious concern and one in which the federal government and private parties have a keen interest. At the time, however, the concept was foreign; the prevailing doctrine of laissez-faire had kept government intervention in the economy to an absolute minimum. Laissez-faire is, perhaps, a workable concept as long as there is some equality of bargaining power among parties. By the late nineteenth century, it was obvious that farmers could not match either the economic strength or knowledge of the merchants with whom they dealt (Miles 1967, p. 10).

To understand better the significance of bargaining power, it is instructive to refer to the perfect competition model. In order for the model to apply, two conditions in particular must hold: (1) no firm is large enough to affect prices by varying quantities bought or sold, and (2) all parties have perfect information on prices and other factors. As a practical matter the model predicts what the turn-of-the-century farmer had already recognized for himself: when economic power diverged from equality, the weaker party is disadvantaged. Subsequently, it fell to the federal government to rectify that situation.

The first action taken was direct intervention. That is, if the economic system was flawed because of deviations from perfect competition, due to divergent bargaining power, the solution was to restore the balance. This involved two distinct approaches. First was the passage of the initial and

FIGURE 10-1. USDA Calendar of Reports for November 1989 (Source: USDA, ERS 1989 *Calendar of Reports*)

NOVEMBER

MONDAY	TUESDAY	WEDNESDAY	THURSDAY	FRIDAY
		1	2 Egg Products N	3 Fruit & Tree Nuts E Poultry Slaughter N
6 Dairy Products N	7 Celery N	8 Vegetables & Specialties Yearbook E	9 World Agricultural Supply & Demand W Crop Production N	10 Holiday
13 World Food Needs & Availabilities E	14 Livestock & Poultry E Turkey Hatchery N Farm Labor N	15	16 Milk Production N	17 Feed E Sugar Market Statistics N Cattle on Feed N
20 Agricultural Outlook E Catfish N	21 Wheat E	22 Foreign Agricultural Trade Update E Cold Storage N Eggs, Chickens & Turkeys N	23 Holiday	24
27 Cotton & Wool E Livestock & Poultry Update E Livestock Slaughter N	28 Outlook Conference Exports E Peanut Stocks & Processing N	29 Outlook Conference	30 Outlook Conference Agricultural Prices N	

225

most formidable of the great anti-trust laws, the Sherman Act of 1890. This Act has gone through numerous reinterpretations by the courts during its 100-year existence, but its initial and obvious intent was to restrict the abuses of the great trusts. These abuses included discriminatory rates, secret mergers with ostensible competitors, and boycotting — all actions that would place a farmer in a disadvantaged bargaining position. (For further information on anti-trust laws, see Neale 1980).

The second action taken was based on a straightforward issue with an identifiable need: the collection and dissemination of market prices. Action came in 1913 with the establishment of the U.S. Office of Markets whose functions were subsequently subsumed by the Agricultural Marketing Service of the USDA. Interestingly, the 1913 Act was seen as benefiting both consumers *and* producers. Consumers suffered from rampant price variability. The more general matter was equity of the division of benefits among producers, processors, and consumers. Producers' concerns, of course, were with distributional equity between themselves and buyers.

While the provision of improved price information could be seen as benefiting both producers and consumers, an inherent conflict exists between the two groups. Essentially, producers gain at consumers' loss, and vice versa, at least in the short run. Faced with this dilemma, the federal government began, in the 1920s and 1930s, to move away from a strict emphasis on improving the efficiency of the marketing system. Instead, increased attention was given to protecting the interests of producers. This has been accomplished through surplus removal, price support programs, and marketing orders (Breimyer 1963). Livestock products have been indirect beneficiaries of these programs — through the stabilization of feed prices (albeit at higher than free market prices for most years) and the stimulation of demand under food stamp and school lunch programs. But the major thrust of these programs has been outside of livestock product markets. Hence, they are not treated in detail here.

SOURCES AND DISSEMINATION OF MARKET INFORMATION

Live Animals and Livestock Slaughter

Reporting data on livestock numbers is the responsibility of the Agricultural Statistics Board, National Agricultural Statistics Service of the USDA. A total of 14 separate reports are prepared for animal numbers, and two for livestock slaughter. Most reports are released several times a year. (Table 10-1 refers specifically to reports released in 1989.) Several

TABLE 10-1. USDA Agricultural Statistics Board Livestock Reports in 1989 (Source: USDA, Ag. Stat. Board, National Ag. Statistics Service, 1989 Calendar)

Livestock Commodity	Months
Hogs and Pigs	Jan., March, Sept.
Livestock Slaughter	Monthly (annual in March)
Cattle on Feed	Monthly
Cattle	Feb., July
Agricultural Prices	Monthly (annual in June)
Meat Animal -- Production and Income	April

additional reports are available on wool production and related matters. It should be noted that the name of this and other USDA divisions changes from time to time.

The series titles are quite descriptive of the data contained therein. Much of this information is useful for projecting short-run supply and, hence, prices. The January sow inventory, for example, may be used to project the spring farrowing and, indirectly, prices. Cattle on feed is a strong predictor of marketings over a six-month period. Slaughter, particularly heifer and gilt slaughter changes, reveals future production potential.

What is not revealed by the data are its sources and, hence, any indication of the level of accuracy. Slaughter statistics are taken from FSIS records (see Chapters 11 and 12) for federally inspected plants and are hence highly accurate (USDA, SRS 1975, pp. 74-75). Non-federal inspection is tabulated from state records and monthly surveys, while farm slaughter data are drawn from annual farm surveys. These data are not as accurate as the FSIS figures but apply to only five percent of total slaughter, or less.

Inventory and production data are derived from farm-level information and cause greater collection problems. Data are collected based on multiple frame survey techniques. The procedures are quite complex, but simplified they amount to relying heavily on mail surveys. Questionnaires are sent to large producers in an area and to a randomized sample of smaller farms. Farms potentially not included on address lists are sampled through area frame surveys in which all farms in a randomly selected area are visited. The errors detected in these farm interviews are used to adjust the raw mail survey information before making final estimates. Through these procedures, the sampling error is estimated to be three to four percent (USDA, SRS 1975, p. 67).

In general, then, Agricultural Statistics Service data are quite accurate.

Several further generalizations may be made about data accuracy. The larger the area enumerated and the larger the numbers counted, the smaller the sampling error is expected to be. That is, national data are generally better than state data, and state data are superior to counties, given for the same level of care in data collection. This is because estimated error is related to the variability across farms, and variability tends to be self-cancelling as numbers increase. Moreover, counts of smaller farms are more problematic as they are more difficult to identify and track. Thus, data relating to small farm enterprises are likely to be less accurate.

Prices

Price data are reported for both live animals and carcasses as well as for selected cuts. Live animal prices are reported by the Agricultural Marketing Service (AMS) and Agricultural Statistics Service (ASS), both of the USDA. AMS reports prices—generally on a cooperative basis with states—from auction and terminal markets. Because these are public markets, the price reporter is free to enter and record. The principal markets reported are as follows: slaughter cattle—Omaha, Sioux City, and Amarillo; feeder cattle—S. St. Paul, Kansas City, Dodge City, Sioux City, Sioux Falls, Oklahoma City, Amarillo (direct), and S. St. Joseph; slaughter hogs—Indianapolis, Peoria, Omaha, National Stock Yards, Sioux City, S. Interior Iowa, S. St. Paul, and Georgia, Florida and Alabama (direct); sheep and lambs—San Angelo, Sioux Falls, and S. St. Paul. Surveyed markets change from time to time.

Prices are reported daily on a graded basis; grades are as estimated by the market reporters. Prices are reported as ranges reflecting both the range of prices paid daily for animals of like grade and quality, and the quality range (and hence value) existing within a grade standard. The midpoint is intended to represent a median value, although it is uncertain just how the reported range is derived from observed transactions. The subjective input for establishing grades and reporting ranges lessen, to a degree, the validity of the reported prices, but they can generally be considered as highly reflective of prices paid in that market that day. The validity of composite prices reflecting those received in several markets, as well as the reflectiveness of public market prices for the bulk of many classes sold privately, are substantial questions which are discussed in Chapter 7.

The second source of live animal prices is the Agricultural Statistics Service (ASS). ASS provides monthly and state-specific prices received by farmers for the major classes, including steers and heifers, calves, hogs, sheep, and lambs. These figures have a possible benefit over those collected by AMS because ASS includes all sales, not just those that occur in public markets. However, the values are averages for all grades within a class and hence are not necessarily indicative of what a particular grade will bring. The usefulness for individual producers is, then, somewhat limited. Moreover, most data are provided on survey forms sent to voluntary reporters including selected farmers, dealers, and packers. Other government figures are also used, and some personal enumeration of cattle prices is carried out (USDA, SRS 1975, pp. 103-4). Nonetheless, with voluntary respondents providing the bulk of the data on unobservable transactions, it is impossible to determine the coverage and accuracy of the reported values. Overall, ASS figures are better suited to their intended use, the estimation of gross farm income, than for use in marketing decisions (USDA, SRS 1975, p. 99).

Meat, or more specifically carcass, prices are available daily on a fee basis from two sources: the "Yellow Sheet" and the "Meat Sheet." The formal name for the "Yellow Sheet" is the *Daily Market and News Service*. It has been published by The National Provisioner, Inc. since 1923. The Meat Sheet is a recent entrant (1974) by the Meat Sheet, Inc. The USDA had been providing data on car lot meat prices from 1916 to July 1990 when they were dropped as a response to market changes decompressing carcass sales. A "carcass equivalent" price report is being considered. The method of reporting prices varies among the reports. In the "Meat Sheet," a High, Low, and Close price are reported. The "Yellow Sheet" typically reports only a single price, interpreted to be the closing price. The USDA reports, when available, contained the full range of prices reported to it without comment on trends or closing levels (USDA, P&SA 1978, p. 23). Perhaps because a single price is more convenient for use, the "Yellow Sheet" is the most commonly used of the three market reports. It has a substantial impact on carcass prices and, indirectly, on slaughter animal prices for cattle and hogs.

Perhaps, too, because of its importance, the "Yellow Sheet" has come under repeated scrutiny, most recently in the late 1970s. Sector participants as well as consumers are wary because the limited number of trades on which the quoted prices are based allows non-reflective prices, and worse, to be reported. Yet, no systematic pattern of error has been detected and a formal-price fixing case brought by a group of feeders was dropped in late 1988 after 12 years. (For a literature review, see Chapter 7.) The two remaining reports, while not scrutinized as closely, use much **the same procedures and data sources as the "Yellow Sheet"** so that the data quality is likely comparable over the group. The procedure of reporting multiple prices (typically in the form of a range) does, however, provide more information so that, strictly speaking, the other news reports are better indicators of observed prices.

Future Prices

Four livestock-related futures contracts are traded, all at the Chicago Mercantile Exchange (CME). The four are fed cattle, feeder cattle, live hogs, and pork bellies. The use of futures markets is discussed in detail in Chapter 14. Here, attention is directed to the sources and quality of reported futures prices.

Futures contracts refer to a well-defined product. For fed cattle it is 40,000 pounds of choice grade steers to be delivered at specified times and locations. This specificity simplifies price reporting because there is no ambiguity over what is being reported. Trades themselves are made on an "open outcry" basis meaning traders must call out their buy and sell offers (Figure 10-2). This din is associated with the trading "pits," and as a practical matter the noise is so intense the buy/sell offers are actually transmitted with hand signals. (Figure 14-4).

FIGURE 10-2. Trading Futures Contracts at the Chicago Merc

The outcry and accompanying hand signals do provide access to price information as is appropriate in these public markets. The Exchanges have employees stationed in and around the pits who record the prices as they evolve. Thus, the reported prices are current and virtually error-free. From the trading floor, price information is transmitted electronically around the world. Many newspapers also publish daily price quotes. Typically, the day's high, low, and closing price are reported.

Dissemination

Private reporting services such as the "Yellow Sheet" sell this information either as hard copy sent through the mail or through some form of wire service. Futures prices, too, can be acquired from several wire services such as *Cattle Fax* provided by the National Cattleman's Association. The sources and forms of information available electronically are evolving rapidly and must be determined at the time of intended use. The placement of microcomputers on many farms makes access to those services more feasible than in the past. Computers can communicate over telephone lines with the aid of a device known as a modem.

The federal government nonetheless remains the principal source of information on livestock and related agricultural commodities. Data bulletins are available on a modest fee basis (many bulletins cost about $5.00), while recorded telephone messages are free. USDA data are also carried by several private electronic data service networks. However, most users receive USDA figures indirectly through newspapers and farm journals as well as over the radio and television. Overall, the available data are reasonably available to interested producers.

EXTENT OF USE OF MARKET INFORMATION

There are a number of ways livestock producers *could* and perhaps should use market information. They could consider (1) what to produce (long-term outlook); (2) when to produce (short-term outlook); and (3) when, where, and how to sell market-ready animals (marketing). Here the last topic shall be considered since it provides the most apparent link between market information and decision making.

In general, knowledge of how producers use market information is both limited and dated. Available studies, however, suggest livestock producers are poor comparison shoppers using few sampling prices in more than one market before making the decision to ship. Indeed, timing decisions are often based on feed availability and market readiness rather than

price considerations. Price information within such a system is more useful for a post-sale evaluation than for determining where or when to sell
(for a literature review see Helmberger, Campbell, and Dobson 1981,
pp. 563-66).

Much of this research dates to the 1960s and even the 1950s. With
increases in farm size as well as price volatility in recent years, it is hoped
that livestock producers are now better marketers than those sampled in
the earlier studies. This issue certainly requires greater research attention
since it is more difficult to improve on current practices when they are not
well documented or understood. Nonetheless, the implication is that more
market information is available than many livestock producers are able or
willing to use.

SYNOPSIS

Precise and timely information on market conditions, particularly volume and prices, is essential for a well-functioning market. Indeed, it is
one of the theoretical requirements for perfect competition. In the U.S.,
most information is provided by the government through the Department
of Agriculture.

Major categories of data are (a) standing herd size, (b) slaughter statistics, and (c) price data. Herd numbers are estimated annually based on a
survey with a good level of accuracy. Slaughter statistics are compiled
from inspection records and are quite accurate. Price data are from two
sources: public markets (where they are observed) and reports of private
transactions. The public market data are valid for the markets observed
but do not necessarily reflect all prices; reports of private transactions
cannot be directly confirmed and are hence less assured. The principal
provider of meat (as opposed to live animal) prices, the "Yellow Sheet,"
has come under considerable scrutiny for at least 15 years but no improprieties or inaccuracies have been substantiated. A greater problem is the
limited price information for boxed meat (the principal product form for
beef).

Volume and price data are most useful for making short-term decisions
on when and how to market. The limited and dated information on how
producers use their data suggest that improvements can be made.

Study Questions

1. The USDA is the major source of price information for live animals as well as meat. How can this activity — at public expense — be justified?
2. Data on herd size, slaughter, and prices are collected in quite different ways. What are the ramifications for the accuracy of these data?
3. To what extent do you use market news in your livestock production or marketing activities? How might better use be made of what is available?

REFERENCES

Breimyer, H. F. "Fifty Years of Federal Marketing Programs." *J. Farm Econ.* 45(1963):749-58.

Helmberger, P. G., G. R. Campbell and W. D. Dobson. "Organization and Performance of Agricultural Prices." Part IV in *A Survey of Agricultural Economics Literature, Vol. 3*, ed. L. R. Martin. Minneapolis: Univ. Minnesota Press, 1981.

Houghton, D. and K. McMahon. "The $50 Million Hog Farm," *Farm Journal* October 1984, pp. 13-15.

Miles, J. G. Jr. "The Impact of the Granger Movement Upon Social Legislation." In *The National Grange, Legal and Economic Influence of the Grange.* Washington, DC: The National Grange, 1967 pp. 10-35.

Neale, A. D. *The Antitrust Laws of the USA*. London: The Cambridge University Press, Second Edition, 1980.

USDA, Ag. Stat. Board, National Ag. Statistics Service, 1989 Calendar.

USDA, Economic Research Service. 1989 *Calendar of Reports*.

USDA, Packers and Stockyards Administration. "Résumé." Various years.

USDA, Statistical Reporting Service, "Scope and Methods of the Statistical Reporting Service." Mis. Pub. 1308, July 1975.

Chapter 11

Regulation of Livestock Marketing

Livestock producers often like to think of themselves in terms of the rugged individualists who made the United States great. That view may be true of their own efforts, but it describes poorly the environment they operate within today. For the truth is that the livestock sector is one of the most heavily regulated in the nation. Prices are not supported to any meaningful extent, nor are production decisions influenced as with many commodity programs. But the pharmaceuticals permitted and the sanitation of both animal and processing facility are all heavily regulated as, are to a slight degree, the trade practices of buyers and marketing agencies. This chapter describes the regulatory activities of the U.S. Department of Agriculture, namely those in inspection and trade regulation. Grading, a related issue, is addressed separately in Chapter 12.

MEAT INSPECTION

This is the way that Upton Sinclair described conditions in 1906 in the meat packing industry:

> And then there was the condemned meat industry, with its endless horrors. The people of Chicago saw the government inspectors in Packingtown, and they all took that to mean that they were protected from diseased meat; they did not understand that these hundred and sixty-three inspectors had been appointed at the request of the packers, and that they were paid by the United States government to certify that all the diseased meat was kept in the state . . . [A] physician made the discovery that the carcasses of steers which had been condemned as tubercular by the government inspectors, and which therefore contained ptomaines, which are deadly poisons, were left upon an open platform and carted away to be sold in the city . . . (pp. 98-99)

Meat Inspection Act of 1906

Subsequently—on March 4, 1907, to be exact—the Meat Inspection Act was passed with the purpose of ". . . preventing the use in commerce . . . of meat and food products which are adulterated . . ." Some attribute the passage of this Act to Sinclair's book, but it seems more likely that like most good writers, the author followed, as well as led, public sentiments. The result of these forces, however, was the enactment of legislation that assures consumers of clean, wholesome meat. It applies to all products sold in interstate commerce or exported, and is enforceable even if only a small portion of a plant's product is sold that way.[1]

Under the Meat Inspection Act, meat may be considered adulterated if, among other things, "it has been prepared, packed, or held under unsanitary conditions . . ." As this wording implies, inspection is directed to both facilities and products. Moreover, stock must be inspected for disease both prior to (ante mortem) and after (post mortem) slaughter. Implementation is the responsibility of the Food Safety and Inspection Service (FSIS) of the USDA. (Prior to June 1981 it was known as the Food Safety and Quality Service.) The FSIS also has the responsibility for inspecting poultry products, but that authorizing legislation, (the Poultry Products Inspection Act) was not passed until 1957.

Viewed as a public service, inspection service under the Act has always been done (during regular business hours) at public expense. In a cost-cutting move, the Reagan administration had proposed at one point that inspection be shifted to a fee basis. Alternatively, it was suggested that more inspection be changed to a statistical sample basis rather than the complete survey (e.g., every carcass) currently used. A law to that effect was passed in 1986. Presently, many processed products are tested on a sample basis, just as the complex residue tests are applied to a sample only. Neither of these proposals had widespread support (see e.g., *Poultry Times* 1985; *The Washington Post* 1982), and in 1990 the so-called Improved Processing Inspection (IPI) proposal was dropped.

---

1. The Meat Inspection Act was not the first such law, only the most influential. In 1890, Congress required that certain meat products destined for export be certified disease-free subsequently, in 1891 and 1895, similar protection was extended to domestic consumers.

Subsequent Legislation

While the Meat Inspection Act of 1906 provided uniform protection for consumers of red meat from plants involved in interstate commerce, those purchasing products from non-federally inspected operations relied on state inspection laws. And since those laws and their enforcement varied from state to state, the amount of protection remained uncertain at best. That matter was rectified in 1967 with the passage of the Federal Meat Inspection Act. This Act, commonly known as the Wholesome Meat Act, mandated that states adopt requirements and practices "at least equal to" those of the federal system. Complying states may have up to half of the costs of operating the system provided by the USDA, while those which are judged to be in noncompliance must be designated by the Secretary of Agriculture as under federal authority. States may also voluntarily select to go under federal authority. Compliance is administered by the FSIS.[2] In recent years, 95 percent or more of all commercial slaughter of the major species has been done under federal inspection.

Following several postponements, the requirements of the Wholesome Meat Act became effective in the early 1970s. Since then, a number of states, generally for financial reasons, have elected to discontinue the state system and rely on federal inspectors. As of 1989, 21 states had no state inspection system for red meats (USDA, FSIS 1990, Table 3-4). Plants operating under state inspection are known as state exempt plants. Talmadge-Aiken Plants meet federal inspection requirements administered by federally trained but state-paid employees.

The implementation of the Wholesome Meat Act undoubtedly led to more uniform and generally higher inspection standards. It also led to the closure of numerous plants, mostly the smaller, older operations which could not justify the investment required to comply with the new "at least equal to" regulations. In the Northeast, more than 50 percent of the pre-Act plants closed during the transition period to the new law (Smalley, Webb, and Lesser 1982, pp. 5-6). That number is probably at the extreme due to the larger number of older plants in that region, but it does indicate the disruptions that occurred because of the Act. New plants are generally made compliant with federal inspection requirements since the cost differential is negligible while the potential market is much greater.

---

2. Federal regulation of poultry slaughter and processing was granted by the Poultry Products Inspection Act of 1968.

Custom Plants

The FSIS is presently authorized to inspect plants involved in international, interstate, and intrastate trade. Not included under this authorization is meat that is not sold, or *farm slaughter*. Under this exemption, anyone owning a live animal may slaughter it, or have it slaughtered, for purely private use in non-federally regulated or "custom exempt" plants. Any inspection that is done is mandated and supervised on a state-by-state basis. Only about half of one percent of total slaughter is "farm slaughter" (USDA, ERS, "Livestock and Meat Statistics," 1989).

Despite the implication of the term, not all farm-slaughtered animals are consumed by the producer. Many are sold live to consumers who have them processed for their own accounts. A number of small — many part-time — operations exist to serve this "freezer" market. Many of these plants also process game animals that are also exempt from federal law. Products processed in custom exempt plants are stamped with "Not for Resale" on the wrapper. Custom plants can be operated on a small scale, part-time basis and are quite numerous in many states. Lists are available from state departments of agriculture.

Function of the FSIS

Inspection, as conducted by FSIS, encompasses seven distinct activities:

1. The examination and certification of plant facilities and meat-processing equipment for sanitation;
2. The inspection of cattle, sheep, swine, goats, horses, and other equines ante-mortem for detectable diseases and abnormalities;
3. The investigation of compliance with humane handling and slaughter regulations (see below);
4. Post-mortem inspection and taking of samples;
5. The inspection of processing operations for sanitation, compliance with required procedures, and use of approved ingredients;
6. The examination of processed products to ensure the proper packaging, marking, and labeling has been carried out; and
7. The control of condemned (unwholesome) and inedible products (those declared inedible for human consumption by law and regulation) to prevent them from entering the human food system. (USDA, FSIS 1981, p. 5)

Products meeting all the requirements are marked as "Inspected and Passed" (Figure 11-1). Any product failing to pass all these requirements

FIGURE 11-1. FSIS Markings Indicating Meat Accepted and Rejected for Human Consumption

Marking of successfully passed carcasses and primals (not in containers). The "38" refers to the USDA establishment number.

U.S. INSP'D AND
CONDEMNED

Rejected product stamps

is tagged "U.S. CONDEMNED" and must be destroyed. Suspect items are so marked until they can be examined further and then either accepted or condemned. In order to qualify for inspection, plants must satisfy stringent regulations as to layout, equipment, sanitary procedures and even operation levels (see, e.g., USDA, FSIS April 1984). If an infringement is subsequently found, it must be rectified to the satisfaction of the inspector. Failure to comply can lead to a withdrawal of inspection services which, of course, would mean the inability to sell products for human consumption (see USDA, *Office Information* 1984, p. 19). Certification for foreign plants shipping to the U.S. market operates similarly.

Cooperative Programs

One of the more complex FSIS responsibilities is the handling of additives and residues in meat products. Under a cooperative program involving FSIS and the Food and Drug Administration (FDA), and the Environmental Protection Agency (EPA), the other agencies establish safe levels of additives and residuals which FSIS enforces. In cooperation with these other agencies, acceptable additives are identified and safe levels determined. Products not in compliance may not be sold for human consumption and, alerted by the FSIS, violators may, in some cases, be prosecuted for criminal violations by the FDA. The labeling divisions of the FSIS see that all products are accurately labeled as to products, producer, volume, and ingredients. Any nutritional claims must be substantiated, and standards describe requirements and label standards for such common products as hot dogs (see, e.g., USDA, FSIS 1981, Bulletin 236).

A number of interrelated changes are under consideration in these areas. One approach currently underway for calves is the "certification" by all handlers that an animal is drug-free. Certified calves receive less intensive testing for drug residues (*Federal Register* 1990). Responsibility necessitates identification and a hog identification rule was adopted in 1988. This was all done despite the FSIS conclusion that "Scientific evidence shows that consumers face very little health risk from chemical residues in meat and poultry . . ." (USDA, FSIS 1990, p. 13). When violations are found, tough new state laws impose heavy penalties on producers and/or handlers. A 1990 California law, for example, allows packers to receive treble damages on condemned carcasses.

Product content legislation is not exempt from this review, especially when mechanically deboned products are involved. Some 10 to 20 percent of non-identified materials are allowed in hot dogs, materials possibly including bone chips and fragments which are potentially harmful to some consumers. At this time, USDA-regulated meat products are exempt from nutritional labeling proposals.

In 1989, FSIS undertook a new approach to food safety control known as the Hazard Analysis and Critical Control Point (HACCP) system. Identifying points where a hazardous product may result if a process goes out of control as a "critical control point," this program is intended to prevent problems from occurring rather than identifying them after the fact (USDA, FSIS 1990, p. 9).

TRADE PRACTICES

Public dissatisfaction with the meat packing industry surfaced partially in response to the health and sanitation issues, contributing to the passage of the Meat Inspection Act of 1906 (see above). But those were not the only issues, especially for producers. They were concerned about the so-called "beef trust." Beginning in the late 1800s, member firms were alleged to have a "monopolistic control of the American meat industry and also were engaged in the production and distribution of many other food products." Thus said the Federal Trade Commission (FTC) in its first major action following its establishment in 1914. To rectify these abuses, in 1920 the FTC reached an out-of-court agreement with the Big Five packers, Armour, Swift, Wilson, Cudahay, and Morris, which at the time handled some two-thirds of the livestock slaughtered under federal inspection. This agreement, the well-known Consent Decree of 1920, forbade these firms from direct involvement in procurement and subsequent sales of their livestock products. In response, these companies had to withdraw from ownership of stockyards, transportation lines, cold storage plants, etc. But their principal activities of slaughtering and processing remained unchallenged (Engelman 1975; for more information see "Report of the Federal Trade Commission on the History and Present Status of the Packer Consent Decree," in a December 8, 1924 letter by FTC Commissioner Nelson B. Gaskill to the President of the U.S. Senate). The Decree was terminated in 1981 when it was recognized in a U.S. District Court that a number of large meat packers established since the 1920s were not bound by similar restrictions.

Apparently, however, Congress as a body was unsatisfied in the 1920s that one legal action by the FTC could control the potential abuses of the few very large packing firms in the critical livestock sector. In response, the Packers and Stockyards Act of 1921 was passed to provide for ongoing regulation of livestock marketing. The Act established the *Packers and Stockyards Administration* under the Secretary of Agriculture. At its formation, it was and remains, one of a very few administrative agencies set up specifically to regulate a particular industry.

The Act is primarily a trade practices act, giving the Secretary of Agriculture specific regulatory powers. All interstate transactions in livestock — cattle, sheep, hogs, horses, mules, and goats — are subject to its provisions. These provisions attempt to safeguard producers, consumers, and firms in the industry from unfair, deceptive, or monopolistic trade practices which deprive them of the true value of their products or purchases. Several of these provisions are of particular value to livestock producers and are worth describing in some detail. It is important to note, however, that the details of the regulations are subject to change. For the most up-to-date regulations, check with the nearest regional office (Figure 11-2).

Provisions of the Packers and Stockyards Act

Regulations can be described as falling under six basic categories: (1) registration and bonding, (2) services and rates, (3) payment, (4)

FIGURE 11-2. Offices of Packers & Stockyards Administration, 1990 (Source: USDA, P&SA 1990)

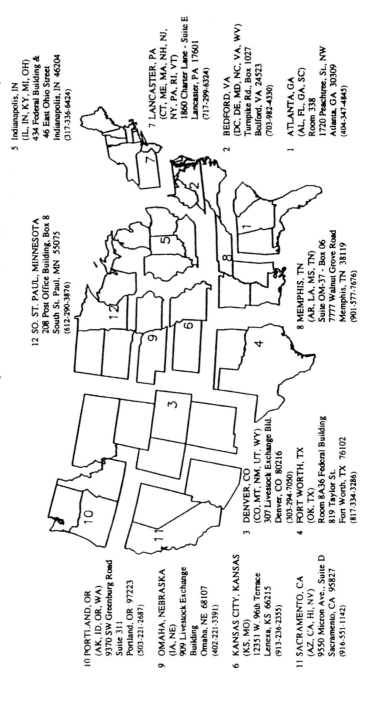

5 Indianapolis, IN
(IL, IN, KY, MI, OH)
434 Federal Building &
46 East Ohio Street
Indianapolis, IN 46204
(317-336-6424)

7 LANCASTER, PA
(CT, ME, MA, NH, NJ,
NY, PA, RI, VT)
1860 Charter Lane - Suite E
Lancaster, PA 17601
(717-299-6324)

2 BEDFORD, VA
(DC, DE, MD, NC, VA, WV)
Turnpike Rd., Box 1027
Bedford, VA 24523
(703-982-4330)

1 ATLANTA, GA
(AL, FL, GA, SC)
Room 338
1720 Peachtree, St., NW
Atlanta, GA 30309
(404-347-4845)

12 SO. ST. PAUL, MINNESOTA
208 Post Office Building, Box 8
South St. Paul, MN 55075
(612-290-3876)

8 MEMPHIS, TN
(AR, LA, MS, TN)
Suite OM-37 - Box 06
7777 Walnut Grove Road
Memphis, TN 38119
(901-577-7676)

3 DENVER, CO
(CO, MT, NM, UT, WY)
307 Livestock Exchange Bld.
Denver, CO 80216
(303-294-7050)

4 FORT WORTH, TX
(OK, TX)
Room 8A36 Federal Building
819 Taylor St.
Fort Worth, TX 76102
(817-334-3286)

10 PORTLAND, OR
(AK, ID, OR, WA)
9370 SW Greenburg Road
Suite 311
Portland, OR 97223
(503-221-2687)

9 OMAHA, NEBRASKA
(IA, NE)
909 Livestock Exchange
Building
Omaha, NE 68107
(402-221-3391)

6 KANSAS CITY, KANSAS
(KS, MO)
12351 W. 96th Terrace
Lenexa, KS 66215
(913-236-2355)

11 SACRAMENTO, CA
(AZ, CA, HI, NV)
9550 Micron Ave., Suite D
Sacramento, CA 95827
(916-551-1142)

scales and weighing, (5) accounting, and (6) business practices (USDA, P&SA 1969, PA-399).

Registration and Bonding: Public markets involved in interstate trade including any which are used by buyers who ship animals or meat products among the states must be "posted" and registered. Posting means literally that notices must be displayed conspicuously in at least three locations stating the market is subject to the regulations of the Act. Posted markets must maintain a surety bond based on its volume of business. The bond is intended to guarantee payment to the producer if the market becomes insolvent or the buyer of the stock defaults on payment.

Dealers and order buyers using those markets for procuring stock must also register and post bonds. Packer buyers must register and their firms (assuming annual purchases exceed $500,000) post bonds equal to about two days' worth of purchases. However, farmers and others not engaged in interstate commerce need not comply with these regulations.

Services and Rates: Public market operators must provide reasonable services and charge reasonable and nondiscriminatory rates. Rates must be announced and adequate notice given of proposed rate changes. Court decisions have mandated that "nondiscriminatory" be interpreted as meaning per head charges rather than percentage fees, the thinking being that costs are related to the class of the animal and not its value.

Payment: Sellers of slaughter stock to packers, market agencies, and dealers have the right to demand immediate payment. The extension of credit or even an agreement to have a check mailed *must* be in writing. For sales not intended for immediate slaughter, payment may be by check without prior agreement. Checks must be in the mail by the end of the business day following the sale. Payment by a draft is considered an extension of credit and must be agreed to in writing. A draft is essentially a legally enforceable form of IOU offered by the seller agreeing to make payment at some future date. Producers and feeders are not subject to these regulations.

Scales and Weighing: Packers and Stockyards requires that all firms and individuals subject to the Act install and maintain accurate scales. All scales must be tested at least twice yearly.

Accounting: Individuals and firms subject to the Act must maintain and file accurate and complete accounts. These records are scrutinized by P&S employees to determine if the financial solvency requirements of registrants are satisfied. Until the rules were changed in 1978, financial records from public markets were used for justifying rate increases. Market fees are no longer regulated so long as they are just and nondiscriminatory.

Business Practices: This is a general category requiring broadly fair and non-deceptive trade practices. Legally, it allows the Secretary of Agriculture to bring cases which might otherwise be the responsibility of the Federal Trade Commission (Engelman 1975). Practically, it has led to a number of specific cases in price fixing, especially at public markets. Two actions require mentioning. First, under the mandate that all exchanges be fair and nondiscriminatory, all sales on a grade and yield basis (see Chapter 5) must be valued on a hot carcass weight basis. That ruling removes much possible misunderstanding, and potential for abuse, when weighing in was done at different points and several adjustment systems used to equate to a hot carcass weight. Second, commissions charged at public markets must be on a per head basis only. Percentage fees, which prevailed previously, neither allowed the seller to judge marketing costs prior to the sale nor did they equate well with actual costs.

Other Functions: The remaining activity of the P&S is the reporting of data on the livestock marketing and packing industry. As part of its information collection responsibility, P&S compiles a great quantity of data not available elsewhere. The principal vehicle for reporting this material is the annual Statistical Report (previously called Statistical Résumé). The Report contains data on numbers of packers and public markets by state as well as numbers of animals by class that move through each type of facility. Other issues such as Vol. XXI No. 10, treat special topics (in this case boxed beef).

Effects of the P&SA

In a 1975 Statement to the House Small Business Committee, Dr. Engelman (then Director of the P&S Industry Analysis Staff) observed that the concentration in livestock slaughter, the proportion carried out by the largest firms, peaked in 1920. (It has since risen above that level, see Chapter 15.) That year was, of course, marked by the Consent Decree and was followed in 1921 by the P&S Act. While he is careful to note that other technical changes were occurring simultaneously that affected the structure of the industry, Engelman clearly attributes much of the effect to the Packers and Stockyards Administration. While undoubtedly valid, his position is probably overstated by a considerable degree. The concurrent technical changes of the 1920s were refinements of mechanical refrigeration and other changes that made it economical to construct smaller plants nearer the source of live animals (see Chapters 3 and 15). Nor is it clear that the actions prohibited by the Consent Decree — and subsequently by P&S regulations — had such a substantial impact on the major firms in the industry.

Rather, the effects of Packers and Stockyards are more likely seen in the protection of individual producers. The key protection allowed is perhaps the opportunity to seek help and, if deemed necessary, redress through the courts with the assistance of a knowledgeable, professional body. The importance of this issue has perhaps diminished with the growing size and sophistication of most livestock producers in the U.S. today. But the support was and remains important for many. Somewhat less important is the bonding provision. Bonding provides protection, but for only two days' worth of purchases or sales. Experience has shown that firms in financial trouble typically have obligations exceeding this amount by many fold. Bonding provides all those creditors with limited protection, although they do benefit from having first claim to the assets of the firm. Many sellers nonetheless feel they are protected from loss by the bonding and extend credit in the form of delayed payment when it is imprudent to do so. The P&S Act does provide an important "safety net" for the livestock seller, but not one that can prevail against major industry trends or counter poor judgment by the individual.

HUMANITARIAN REGULATIONS

Regulation of the livestock sector, while extensive, is predominantly directed to protecting consumers or promoting commerce. But people are not the exclusive domain of the regulators. Laws exist—two in particular—to protect the well-being of the animals themselves. These laws stipulate handling practices during transit and slaughter.

Humane Slaughter Act

Several decades ago, it was common prior to "sticking" to stun animals with a sledgehammer, club, or pipe. While this did render the animal insensitive, the time and effort involved was not considered humane for the animal. Alternatively, a small caliber gun was used, but that proved dangerous to the workers. In 1958, in response to mounting concerns about animal welfare, the Humane Slaughter Act was passed as a means of stipulating the acceptable handling and slaughter practices. Three procedures were approved for stunning prior to slaughter. The most common is a captive bolt gun powered by a small explosive charge. Electrical stunning or use of a carbon dioxide chamber is also acceptable. Ritual religious slaughter is exempted. Kosher slaughter practices, for example, require the animal to be conscious prior to slaughter (Westervelt and Kinsman 1973).

The Act is an amendment of the Meat Inspection Act and as such is administered by the Food Safety and Inspection Service of the USDA. The original law carried no mandate, requiring only that any firm selling meat to the U.S. government be in compliance. In 1978, the Humane Methods of Slaughter Act required compliance as a component of federal inspection. Presently, any firm selling red meat (including foreign plants exporting to the United States) must be in compliance. Regulations associated with the Act mandate such things as having water available during overnight stands. Regulations are set administratively and as such are subject to more rapid change.

Transportation

The so-called 28-Hour Law, passed in 1906, mandated a periodic rest, feeding, and watering stop for stock in continuous transit. This law, which replaced an 1873 statute, provides detailed stipulations of handling methods, holding conditions, and loading/unloading procedures. If sufficient space is available for animals to lie down, then unloading is not required; otherwise, animals must be off-loaded into a more spacious holding area. Despite the title of the law, an eight-hour extension (to 36 hours) is available on request.

As might be expected from the date of passage, the Act applies to rail transit. Several unsuccessful efforts have been made to extend coverage to trucks. The decline in use of trains, combined with the relative rapidity of transport, has diminished the specific importance of this law. Moreover there are commercial benefits to resting, feeding, and watering during transit, especially for animals to be returned to feed (see Chapter 7). The spirit of the Act undoubtedly still prevails and may restrain some potentially damaging actions when short-term needs would otherwise prevail.

SYNOPSIS

The livestock sector is regulated in numerous ways. The major ones are: (a) meat inspection, (b) trade practices, and (c) humanitarian factors. Meat inspection, still based on the major Act of 1906 and paid for by taxes, assures wholesome meat and products through ongoing inspection of plants and animals (ante and post mortem). A proposal to move to the inspection of carcass samples received sharp criticism and was dropped. Extension of requirements to state-inspected plants was made in 1967. Foreign plants must meet similar requirements.

The responsible division of the USDA, the Food Safety and Inspection Service, also administers residue and product content standards established by the Food and Drug Administration and the Environmental Protection Agency.

Trade practices are regulated by the USDA Packers and Stockyards Administration. It controls payment timing and bonding requirements for packers and public market operators, as well as other practices and procedures such as checking scales.

Humane practices regulations are limited to the proper stunning of animals before bleeding (ritual slaughter is excluded) and to rest stops for livestock on rail cars. Truckers are technically exempt, but the indication is that compliance levels are high anyway.

Study Questions

1. What conditions led to the passage of the Packers and Stockyards Act of 1921? What can be said about how effective it has been in rectifying those issues and what might occur in its absence?
2. Suppose you wanted to construct and operate a meat-processing plant. What regulations would you have to comply with in both the planning and operation of that plant? How would you determine who had authority to regulate your operation?
3. Suppose you were hired by the USDA and put in charge of a cost-benefit study to determine if the government should move from continuous carcass inspection to sampling or "self inspection." What is implied by those terms, and how would you organize the study?
4. To what does the term "at least equal to" apply and what does it mean?
5. What are the benefits – and possible dangers – of ending provisions?

REFERENCES

Engelman, G. "Trends in Livestock Marketing Before and After the Consent Decree of 1920 and the Packers and Stockyards Act of 1921." Statement before the House Small Business Committee, June 23, 1975.

Federal Register. "FSIS, Sulfanamide and Antibiotic Residues on Young Calves; Certification Requirements." March 2, 1990, pp. 7472-75.

Poultry Times. "User Fee Plan Takes Hot Seat on Capitol Hill." February 18, 1985, p. 1.

"Report of the Federal Trade Commission on the History and Present Status of the Packer Consent Decree." In a December 8, 1924 letter by FTC Commissioner Nelson B. Gaskill to the President of the U.S. Senate.

Sinclair, U. *The Jungle.* New York: Signet Classics, 1960.

Smalley, H. R., T. F. Webb, and W. Lesser. "An Economic Evaluation of Cattle

Supplies and Slaughter Plant Capacity in New York and the Northeast Region." Dept. Agr. Econ., Cornell Univ., A. E. Res. 82-12, December 1982.

USDA, Economic Research Service. "Livestock and Meat Statistics." Various issues.

USDA, Economic Research Service. "Livestock and Poultry: Outlook and Situation Report." Various issues.

USDA, Food Safety and Inspection Service. "Meat and Poultry Inspection, 1989: Report of the Secretary of Agriculture to the U.S. Congress." March 1, 1990.

USDA, Food Safety and Inspection Service. "Meat and Poultry Products: A Consumer Guide to Content and Labeling Requirements." Home and Garden Bull. No. 236, July 1981.

USDA, Food Safety and Inspection Service. "U.S. Inspected Meat and Poultry Packing Plants: A Guide to Construction and Layout." Ag. Handbook 570, April 1984.

USDA. *Office Information.* August 17-24, 1984.

USDA, Packers and Stockyards Administration. "The Packers and Stockyards Act: What It Is—How It Operates." PA-399, Washington, DC, August 1969.

The Washington Post. "USDA Wants to Slice Its Meat Inspectors." August 18, 1982, p. A21.

Westervelt, R. G. and D. M. Kinsman. "Humane Slaughter Methods." College of Ag. and Natural Resources, Univ. Conn. 73-61, 1973.

Chapter 12

Grades and Grading

Carcass grading is done in the packing plant by USDA Food Safety and Inspection Service (FSIS) employees, and the grade is stamped on the carcass with the same color ink as the inspection stamp (Figure 12-1).[1] There are very obvious differences between inspection and grading, among the most apparent is the fact that grading is both optional and charged to the packer. The two services are nonetheless confused on occasion, especially by consumers. In fact, although inspection and grading can be equated on a conceptual level, the justification and implementation of a grading system is the far more complicated of the two.

PURPOSES OF GRADING

The value of grading can be conceptualized by considering a shopper attempting to locate a package of lean bacon in a supermarket. Suppose there is only "lean" and "fatty" bacon available but they are unmarked. Further assume the fatty product is more abundant so that, on average, only every third package will be acceptable to Mr. Sprat, our shopper. To satisfy his tastes (if not his wife's) he must sort through the packages until a lean one is found. This involves a time or search cost that would be saved if the bacon were graded and sorted into separate piles. Grading, by saving time for consumers, can then make the market more efficient.

More about efficiency in a moment. Now let us examine what Jack did. He sorted through the packages until a suitable one was found. That is exactly what grading is: sorting into groups according to selected characteristics. Grading is applied to products of nature, like fresh meat, which are not subject to precise control. This is simple enough. What is complex

---

1. Prior to 1977, meat grading was undertaken by the Agricultural Marketing Service of the USDA.

FIGURE 12-1. USDA Grade Stamps: Quality Grade and Yield Grade (Sources: USDA, published in U.S. Congress, OTA 1977; USDA, AMS 1985)

is the choice of criteria and how to delineate grade boundaries in what is essentially a continuum of fattiness from none to all.[2]

For the retail buyer, the sorting of bacon by fattiness would be only a modest service. But it does fulfill one essential requirement of grading: the standards are meaningful to users. Thus, to be viable, grading must be a genuine aid to marketing. Selection on an obvious characteristic, like fattiness, is of only limited utility for the shopper. More valuable is the use of unobservable criteria for grade standards. The healthfulness of meat is not necessarily apparent from visual inspection. Health standards imposed by FSIS are, in a sense, a grading; products are sorted into those which are judged safe for human consumption and those which are not. That form of grading is considered so essential in this country that the requirements are mandatory.

Beyond safety, consumers generally select products on their edible characteristics that can be described as sensory components of flavor, juiciness, and texture. Since those aspects are not readily observable, they constitute a basis for a grading system. Suppose, for example, Mrs. Sprat liked juicy meat. Wouldn't it be helpful if the government could grade

---

2. This and other points in this section were derived from Breimyer's talk of November 14, 1960.

meat as "juicy," "less juicy," "not so juicy," etc.? That precise measurement is not impractical; however, one aspect of the grading system, known as *quality grade*, attempts to do just that. In that way, grades are not so much a source of market distinctions as they are a reflection of them. Moreover, grades become a language for expressing those distinctions.

Something, though, is faulty with this explanation of grades as a benefit to consumers. If, in fact, grades are so valuable, why then are so few shoppers familiar with the grading system? A survey done in Texas showed that only 25 percent of shoppers had even minimal knowledge of federal beef grades, grades that have been in effect since 1926 (Branson and Rosson 1981, p. 10). In one supermarket test, "generic" and choice beef were placed side-by-side, but "sales seemed to be a function of space and nothing else . . . which tells me consumers' awareness of grades is low" (Zwiebach 1985).

Of course, consumers need not be informed on the technical aspects of grades so long as the storekeeper selects by those grades and provides the shopper with consistency. Stores do tend to provide consistency, but it need not be based on USDA grades. Using grades poses two problems. First, buyers tend to equate price with quality so that the higher priced product may be selected on that basis irrespective of grade (see, e.g., Brunk and Darrah 1955, pp. 287-88). Second, and more substantially, quality grades are heavily influenced by marbling (see below) but the degree of marbling is not clearly related to rating qualities (Fox and Black 1975). A USDA-commissioned study of lamb found the overall quality grade accounted for only 6.8 percent of the palatability variation of primal cuts (*Federal Register* Sept. 13, 1982, p. 40142).

In fact, what appears to happen is that stores carry out the sorting process for the shopper. A few use and promote an all-choice policy for beef (Figure 12-2). Many, however, use their own selection, which may or may not be derived from federal standards, for at least some of their red meat products. Stores do this consciously, and among careful shoppers different chains have varying reputations for the quality of their meat counter. This becomes an important aspect of the store image which, in turn, affects the competitive environment. The sensitivity of shoppers to meat can be inferred from the frequent specialing of this product. Indeed, specialing can be so persistent that the meat department in many supermarkets break even, at best (NCFM 1966, Tech. Study No. 7, pp. 175-77). Through food chain "grading" of fresh meat, the need for differentiation by advertising is reduced at a likely savings to consumers (U.S. Congress, OTA 1977, p. 7).

FIGURE 12-2. Some Supermarkets Feature USDA Choice Beef

Missing from the above is a clear justification of the value of meat grades to consumers. And, indeed, the grading system has always been more of a mechanism for facilitating wholesale transactions (U.S. Congress, OTA 1977, p. 3). Grading facilitates trading by permitting exchange on a standards, or sight-unseen, basis. This can be a considerable savings to the small seller and buyer who lack the resources to establish private grades and/or the select product directly. Grades further permit price reporting by establishing known classes for which prices can be quoted. Formula trades based on "Yellow Sheet" prices, for example, would be impossible without a widely understood carcass grading system (see Chapter 7).

By reducing transaction costs, grades enhance the efficiency of the marketing system. But efficiency is improved in another way related to the harmonizing of the production, processing, and retailing sectors. To see how it is helpful, return to the bacon example. Many consumers who prefer lean bacon will pick it out first. If this product sells better at the same price, the store would like to order more lean and less fatty product. With no criteria available, the store will be frustrated in its efforts. And since there is no way of measuring consumer preferences, producers will not be paid more to produce leaner hogs and will continue to produce a less desirable product. A sharp entrepreneur could enter and pick out the leaner bacon and sell it under a trade name that would connote leanness to shoppers.

As an alternative, pork could be graded according to leanness. If bacon were packed by grade (whether consumers could identify the grades or not), then store operators would be reordering the popular lean grades more frequently. To meet this need, packers, too, would seek lean hogs, paying higher prices in line with consumer preferences. The clear demand for lean hogs would change producer preferences to this type of animal so that supply and demand could be harmonized. In so doing, the grade system has contributed to the equity of the marketplace. It is through the facilitation of signalling from consumers to producers that grades serve a most important function.

At the same time, grade continuity from feeder animals to slaughter stock presents the greatest challenge to devising the system. What is required is a system that identifies economically meaningful characteristics at all levels of the system. Yet these values are not the same. The feeder **seeks calves with the potential for fast gain. The packer** values the volume and type of meat producible from the carcass, while the retailer wants a carcass that will yield the maximum value in retail cuts. Not all of these needs can be satisfied within a single grading scheme. In recent years, an attempt has been made to satisfy diverse requirements for beef by separating the quality from the cutability or yield characteristics (see below). But, in general, a grading system will not and can not be perfectly satisfactory to all users.

The need to balance the practical requirements determining the number of grades adds to the problems of devising a satisfactory system. There must not be so many as to confuse users nor so few that the range within a grade is so large the grade designation is not meaningful. Because grades balance a number of disparate factors, they are inherently judgemental. Age, for example, is seldom known with precision and must be estimated from certain carcass characteristics. Thus, the human factor introduces additional variability into the system. Despite these factors, grading serves important ends and has survived and evolved for over two-thirds of a century. In the sections below, the current systems are described in some detail.

THE HISTORY AND PREVALENCE OF GRADING

The Food Production Act of 1917 encouraged the development and use of grades as part of the effort to supply U.S. and allied troops during the war. Tentative live animal standards were developed that year. Carcass beef standards, still tentative, were issued in 1923 and expanded in 1925. Around that time, producer groups recognized the value of preserving goodwill with consumers by preventing shoppers from being misled and buying poor quality beef. This was the "truth-in-meats" movement that culminated in the formation of the Better Beef Association in 1927. The Association petitioned the USDA to set up a grading and stamping program for beef. Begun in 1927, that program continues to this day.

Specific authorization came in 1946 with the passage of the Agricultural Marketing Act which directed the Secretary of Agriculture to ". . . inspect, certify, and identify the class, quality, quantity, and condition of agricultural products when shipped or received in interstate commerce . . ." (Section 203, 7 U.S.C. 1621-1627). Grading, which was generally voluntary, caught on slowly. By 1940, only eight percent of commercially produced beef was graded. Following mandatory grading during World War II and the Korean conflict, use again declined, but only to about 25 percent (Breimyer 1960). By 1977, the percent had increased to over 50 percent of beef (U.S. Congress, OTA 1977, p. 31). There is reason to believe the proportion had been declining in recent years as government purchases became less important and as packers and retailers grew larger and could substitute private grading schemes. However, the amount of officially graded "USDA Good" beef jumped 700 percent when it was renamed "USDA Select" in November 1987 (USDA, News Div. 1990), and the amount of all beef graded edged up to 54 percent by 1988 (AMI 1989, p. 22). Pork grades are less used.

Since their inception, alterations in the grade systems for the three major species have been made some 10 times. Standards for lamb and lamb carcasses were changed in 1982 and for slaughter animals in 1984, while a simplification of barrow and gilt carcasses and live animals was adopted in January 1985. The proposal to revise the beef carcass and slaughter cattle standards was subsequently withdrawn. A major change for beef had been made in 1965 when conformation was removed from the quality standards and treated as a separate "yield" component. In 1975, yield grading was made mandatory with quality grading along with a few other relatively minor changes (see below). Then, in 1989, the grade was permitted to consist of the quality grade only, the yield grade only, or a combination of the two.

CURRENT STANDARDS

In general, grade standards are designated for the carcass. Slaughter and feeder stock standards are such that a U.S. No. 1 barrow or gilt, for example, is expected to produce a carcass of the same grade. Similarly, a No. 1 feeder pig should mature into a No. 1 slaughter hog. Carcass grading is carried out by the Food Safety and Inspection Service while the Agricultural Marketing Service is responsible for the live animal grades. Live grades are not officially applied but they are used by the USDA Market News Service when reporting prices.

When setting standards, an obvious distinction needs to be made between the species. Further designations are required for the age and intended use of animals. Clearly, feeder calves would not be treated similarly to slaughter cattle and vealer calves should not be directly compared to those going back to feed. These subgroupings of the species are referred to as classes. For the purposes of grading, 10 classes of animals are recognized, as follows:

Cattle: Live, Slaughter Cattle, Slaughter Calves, and Feeder Calves.
Hogs: Live, Slaughter Hogs, and Feeder Pigs.
Sheep: Live, Slaughter Sheep, and Feeder Lambs.[3]

---

3. Wool is also graded, but this product is not considered here.

Grades for each of these classes are treated below.

Cattle

Beef Carcasses

A 1981 proposed revision in the standards and corresponding changes in live animal grades was abandoned the next year because of widespread concern that palatability would suffer. The justification for the revision was lowering of costs of finishing animals to desired levels. The cost competitiveness of cattle compared to other red meats and poultry is an ongoing issue, but taste is apparently the overriding concern with the industry (USDA, News Center 1982). Current grade standards, therefore, date to the mid-1970s, with some minor additional modifications adopted in 1980, 1987, and 1989.

In addition to the live grades, there are carcass grades for each slaughter class. The major grade standards are as follows. For each species, the carcass grade is described first since it forms the basis from which the other grades are derived.

Two bases for grading are recognized: palatability or "quality" grades, and cutability or "yield" grades based on the major retail cuts (round, sirloin, short loin, rib and square-cut chuck). When officially graded, both designations may, at the option of the packer, be applied to the carcass (except for bull meat which is yield graded only). The yield grade designation, however, may be removed if the external fat does not exceed 3/4 inches. In total, eight *quality grade* standards (prime, choice, select,[4] standard, commercial, utility, cutter, and canner) and five yield grade standards (1 through 5 with 1 designating the highest cutability) are recognized. Finally, five types of carcasses are designated: steer, heifer and cow, and bullocks and bulls. Bullocks and bulls are treated separately from the other group so that grades are not necessarily comparable between the groups.

Yield grades. Yield grades are based on external fat thickness, proportion of internal fat (kidney, pelvic, and heart) and carcass weight/size according to the following formula:

Yield grade = 2.50 + (2.50 × adjusted fat thickness) + (.20 × % kidney, pelvic, and heart fat) + (.0038 × hot carcass weight in pounds) − (.32 × area ribeye, square inches).

---

4. Known as "Good" prior to November 1987.

The grade is reported as a whole number with any fractional part dropped (not rounded to the nearest whole number). That is, a 3.9 becomes a 3, not a 4.

The external fat is measured over the ribeye muscle, although this is not a rigid standard if the fat covering is irregular. Internal fat quantity is determined subjectively while the ribeye area is that at the 13th rib. This area is generally estimated and, indeed, it is generally believed the yield grade can be accurately determined based on a visual appraisal. This approach is facilitated by a procedure that uses external fat thickness and adjusts for the other characteristics (Figure 12-3).

Quality grades. The quality grade is based on the composite of the fat factor of streaking (marbling), firmness, and maturity according to the system shown in Figure 12-4.

Marbling is recognized in seven levels: slightly abundant, moderate, modest, small, slight, traces, and practically devoid (Figure 12-5). All assessments are visual and are to be made on a properly chilled carcass ribbed between the 12th and 13th rib. Except for the maturity group, the required degree of marbling increases with apparent age within each quality grade designation.

Maturity and firmness are recognized in five (A to E) and nine classes (firm, moderately firm, slightly firm, moderately soft, slightly soft, soft, soft and slightly watery, soft and watery, and very soft and watery), respectively. Firmness refers to the condition of the ribeye muscle, the same one used for determining marbling. Maturity is more complex but is based predominantly on the ossification of the split chine bones (vertebrae) and the shape of the ribs. As secondary factors (meaning whenever the preliminary designation is near the boundary of a maturity class), the color and texture of the lean may be used for a final decision. Beef tends to grow darker and grainier with increasing age. Color, however, refers to that related to age and not "dark cutting" meat. The maturity and firmness characteristics by quality grade are summarized in Table 12-1.

Veal and Calf Carcasses

Young beef-type animals are graded based on a combination of quality grade and conformity. Conformity refers to the thickness and fullness of the carcass, with superior conformity implying a higher proportion of meat to bone. Quality standards are similar to those used for more mature animals and particular attention is given to marbling as well as fat feathering. The marbling classifications are identical to those for steers and heifers (see Figure 12-5). Like cattle, evaluations are made at the 13th rib, and the firmness of the ribeye is also important.

FIGURE 12-3. Determination of Yield Grades on Beef Carcasses (Source: USDA, AMS 1984)

1. The preliminary yield grade is determined from the following table based on the estimated adjusted thickness of fat over the ribeye:

Fat Over Ribeye	Preliminary Yield Grade	Fat Over Ribeye	Preliminary Yield Grade
.2 inch	2.5	.8	4.0
.4	3.0	1.0	4.5
.6	3.5	1.2	5.0

2. The adjustment for area of ribeye is based on the area of ribeye-carcass weight relationships in the following table. For each square inch by which the area of ribeye is estimated to exceed the area shown for the estimated carcass weight, subtract .3 of a grade. For each square inch less than the area shown for the estimated carcass weight, add .3 of a grade.

Warm Carcass Weight	Area of Ribeye	Warm Carcass Weight	Area of Ribeye
350 lb.	8.0 sq. in.	650 lb.	11.6 sq. in.
400	8.6	700	12.2
450	9.2	750	12.8
500	9.8	800	13.4
550	10.4	850	14.0
600	11.0	900	14.6

3. The adjustment for estimated percent kidney, pelvic, and heart fat is made from the following table:

% Fat	Adjustment	% Fat	Adjustment
.5	− .6	3.0	− .1
1.0	− .5	3.5	0
1.5	− .4	4.0	+ .1
2.0	− .3	4.5	+ .2
2.5	− .2	5.0	+ .3

4. Combine the preliminary yield grade and the adjustments to obtain the final yield grade.

FIGURE 12-4. Relationship Between Marbling, Maturity, and Beef Carcass Quality Grades* [Note: Bulls are not quality graded, while for bullocks only maturity applies.] (Source: USDA, AMS 1989, p. 10)

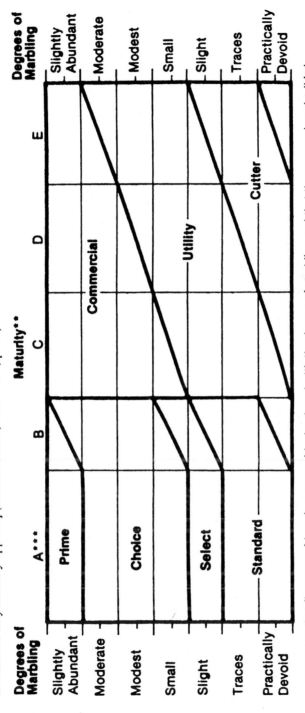

*Assumes that firmness of lean is comparably developed with the degree of marbling and that the carcass is not a "dark cutter."

**Maturity increases from left to right (A through E).

***The A maturity portion of the Figure is the only portion applicable to bullock carcasses.

FIGURE 12-5. Marbling Levels in Cattle Carcasses (Source: USDA, AMS 1976)

ILLUSTRATIONS OF THE LOWER LIMITS OF CERTAIN DEGREES OF TYPICAL MARBLING REFERRED TO IN THE OFFICIAL UNITED STATES STANDARDS FOR GRADES OF CARCASS BEEF

Illustrations adapted from negatives furnished by New York State College of Agriculture, Cornell University

1—Very abundant	4—Slightly abundant	7—Small
2—Abundant	5—Moderate	8—Slight
3—Moderately abundant	6—Modest	9—Traces

(Practically devoid not shown)

UNITED STATES DEPARTMENT OF AGRICULTURE
AGRICULTURAL MARKETING SERVICE
LIVESTOCK DIVISION

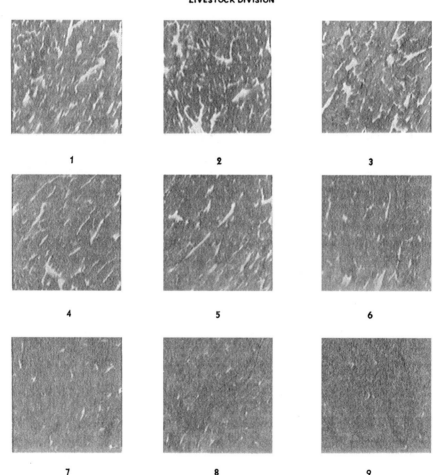

1   2   3

4   5   6

7   8   9

TABLE 12-1. Maturity and Firmness of Beef Quality Grade Standards (Source: USDA, AMS 1985, pp. 27-31)

Quality Grade	Maturity Group	Chine Bones	Ribs	Firmness
Prime	A	slightly red and soft	slightly wide and flat	moderately firm
	B	tinged with red	slightly wide and flat	firm
Choice	A	slightly red and soft	slightly wide and flat	slightly soft
	B	tinged with red	slightly wide and flat	slightly firm
Select	A	slightly red and soft	slightly wide and flat	moderately soft
	B	tinged with red	slightly wide and flat	slightly soft
Standard	A	tinged with red	slightly wide and flat	soft
	B	slightly red and soft	slightly wide and flat	moderately soft
Commercial*	C	moderately hard and rather white	moderately wide and flat	slightly firm
	E	hard and white	wide and flat	firm
Utility*	B	slightly red and soft	slightly flat	soft and slightly watery
	D	moderately hard and rather white	moderately wide and flat	moderately soft
	E	hard and white	wide and flat	slightly firm
Cutter	C	moderately hard and rather white	slightly wide and flat	very soft and watery
	D	moderately hard and white	moderately wide and flat	soft and watery
	E	hard and white	wide and flat	soft and slightly watery
Canner	E	Inferior to listed standards		

Note: Excluded maturity groups are averaged (interpolated) from the available categories.

Maturity is judged largely, but not exclusively, from the lean color and indeed is used to distinguish maturity levels within veal and between vealers and calves. In general, veal carcasses are no darker than grayish pink in color while calves may range from grayish red to moderately red. The relationship between maturity and quality grades are shown in Figure 12-6. In general, this is the same procedure as for steers except for the "Maximum Prime" designation. That simply means no credit is given for marbling in excess of designated amounts.

The final grade must include conformity as well as the quality grade developed as from Figure 12-6. Within limits, superior quality can compensate for limitations in conformity. The reverse is limited to one-third of a grade except for prime and choice where it is prohibited. Conformity is the judged width and thickness of the carcass, in relation to length, and the appearance of fleshiness and plumpness (see USDA, Ag. Mkt. Serv. 1985, pp. 36-38).

Slaughter Cattle

The current standards went into effect in 1989, although changes since 1985 have been relatively minor. The current standards supplanted previous systems adopted in 1918, 1925, 1928, 1950, 1966, 1973, 1975, and 1987. The 1966 revisions introduced yield grade standards (in addition to quality); in 1975 the two systems were mandated for graded animals. Other major changes in 1975 were an increased emphasis on marbling for the youngest (of maturity) class combined with a general reduction in the minimum marbling levels required for slaughter steers, heifers, and cows about 30 to 42 months of age. These changes were challenged in court and did not become effective until February 1976. The 1987 change was limited to a renaming of "Good" to "Select." The standards recognize five classes of slaughter cattle (steers, bullocks, bulls, heifers, and cows), eight quality grades (prime, choice, select, standard, commercial, utility, cutter, and canner), and five yield grade levels (1 to 5, with 1 signifying the highest cutability). These designations parallel those for carcasses as, indeed, they must.

Yield grades. Yield grades are based on the estimated cutability or yield of major retail cuts. This is determined by simultaneously observing muscling and fatness in relation to skeletal size. Particularly indicative of muscling are the round and forearm, while those areas where fat tends to be deposited—especially the back, loin, rump, flank, cod or udder, twist and brisket—are good places to identify the fattiness. At the extremes, lean, thinly muscled cattle generally have back widths exceeding that through the center of the round. Because of the lower proportion of bone to lean

FIGURE 12-6. Relationships of Maturity and Marbling for Veal and Calf Carcasses (Source: USDA, Form LPGS-32 (11-80))

QUALITY GRADE COMPENSATIONS

I   In each grade, superior quality is permitted to compensate for deficient conformation, with out limit.

II   In the prime and choice grades, superior conformation is not permitted to compensate for deficient quality. In all other grades, such compensation is permitted, on an equal basis, but only to the extent of 1/3 of a grade of deficient quality.

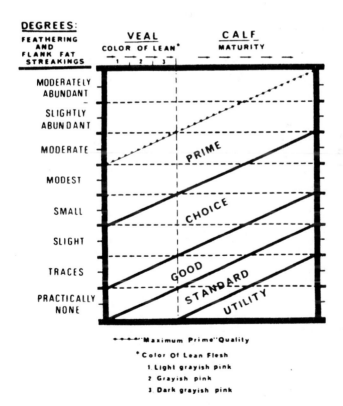

QUALITY GRADE EQUIVALENT OF VARIOUS
DEGREES OF FEATHERING AND FLANK FAT STREAKINGS
IN RELATION TO COLOR OF LEAN (VEAL) OR MATURITY (CALF).

breast on these cattle, they are almost never classified as Yield Grade 1. In contrast, very fat cattle tend to be wider through the back than through the round (Figure 12-7). Within a yield grade, the quality grade can vary by three or more levels.

Quality grades. The amount and distribution of external finish (as evident on the chops, back, ribs, loin and rump) are the major visible indices of quality grades. Animals judged to be between 30 and 42 months of age must exhibit increasing amounts of finish and firmness of muscling to qualify for a given quality grade. Below 30 months, a progressive escalation of quality-related factors is not required. Forty-two months is the approximate division between prime through standard and the inferior grades. Specific requirements for steers, heifers, and cows are shown in Table 12-2 but are perhaps better expressed by reference to Figure 12-8.

Other factors. A large number of other factors beyond those identified above contribute to the grade of slaughter cattle. Principal among these are breed and management practice. While the first is visible, buyers also frequently like to know the source of slaughter stock as a means of enhancing their ability to project the carcass grade.

Bullocks and bulls. Bullocks are very young bulls—up to about 24 months of age—when maturity characteristics begin to appear. Quality grade standards for bullocks are comparable to steers in the youngest (less than 30 months) age class. Bull carcasses are not quality graded, but a yield grade will be affixed (USDA, Ag. Mkt. Service 1989).

Feeder Cattle

Feeder cattle standards, adopted in 1979, reflect major changes over previous systems. The present arrangement attempts to measure the value (merit) of the animals in three dimensions: frame size, thickness of muscling, and thriftiness. Frame size is the skeletal size (height and body length), thickness is the development of the muscular system in relation to frame size, while thriftiness actually refers to the apparent health of the animal. Health (disease, parasitism, emaciation) relates to the ability to grow and fatten normally and may change over time. The term feeder applies only to animals less than 36 months old.

Three frame sizes (large, medium, and small) are recognized, as are three degrees of muscling (No. 1, No. 2, and No. 3) for a total of nine possible classes of thrifty cattle. Seemingly unhealthy feeder cattle are classified as unthrifty regardless of their other characteristics. In general, the larger the frame size, the higher the rate of gain, the longer the period required to fatten, and the greater the live weight necessary to attain a given slaughter quality grade. Frame size and slaughter weight for choice

FIGURE 12-7. Slaughter Steers: Yield Grade (Souce: USDA, AMS, Livestock and Seed Div. 1990)

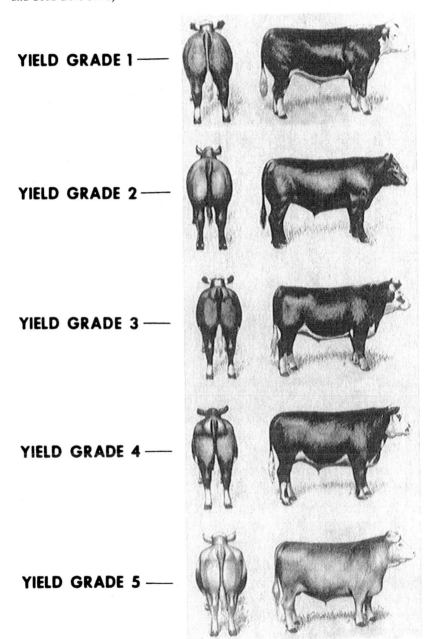

YIELD GRADE 1 ——

YIELD GRADE 2 ——

YIELD GRADE 3 ——

YIELD GRADE 4 ——

YIELD GRADE 5 ——

TABLE 12-2. Quality Grade Standards for Slaughter Steers, Heifers, and Cows (Source: USDA, AMS 1985, pp. 4-5 and 8-12)

Grade	Age/Months	Fat Covering[a]	Fullness[b]	Firmness[c]
Prime	LT 30	moderately thick	marked	firm
	30 - 42	tends to be thick/smooth	upper full	very frim
Choice	LT 30	slightly thick	moderately full	moderately firm
	30 - 42	moderately thick	marked	firm
Select	LT 30	thin	slightly	slightly
	30 - 42	slightly thin	some	firm
Standard	LT 30	very thin	-	-
	30 - 42	very thin	-	-
Commercial	Younger	slightly thick	-	moderate
	Mature	moderately thick	moderately full	firm
Utility	Younger	very thin	slightly thin	
	Mature	slightly thick	slightly full	
Cutter	Younger	practically none	-	-
	Mature	very thin	-	-
Canner	--- Inferior to above ---			

LT = less than

___

a/ Over chops, back, ribs, loin and rump. In lower quality grades applies only to back, loin and upper ribs.

b/ Muscling in the brisket, flanks and cod or udder.

c/ Firmness of muscling.

FIGURE 12-8. Slaughter Steers: Quality Grades (Source: USDA, AMS, Livestock and Seed Div. 1987)

PRIME —

CHOICE —

SELECT —

STANDARD —

UTILITY —

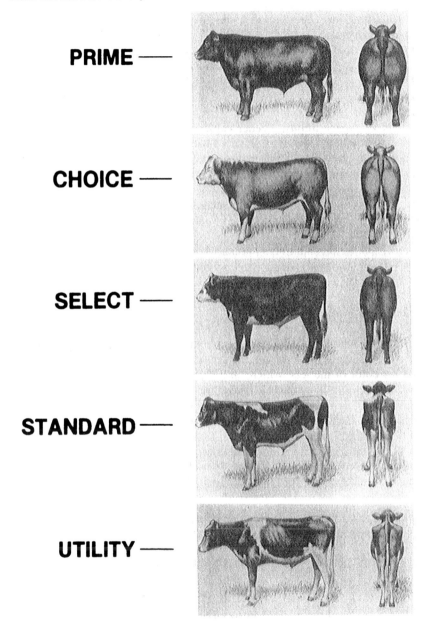

grade can be related as in Table 12-3. Frame size is based on the relationship of height and length (Figure 12-9).

Muscling, too, correlates to mature weight for a given grade and particularly to the carcass yield grade. Because apparent muscling is influenced by the degree of fat covering, a slightly thin cover is assumed for the standards; in practice, adjustments must be made for variation around this level. No. 1 feeder cattle are slightly thick throughout, while No. 2 cattle are narrow through the forequarter and middle of the rounds. Inferior muscling is classed as a No. 3 (Figure 12-10).

Other considerations. Feeder cattle grades, especially with regard to the frame size, are heavily influenced by age and breed. That is, a medium could be a large Angus or small Charolais. Clearly, the feeder needs to know the breed, or mix, in addition to the grade. Frame size is also highly correlated to age so that a buyer must be a good observer of age. Maturity is most evident in head size which increases proportionally with age while ears become relatively smaller. The tail and feet expand proportionally while the tail has a more pronounced switch. These factors are, however, highly qualitative and make purchases of feeder cattle, particularly sight-unseen, more complicated and the reliance on a known source correspondingly greater.

Slaughter Vealers and Calves

Calves are classified primarily by maturity and are generally eight months of age or less. The distinction between a calf and a vealer is based on the color of the lean which, in turn, is determined by the diet (roughages or milk/milk replacer). Thus, for calves older than three months, designation as a vealer requires evidence of the diet. Slaughter calves, like cattle, are rated on a quality basis (prime, choice, select, standard and utility) with no yield grade estimate. Grades are based on quality (fatness, maturity, and related factors) and conformation. Superior quality is permitted to compensate without limit for deficient conformation (compensation of conformation for quality is limited in the two top grades).

TABLE 12-3. Relationship Between Frame Size and Slaughter Weights for Feeder Calves (Source: *Federal Register* 1979, p. 45322)

Approx. slaughter weight for choice (lbs.)

Frame Size	Steer	Heifer
Large	above 1,200	above 1,000
Medium	1,000-1,200	850-1,000
Small	below 1,000	below 850

FIGURE 12-9. Feeder Cattle Frame Sizes Classes (Source: USDA, AMS 1980)

Frame Size

Large

Medium

Small

Large and medium frame pictures depict minimum grade requirements. The small frame picture represents an animal typical of the grade.

FIGURE 12-10. Feeder Cattle — Muscling Classes (Source: USDA, AMS 1980)

Thickness

No. 1

No. 2

No. 3

No. 1 and No. 2 thickness pictures depict minimum grade requirements. The No. 3 picture represents an animal typical of the grade.

Quality standards are determined primarily by viewing the overall degree of muscling with special attention paid to the width of the back and loin, and shoulders and hips. Prime is moderately thickly muscled throughout and moderately wide over the back, shoulders, and hips. Standard ranges down to very thinly muscled and narrow over the back, shoulders, and hips. Utility grades are inferior to other standards (USDA, Ag. Mkt. Service 1985).

Hogs

Pork Carcasses

The grading standards provide for three carcass classes: barrows and gilts, sows, and stags and boars. No provisions exist for grading the final one, while the first two are treated separately.

Barrow and gilt carcasses. A revised grade standard was adopted in January 1985, replacing the system that went into effect in 1968. (Previous standards date to 1955 and 1952, the original). The revisions reflect the efforts by swine feeders to raise longer and, most importantly, leaner hogs since the 1968 standards went into effect. By 1980, over 95 percent of slaughter barrows and gilts were U.S. No. 1 or 2, compared to 50 percent in 1968. Whenever such a large portion of production is classified within a few grades, the value of the grade standard declines. The 1985 revisions essentially rename No. 1 as No. 2, and so on, through No. 4. Any carcass not satisfying these requirements is graded utility (Parham and Agnew 1982). Boars and stags are not graded under this plan. Grades are based on the two considerations of *quality* and *expected yield* from four lean cuts: ham, loin, picnic shoulder, and Boston butt. (These standards are reproduced in the *Federal Register* Dec. 14, 1984). Quality is either acceptable or unacceptable, with unacceptable carcasses receiving the U.S. utility grade. In determining quality, graders look for minimally acceptable standards in the (1) firmness of the fat and lean, (2) feathering between the ribs, and (3) color of the lean. Ideally, evaluation is judged from the condition of the loin eye muscle above the 10th rib. The carcass must satisfy the additional requirement of a belly with sufficient thickness for bacon production.

The remaining criteria are based on the backfat thickness and the degree of muscling. These measures are used as approximations for the yield of the four lean cuts stipulated above. Backfat thickness is measured at the last rib, giving preliminary grades as follows:

U.S. No. 1:   < 1.00 inches backfat thickness
U.S. No. 2:   1.00 to 1.24 inches thickness
U.S. No. 3:   1.25 to 1.49 inches thickness
U.S. No. 4:   1.50 inches and over

Adjustment is then made for degree of muscling based on a subjective evaluation, with special attention given to the ham. Three degrees are recognized: thick (superior), average, and thin (inferior) (Figure 12-11). In general, more heavily muscled carcasses have less fat. Applying a score of 1 for thick muscling, and 2 and 3 for average and thin, respectively, the grade can be determined from the formula:

Carcass grade = (4.0 × backfat thickness) − (1.0 × muscling score)

Note, however, that carcasses with over 1.75 inches of backfat thickness cannot be graded as No. 3 even if the muscling is superior. All other carcasses are graded utility.

These revised grade standards reflect a general movement toward valuing lean carcasses more highly and fatter ones lower. One Wilson Foods Co. executive commented, "If we don't have to trim as much fat off the lean hog—we can afford to pay more for that hog." A Wilson premium/penalty scheme emphasizes the backfat thickness, with premiums paid for average weight carcasses (160-174 lbs.) with 1.3 inches and less of backfat (Fleming 1984).

The revised grade standards are intended to provide ongoing economic incentives for hog producers to provide leaner and, hence, more valuable animals. However, the practical minimum is being approached because packers need some fat cover to protect the lean. Future improvements—and grade standards—are likely to emphasize carcass length (Parham and Agnew 1982).

Sows. Sow carcass standards were excluded from the 1984/85 revisions so that they date to 1968. While the criteria are similar to those for barrow and gilt carcasses, the recognized grades are slightly different (U.S. No. 1, No. 2, and No. 3 plus Medium and Cull). The principal standard is backfat thickness which, unadjusted, allows heavier coverings than for comparable younger animals. The standards for preliminary grades are as follows:

U.S. No. 1:   1.5-1.9 inches backfat thickness
U.S. No. 2:   1.9-2.3 inches backfat thickness
U.S. No. 3:   2.3 and over inches backfat thickness
Medium: 1.1-1.5 inches backfat thickness
Cull: below 1.1 inches backfat thickness

Final grading also includes a consideration of the thickness of muscling, conformation, uniformity of fleshing, and finish firmness (USDA, AMS 1985, pp. 55-57.)

Slaughter Barrows and Gilts

Slaughter animal standards are intended to reflect directly the condition of the carcass as it relates to the aforementioned carcass standards. The date of adoption is therefore the same as the preceding carcass standards. To accomplish that objective, the same three criteria are used as for carcasses (acceptable belly thickness, acceptable quality, and backfat thickness). And, in fact, these standards were revised simultaneously with the new carcass standards. However, for live animals these characteristics must be estimated rather than measured. Only general factors may be used to judge lean quality. The fatness and muscling evaluations are typically made simultaneously. The ham, little affected by fat, is an indicator of muscling. Fatness is evident in several ways since fat deposition is less regular than muscling. Observed areas should include the backfat, shoulder, edge of the loin, rear flank, and the belly. Determination is relatively straightforward for the experienced judge (Figure 12-12). Once the estimates are made, the same grade-determining formula used for carcasses may be applied (USDA, AMS 1985).

Feeder Pigs

Feeder pigs grades have not (as of this writing) been revised to reflect the new carcass standards so that the current standards date to the 1970s. An adjustment bringing them into agreement can be expected at some point. Two classes of feeders are recognized, thrifty and unthrifty. Thrifty pigs may be graded U.S. No. 1 through 4, while unthrifty animals are utility or cull depending on the severity of the impairment to their ability to gain weight. Cull pigs will require a long period to recover from disease and/or poor care but, once restored to thriftiness, a utility or cull pig may be classified as No. 1 through 4.

Thrifty pigs are graded entirely on the basis of their slaughter potential at a market weight of 220 pounds. No. 1 pigs, due to superior musculature and skeletal systems, reach that weight with a minimum degree of finish. Higher numerical grades (2, 3, and 4) will have higher finish levels at the target 220 pounds. This is another way of saying the frame size and muscling decline from 1 to 4.

No. 1 feeder pigs are long and have thick muscling throughout, which

FIGURE 12-11. Barrow and Gilt Carcass Muscling [Note: Under the 1985 standards, three degrees are recognized: Thick, Average, and Thin. These correspond in the illustration to Very Thick, Moderately Thick, and Thin.] (Source: USDA, AMS)

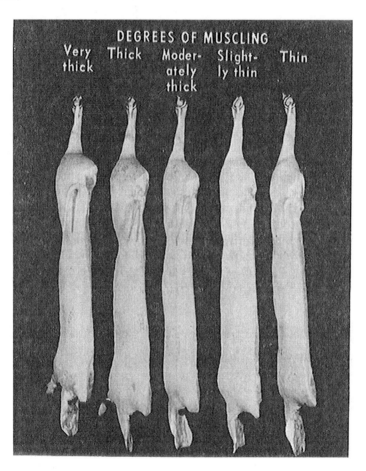

is most evident in full hams and shoulders. Other attributes may partially compensate for muscling, but a No. 1 must be at least moderately thick-muscled. These standards apply equally to barrows and gilts. (Sows and boars are seldom fed and the standards do not apply.) At the other extreme, No. 4 pigs are short and thinly muscled throughout, with thin and rather flat hams (Figure 12-13) (USDA, AMS 1974).

FIGURE 12-12. Slaughter Barrow and Gilt Grades (Source: USDA, AMS 1988)

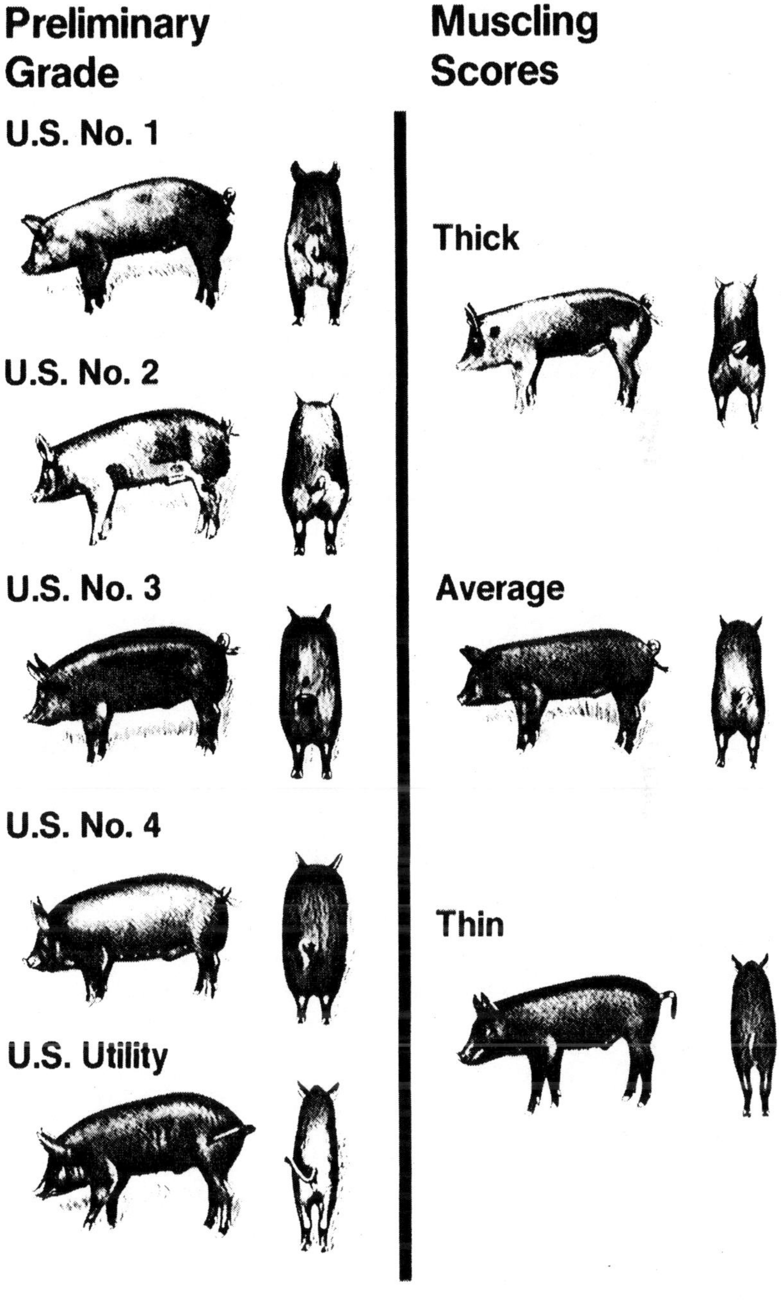

Preliminary Grade

U.S. No. 1

U.S. No. 2

U.S. No. 3

U.S. No. 4

U.S. Utility

Muscling Scores

Thick

Average

Thin

FIGURE 12-13. Feeder Pigs — U.S. Grades (Source: USDA, AMS 1988)

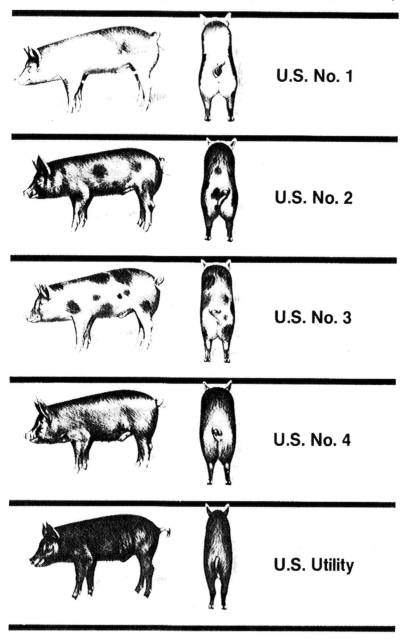

Sheep

Lambs, Yearling Mutton, and Mutton Carcasses

Revised standards for these three carcass classes recognized for bovines by the USDA became effective in October 1982, replacing the 1960 standards. Previous standards were effective in 1931, 1951, and 1957. These standards refer to quality characteristics only. A yield grade system was applied in 1969 as a voluntary addendum to quality grades, but the industry has not made use of the yield grades. The yield grades remain in place but were unaffected by the 1982 quality grade revisions. Quality grades recognize five levels (in descending order): prime, choice, good, utility, and cull, although cull applies to mutton carcasses only (see *Federal Register* Sept. 13, 1982).

The carcass *class* is based on age as interpreted from the condition of two joints: the "break joint" and "round" or "spool joint." The round or spool joint is the end of the cannon bone and is visible when the foot and pastern are removed at the ankle. Directly above the spool joint, at the point of bone growth, is an area of cartilage in young animals that is subsequently replaced by bone during the aging process. This is the point, or break joint, at which the foot can be removed in young animals.

Under the current standards, a carcass with only one break joint may be classified as lamb or yearling mutton. The distinction is based on other evidence of maturity (the rib shape and color of the lean). Lamb carcasses have slightly wide and moderately flat rib bones and a light red color and fine texture of the lean. Carcasses with two spool joints or those with missing cannon bones ("trotters") are yearling mutton or mutton. Again, shape of the ribs and color of the lean are used for the final classification.

Quality grade standards were simplified considerably in this most recent revision. Feathering and flank fullness and firmness were eliminated. Three standards remain: (1) conformation, (2) quantity of fat streakings, and (3) firmness of fat and lean. These factors are taken as indices of the palatability of the lean. Rather than establishing fixed standards, the requirements vary with the maturity (class) of the carcass (Figure 12-14).

Conformation refers to the relative development of the muscular and skeletal systems and is determined by the thickness of muscling as well as the degree of thickness and fullness of the carcass. Prime carcasses tend to be thickly muscled throughout and are moderately wide and thick in relation to length. Those carcasses graded choice are slightly muscled and slightly wide and thick, while good carcasses may be slightly thin muscled and moderately narrow in relation to length.

There are ten degrees of marbling (flank fat streakings): abundant,

FIGURE 12-14. Quality Grades of Lamb Carcasses: Relationship Between Maturity and Marbling (Source: *Federal Register* 1982, p. 40145)

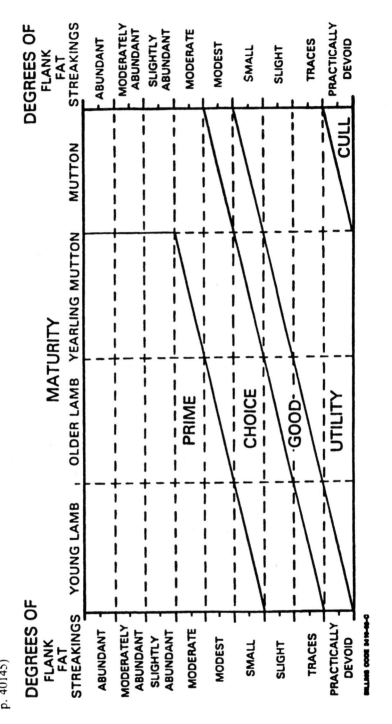

278

moderately abundant, slightly abundant, moderate, modest, small, slight, traces, practically devoid, and devoid. Evaluation is done on the inside flank muscle. Firmness is also measured in degrees, in this instance, 14 of them: extremely firm, tends to be extremely firm, firm, tends to be firm, moderately firm, tends to be moderately firm, slightly firm, tends to be slightly firm, tends to be slightly soft, slightly soft, tends to be moderately soft, moderately soft, soft, and very soft. A prime lamb carcass must rank at least "tends to be moderately firm." Choice carcasses must be at least "tends to be slightly firm" while good carcasses achieve the status of not less than "slightly soft."

The final carcass grade characteristic is the *yield grade*, a referral to the estimated percent of closely trimmed, boneless, major retail cuts to be taken from the carcass. As noted, quality and yield grades are determined independently, although, to date, the industry has not made use of the yield grades. Yield grades are based on the amount of external fat, the amount of kidney and pelvic fat, and the conformation grade of the legs. In 1982, the yield grade standards remained unchanged from those first established in 1969.

Conformation for yield grade purposes is measured in the same way as for the quality grade. A number system is used, with a 1 for low cull up to 15 for high prime. The kidney and pelvic fat quantity, assessed judgementally, is expressed as a percentage of the carcass weight. Finally, the external fat cover is the actual thickness of the fat between the 12th and 13th ribs. These factors are combined by using the formula:

$$1.675 + (.33 \times \text{No. fat thickness increments (in .05 inch increments)} + (.25 \times \% \text{ pelvic and kidney fat}) - (.05 \times \text{conformation score } (1-15))$$

Any fraction is dropped from the final yield grade designation so that a 3.8 is reported as a 3.

Slaughter Lambs, Yearlings, and Sheep

Revised live-animal standards became effective in November 1984. These revisions were necessary to conform to the new carcass grade standards, but no basic changes were made. Thus four quality grades are used—prime, choice, good, and utility, except for sheep, where the cull grade still applies.

Paralleling the carcass standards, live animal grades are based on conformation and quality, principally the fatness and maturity. Conformation refers to both the general body proportions and ratio of meat to bone. Quality, too, includes a measure of thickness of the fat covering and its firmness and smoothness. Variations in the fleece make some handling necessary to augment the visual inspection, and considerable judgement and experience are needed for establishing accurate grade standards (*Federal Register* October 26, 1984).

Prime animals are thickly muscled as well as moderately wide and thick. The body is smooth and symmetrical. Choice requires slightly thick muscling and a slightly wide and thick body in proportion to length. The appearance is moderately refined. Those animals grading good must be at least slightly thinly muscled and moderately narrow in relation to length.

The yield grade standards use the same criteria and, indeed, the same criterion as for the carcass. Assessing the percentage of kidney fat and other factors on a live animal is, however, a speculative effort. A more practical approach uses but two factors, the degree of fatness and leg conformation. In judging the relative fatness, the areas of rapid fat deposition (including the back, loin, rump, flank, breast, and cod or udder) are given particular attention.

Feeder Lambs

At this time, the USDA has no feeder lamb standards, and none appear imminent. However, the Livestock and Grain Market News Service does use descriptive terminology such as fancy, choice, and good in its market reports (Ray, USDA, AMS 1990, personal communication).

SYNOPSIS

Grading, which has its origins in wartime at the beginning of this century, did not receive formal authorization until the next war in 1946. Even then, it was adopted slowly (except in wartime) as a voluntary system paid for by the packer. Using beef as an example, it has exceeded 50 percent only quite recently. Grading is done by the Food Safety and Inspection Service (FSIS) of the USDA.

Grades serve several, quite different, purposes. They enhance price reporting and trading on a sight-unseen basis. And they provide a means for the system to identify products in relatively high demand and signal those products back to producers. Satisfying those goals requires that grades' present characteristics be meaningful to consumers (even if they are unaware of them) and to producers, packers, and processors. At the same time, there must be neither too many nor too few grades and the standards need to be relatively easily describable. All this makes for a complex system that must evolve over time. However, no system will satisfy all users.

While grades vary by species and animal class, they are all based on carcass grades. Typically, there are two parallel systems, one measuring quality (palatability) and one measuring yield (cutability). Slaughter stock and feeder grades are all based on the carcass system, so that a choice slaughter steer should yield a choice carcass. Feeders are graded somewhat differently, so that a low number will reach a target slaughter weight with a lower degree of finish (fat).

Within quality and yield standards, there are a number of subcomponents that are related in a complex fashion, often by a formula. In the final analysis, many of the decisions are a judgement call by the graders.

Study Questions

1. Justify the use of public funds for mandatory health inspection while grading is voluntary and paid for by the packer.
2. How can grades serve a purpose when only a portion of consumers understand what they mean?
3. Devise a system of standards for feeder lamb grades.
4. Since slaughter and feeder stock grades are not formally applied, how are they used?
5. In your judgement, why is grading used less for pork carcasses than for beef?

REFERENCES

American Meat Institute. "Meat Facts." 1989.

Branson, R. E. and P. Rosson. "Consumer Attitudes Regarding Leanness in Beef, Animal Fats, and Beef Grading Systems." Texas Ag. Exp. St., MP-1495, September 3, 1981.

Breimyer, H. F. "The Purpose of Grading." Talk before the Inter-Industry Beef Grading Conference, Kansas City, November 14, 1960.

Brunk, M. E. and L. B. Darrah. *Marketing of Agricultural Products*. New York: Ronald Press, 1955.

Federal Register. "Standards for Grades of Slaughter Lambs, Yearlings, and Sheep." Vol. 49, No. 209, October 26, 1984, pp. 43035-9.

Federal Register. "Standards for Grades of Barrow and Gilt Carcasses and for Slaughter Barrows and Gilts." Vol. 49, No. 242, December 14, 1984, pp. **48669-675.**

Federal Register. "Standards for Grades of Lamb, Yearling Mutton, and Mutton Carcasses." Vol. 47, No. 177, September 13, 1982, pp. 40141-9.

Fleming, B. "The New Wilson Carcass Program: Lean Content Outranks Weight," *National Hog Farmer* August 15, 1984, pp. 31-35).

Fox, D. G. and J. R. Black. "The Influence of Cow Size, Crossbreeding, Slaughter Weight, Feeding System and Environment on the Energetic and Environmental Efficiency of Edible Beef Production." Proc. 28th Ann. Reciprocal Meat Conf., 1975, pp. 90-115.

National Commission on Food Marketing, Organization and Competition in Food Retailing. Tech Study No. 7, June 1966.

Parham, K. D. and D. B. Agnew. "Improvements in Grades of Hogs Marketed." USDA, Economic Research Service, ERS-675, February 1982.

United States Congress, Office of Technology Assessment. "Perspectives on Federal Retail Food Grading." Washington, DC: U.S. Govt. Printing Office, June 1977.

USDA, Agricultural Marketing Service. "Official United States Standards for Grades of Carcass Beef." Washington, DC: U.S. Gov. Printing Office, 241-790/80702, 1989.

USDA, Agricultural Marketing Service. "Official United States Standards for Graders of Slaughter Cattle." Washington, DC: U.S. Govt. Printing Office, 241-790/80854, 1989.

USDA, Agricultural Marketing Service. "Official United States Standards for Grades of Slaughter Swine." 1985.

USDA, Agricultural Marketing Service. "Meats, Prepared Meats and Meat Products (Grading, Certification and Standards)" 1985.

USDA, Agricultural Marketing Service. "Facts About: U.S. Standards for Grades of Feeder Cattle." AMS-586, April 1980.

USDA, Agricultural Marketing Service. "USDA Grades for Slaughter Swine and Feeder Pigs." Mkt. Bulletin No. 51, July 1974.

USDA, News Center. "NEWS" September 20, 1982.

USDA, News Division, News Feature. "Consumers Like New Name for Lean Meat." April 24, 1990.

Zwiebach, E. "Meat Scanning Aids Targeting of Consumer Needs,"*Supermarket News*, May 13, 1985, p. 11.

Chapter 13

Marketing Alternatives

A MARKETING STRATEGY

Most livestock producers have several market outlets available. Perhaps there are several packer or order buyers willing to call at the farm, or animals may be shipped to the nearest auction market. Still other choices may exist in some areas, possibly in the form of dealers or terminal markets (see Chapter 9 for a description of these terms). Which, then, is the best alternative?.

One obvious way to evaluate these several markets is to compare (expected) price. From that must be deducted transport costs and shrink (actual or pencil) paid by the seller so that prices can be compared on an equivalent basis (Chapter 7). Now you go ahead and sell to the highest-priced outlet, correct? Well you would if you were *selling* livestock, but not necessarily if you were *marketing* it. The key distinction between selling and marketing is that marketing focuses on the needs of the customer, not those of the producer (see Chapter 1).

In a more applied sense, the producer *marketing* his/her stock would consider numerous factors other than the differences in (expected) net prices among available market outlets. Certainly the services required are important. Is it necessary to do the trucking yourself? Will the buyer help you with deciding when to sell? Then, you must consider when you are paid and what the likelihood of default is. Moreover, can you, at a reasonable cost, withhold animals if prices are unexpectedly depressed or, alternatively, sell all types and grades of your stock at this market? Or is demand highly specialized? It is necessary to determine how accurately you can project prices at these markets. If the price is a firm quote from a buyer, accuracy is high, probably far higher than using the judgement of the manager of a public market, for example. Finally, one must look beyond the current sale to determine any possible impacts on subsequent sales. If you are indeed in the livestock business for the long term, your marketing system should reflect that long-term outlook.

Considering these, and related factors, as they apply to the available markets is what separates the marketer from the seller. The seller attempts to maximize price on a sale-by-sale basis. The marketer attempts to maximize the multiple long-term prospects of his or her livestock enterprise. Does not everyone have the same prospect in mind — the maximization of profits? That may be true at the most general level. More specifically, each of us has other goals or objectives that must be considered simultaneously. Some like to get on the telephone and deal, while others detest it. Some are willing to accept large risks, while others are content with the steady, if possibly more limited, profit.

Marketing, then, requires the development of a *strategy*. This includes several steps, the first of which is to establish your objectives; only *you* can do that. It is important to be realistic, taking into account your abilities, likes, and dislikes. Remember, nobody can do everything, or at least everything well. Nor can all objectives necessarily be maximized simultaneously. A set of basic objectives could incorporate the following:

- Limited time available to arrange each sale;
- Stable returns necessary to meet loan obligations (e.g., occasional low prices, delayed payments, etc., must be avoided to the degree possible);
- A mixed livestock enterprise requires a market or markets that accept a wide range of types and sizes;
- Due to an isolated location, transport-related costs are high and must be considered carefully.

Objectives should always be written down so they may be referred to for consistency with actual marketing arrangements. Seen in print, the farmer whose objectives are described above may recognize that he is imposing a strict limit on his marketing options by allowing little time for evaluating them. Perhaps easing the time restraints, even if it means reducing the number of head slightly, would be a profitable decision.

Once the objectives are established, the next step is deciding how best to achieve them — *the plan of action*. The objectives, plus the action plan, constitute the marketing strategy. Action, however, requires knowing the attributes of each marketing alternative so that they may be matched. For example, no one who is uncertain about the grading of his/her stock should ship to a buying station that receives a limited range of grades. Nor should anyone with a real concern for defaults on payment sell on a non-cash basis to an unknown farmer; there are no formalized remedies for nonpayment short of a legal suit.

The purpose of this chapter, then, is to describe the attributes of the major marketing alternatives as they affect strategic marketing planning. The material in this chapter is limited to Cash Markets. Chapter 14 is directed to Futures Markets. Cash markets include the traditional outlets — public markets, buying stations, and direct sales — as well as integrated outlets in that the producer typically has a contracted selling agreement that may include an equity interest (investment). This material expands on the descriptions of the functionaries of the system presented in Chapter 9.

Futures markets (Chapter 14) are not markets proper. Certainly, some livestock does exchange hands (the contracts mature to delivery), but typically less than one percent. Rather, these markets are an institutionalized means of limiting (actually shifting) price risk for future sales. Using the future markets, then, is an adjunct to some other form of cash market.

MARKETING ALTERNATIVES

The major cash markets included here are described according to the following characteristics:

- *Description:* A description of the physical operation and layout of the markets.
- *Livestock Types and Grades Accepted:* Markets typically differ on the range of types of livestock handled; some are very specific while others provide buyers for nearly everything.
- *Marketing Costs:* The costs of using a market, net of shrink and transport.
- *Convenience:* The markets are ranked according to the amount of time and effort required to use them.
- *Cost of Cancelling Sale:* Sometimes, especially if prices appear to be temporarily depressed, it is prudent to withdraw from the sale. This category considers the mechanism for withdrawal and the associated costs such as trucking.
- *Price Uncertainty (Risk):* Unless the price is guaranteed before the sale, there is always an uncertainty about the eventual price. To the extent price uncertainty is related to a market type, it is discussed.
- *Duration of Relationship:* Does the sales arrangement encompass only the pending sale, or are there longer term components to it?
- *Payment Risk:* Following a non-cash sale, how confident can you feel about receiving full payment in a timely manner?
- *Equity Requirements:* Does participation in the market necessitate any investment by the producer in the marketing organization?

- *Comments/Suggestions:* A listing of other market aspects that is useful in understanding or utilizing them.
- *Options:* Any variants of the basic operational practices are described.

Public Markets

Description

Public markets include both *auction markets* and *terminal markets* (Figure 13-1). Terminal markets are named for their location at railroad terminals. They began a decline in numbers during the 1930s along with the decentralization of packing plants (see Chapter 4) and the shift to truck transport. From a high of 80 terminal markets in 1937 there were only 21 in mid 1980. At the same time there were 1,564 auction markets, but this, too, is down from a peak of 2,400 in 1949 (USDA, P&SA, Resume 1986, Table 20). These markets tend to be concentrated in the Midwest (Figure 13-2). Today, public markets handle between 11 and 20 percent of livestock sales by number for cattle, hogs, and sheep (USDA, P&SA, Resume 1987, Table 2). For calves, the total is closer to 40 percent. Public markets, however, are disproportionally important to pricing because they are the source of the major price reports (see Chapter 10).

Other than their historical origin, terminal and auction markets differ in the use of a commission agent. A terminal market's sales are typically made by the commission agent who receives a payment from the seller (see Chapter 9). Sellers may choose to act as their own agent, but seldom do so. The actual sale may be made by negotiation (private treaty) or by auction. Auction markets lacking commission agents and use the auction method exclusively.

Livestock Types/Grades Accepted

Most markets will accept a wide range of types and grades. This is especially true of auction markets where virtually anything can be run through the ring. The ability to cash in stock quickly is a prime attribute of public markets and helps explain their prevalence in areas of cull stock whose sale is often unplanned. Prices for unusual types and grades may, however, be depressed due to high marketing costs or use for lower value purposes. The author, for example, has seen good feeder stock going for veal slaughter due to the lack of a feeder calf buyer that day.

FIGURE 13-1. Typical Public Livestock Market (Source: Torgerson 1978)

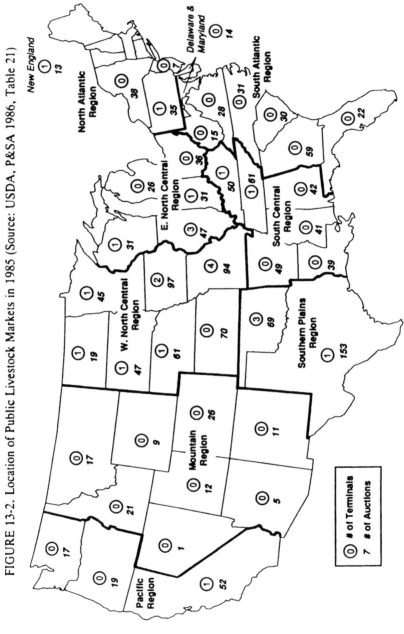

FIGURE 13-2. Location of Public Livestock Markets in 1985 (Source: USDA, P&SA 1986, Table 21)

288

Cost

Public markets are costly; 1972 fees of about $2.60 per head for feeder calves and 95 cents for slaughter hogs rose to over $9.00 for feeder calves by 1980 (Stoddard 1975, Table 4; Sullivan and Linton 1981, Table 2). The major explanation for these high costs—this is typically the most costly marketing alternative—is the considerable labor involved in receiving and sorting the animals and the utilization of the market only one or two days per week. In addition to these fees, the seller is responsible for transportation and shrink loss (both in transport and in the market up to the sale). Moreover, some markets assess the buyer, and a portion of that charge may be passed back, in the form of a lower price, to the seller as an indirect charge. So, altogether, public markets are a costly way to market. This probably explains, to a large degree, their decline in use.

Convenience

Public markets are convenient; it is convenience that one is buying at a rather high price. All that is *necessary* is a telephone call to the market. Transport, too, can be arranged that way. Some producers are not even on the farm when the pickup is made, although this is not advised.

While public market sales can be quick and easy, it is seldom the way to get the best price, especially if you are unknown and/or are selling a type or grade that is uncommon in that market. Under those circumstances, certain preparatory steps described below under Comments/Suggestions are beneficial.

Cost of Cancelling

There are several means of retreating from a public market sale if prices are inadequate. Many markets will permit the placement of a reservation, or "no sale" price, prior to the sale. This is especially common for feeder stock and for commission agent sales. Alternatively, the producer may be able to signal "no sale" from the stands or, finally, bid on his own stock and "buy" them back if necessary.

Once stock are delivered to the market, the costs of these options are all high. A fee (probably the full commission) will be levied. Also, the producer is responsible for shipment and/or additional yardage fees. Weight, and possibly even vigor, suffer from the stress, especially for young animals. In general then, livestock shipped to one of these markets should be viewed as sold except under unusual circumstances.

Price Uncertainty

Prices on public markets can be expected to vary more than those in private treaty sales. This may occur because public markets can be affected very rapidly by supply and demand changes. Packer buyers, especially, move in and out of these markets to balance their slaughter requirements. Producers can redirect supplies as prices move up and down. These short-term supply and demand changes can lead to price volatility and, hence, to price uncertainty when sales are made there (see also the explanation in Chapter 7). On the down side, price uncertainty may be controlled by setting a minimum acceptable (or reservation) price, but that risks a no sale and its inherent costs and complexities. Certainly the reservation price should not be set unrealistically high.

Under Packers and Stockyards regulations (see Chapter 11), public markets are prescribed to engage in market support activities. This usually means a bid, and possible purchase, on the part of the market as a means of strengthening prices. Ostensively, this practice protects the seller from very low prices and, on occasion, it is undoubtedly effective. In general, however, market support is undertaken infrequently so that it is nothing to be relied upon. Realistically, through limited market intervention, it is not possible to achieve more than a temporary price smoothing and possibly an increase in the momentum of the sale.

Duration of the Relationship

Sales are made on a shipment-by-shipment basis with little carryover between sales. With each exchange an essentially independent exchange, price per sale should be maximized with little consideration for market development and goodwill interests. However, a regular seller with consistent quality stock can become known and command a higher price (see Chapter 7). So, too, will the producer who disguises a defect in his/her animals. Even in these markets, then, the producer should proceed with an interest in developing a reputation for consistency. Quality should be consistent so as to be predictable from sale to sale, but need not necessarily be high.

Payment Risk

Public markets are required by the Packers and Stockyards Administration (see Chapter 11) to make payment by the close of the next business day (unless credit has explicitly been extended) and to be bonded should the buyer forfeit. Payment, then, is quite secure although the potential of a major forfeit, exceeding the amount of the bond, always exists and should be considered.

Equity Requirements

Privately owned, public market sales are on a fee basis with no investment in the market required. However, a number of the existing markets (handling about 13 percent of domestic livestock in 1974, but with 40 percent of livestock producers as members) are operated as cooperatives owned by the member-users, possibly combined as a subsidiary operation of a larger cooperative providing feed and supplies (Torgerson 1978, p. 1). By mid-1980, cooperative markets numbered 67, about four percent of the total (Hogeland 1987, p. 3). These operations require some ownership, or equity investment, by producers/users. Typically, producers entered this sector as a means of protecting a market outlet rather than as profitable enterprises in their own right, although cooperatives tend to provide additional services.

Comments/Suggestions

If you have the luxury of several public markets within a reasonable distance from the farm, the first decision is to choose among these markets. A series of price data will help determine if there is any regular price advantage in going to one of these markets. Local auction prices are typically carried by newspapers or farm journals; regular reports should be available from the Market News Service of your state department of agriculture. This, of course, applies only to those markets for which prices are collected and reported, perhaps half of all markets. For nonreported markets, the manager should be willing to supply information on past prices, but these will not be accompanied by an objective judgement of the grade of animals sold. In general, there is no reason to expect major systematic differences in prices across similar markets unless the markets differ greatly in size or in principal species and classes handled. But it is advisable to make periodic checks anyway. When comparing net costs, be certain to adjust for differences in transport charges, commission fees, and shrink (Chapter 7). Fees vary substantially.

Once the market has been selected, it is necessary to determine the optimal day to ship to market. In general, it is unadvisable to advance or retard the sale date by more than a few weeks from the optimal weight range. By delaying the sale date, in particular, additional holding and feeding costs are incurred, and possibly a discount for overly fat slaughter stock. Look to the market manager or your commission agent to help you decide on the actual sale date. Market personnel are likely to have a good feel for the daily market and will have information not available to you on daily supplies and requirements. You are paying for this expertise through the commission charges and ought to take advantage of it.

As the sale day approaches, notify the market of your plans. This will allow the manager to make the necessary arrangements, including notifying buyers. But once you commit yourself to ship to that market, it is important to follow through. Unreliable suppliers should not expect assistance from the market in the future. Follow the manager's suggestion on feeding and watering before the sale. (See Chapter 7 for suggestions on what time of the sale to ship.)

Many producers find it advantageous to attend the sale. Often, it is felt, the commission agent or auctioneer will try just a little harder if the owner is in the audience. Be certain the animals are identified with your farm so that you can establish a reputation among buyers. Also, the market will give you an idea of what other farmers are producing and what the buyers are looking for. For part-time producers, attending a market regularly is often impractical, but should be attempted on occasion.

Variations – Pooling

Pooling, also known as commingling, is not so much a distinct marketing process as a variation on selling at public markets. Within a pooling arrangement, animals are sold in lots (often as much as a truckload) and sorted for uniformity. Among slaughter stock, the market employee would strive for similar weights and quality and yield grades, as well as identical breeds and sex, when feasible. Feeder stock would be sorted on the basis of grades, sex, weight, and age as well as breeds and general vigor. Sorters also must consider the degree of fill of the animals (Figure 13-3).

Highly uniform lots help the buyer interpret the value of the animals purchased as well as facilitating handling, especially with feeder animals. Due to the more predictable value of the lots, pooled animals are generally considered to raise the bid price. Some observers even feel that the more uniform appearance itself enhances the value. Finally, by reducing the time required to sell the animals compared to individual sales, the marketing costs are reduced.

While these sources of benefit from pooling are widely recognized, substantiating them quantitatively is another problem. In one study of veal calves in Maryland that covered two auction markets over five days, the average price of the pooled calves exceeded that of those sold individually. However, there were instances of individual calves selling for more

FIGURE 13-3. A Pooled Lot of Homogenous Animals Moving Through an Auction Market (Source: Dept. of Animal Science, Cornell University)

than the pool price (Smith and Smith 1960). Properly measuring the impact of marketing in lots requires selling identical animals both individually, and as lots sorted according to various standards. Otherwise, price differences can be ascribed to differences among the individual animals. Clearly, it is impossible to replicate exactly a sale in two different ways, but several efforts have been made to sell the same stock twice at the same market, once individually and once as part of a lot. In a few cases, when this was attempted for veal calves, the price of the second selling of the commingled calves exceeded the individual price by about $1.00 to $3.30 per cwt (reported in Haas 1963, pp. 28-29). For feeder calves, commingling was found to speed up the sale and result in higher average prices, although the differential was small (Chambliss and Bell 1974). Schroeder, Jones, and Nichols found feeder pig prices averaged $13 per cwt higher for lots of 60 head compared to single head sales (1989, p. 257).

Several examples do not constitute proof, but these results are consistent with impressions and lead to favoring pooled sales as a general matter. Complicating the issue is the impact on individual producers who make up lots. As everyone receives the (average) lot price, sub-average quality producers can receive a benefit at the expense of suppliers of above-average quality. Recognizing there is typically a difference in the perceived quality among a group of producers, the potential for a disagreement — and breakdown of the arrangement — is great. A well qualified and scrupulously impartial grader is essential.

Other suggestions for running successful livestock pools are:

- Keep weight lots within a narrow range, and consistent from week to week;
- Tailor lot sizes to buyers' needs, being particularly careful not to exclude smaller buyers from at least some of the bidding;
- Do not split pools (especially by allowing the successful bidder to pick over the lot); the remaining animals are then viewed as inferior and typically bring a lower price;
- Maintain a good accounting system to ensure that each seller is paid according to the price received by his/her animals (Haas 1963, pp. 31-39).

Buying Stations

Description

Buying stations are assembly points typically operated by an individual packing plant, although a few livestock marketing co-ops also operate these facilities. Buying stations are an intermediary form between public markets and on-farm sales, best serving the medium-sized producer in a moderately dense production area (Chapter 9). With escalating costs and especially larger, specialized producers, buying stations have begun to disappear, but there are no good figures on those presently in operation.

Livestock Types/Grades Accepted

With a station typically supplying a single packer, the animal classes purchased are limited to what is handled by that firm. In most larger plants, the range is quite limited so that other outlets will have to be found for different species and grade/weight ranges.

Cost

No specific marketing charge may be made at buying stations, although the cost of the service is likely to be reflected in the offer price. When there is a charge, it has been in the range of $1.60 to $2.60 a head (Johnson 1972, pp. 21-22).

Convenience

The seller is responsible for transport to the station. Since the negotiation over price is also done there (following the evaluation by the buyer) the producer typically is the one who does the transportation. This involves an investment in transportation equipment as well as availability of the seller during the hours the station is operating. Stations may operate one or possibly two days a week. Thus this option is less convenient than public markets.

Cost of Cancelling

A cancelled sale at a buying station really means the seller and buyer cannot agree on exchange terms. Perhaps this could better be called a "no sale." The seller incurs no penalty other than the two-way transport cost, stress on the animals, and any personal time involved. Together these are not trivial costs, although the value of one's own time is often difficult to gauge. Clearly, though, numerous "no sales" would make this option unprofitable.

Price Uncertainty

At buying stations, sellers do not relinquish title until a set price has been agreed on. Thus, there is no particular price risk, only the risk of having to make a return haul. Even that risk can be minimized by asking for an approximate quote prior to shipping. Once the buyer is familiar with the performance of your stock, it should be possible to get a near quote over the telephone.

Duration of the Relationship

Each sale is, in itself, complete so that there is no need for an ongoing relationship. Nonetheless, with few outlets convenient to most producers, a buying station can become part of a regular marketing program. Thus, it is advisable to consider the future ramifications of each sale.

Payment Risk

Under Packers and Stockyards Administration requirements, the packer buyer must make payment by the end of the next business day, unless the seller expressly grants an extension. Moreover, the packer is required to carry a bond to cover outstanding obligations to suppliers (see Chapter 11). Despite these safeguards, packers sometimes do default on payments to producers, especially if the firm fails. To minimize exposure, it is advisable not to extend terms, and to watch for danger signs such as requests for payment extensions and delayed checks. If you are suspicious, check with neighbors to get their experiences.

Equity Requirements

Buying stations are generally packer-owned and do not involve an investment by producers. Co-op stations typically require some equity ownership.

Comments/Suggestions

It is always advisable to call before shipping, and to get as much price information as possible over the telephone. But do not rely on the buyer for all your market information. The more knowledgeable you are of current prices, the stronger your bargaining position. In order to be credible in your "no sale" threats, at times it may be necessary for you to haul a load of animals back home. Think of this as a long-term cost and not one that relates solely to that sale.

The value of your slaughter stock is closely related to your carcass yield percentage. See if the packer will cooperate by giving you this information, and use it as part of your negotiation position. The packer should recognize better yielding, pre-shrunk animals, if that is what you are supplying.

Direct Sales

A significant portion of livestock marketing costs is attributable to loading and shipping charges. If the market serves, in a figurative sense, as a halfway point between seller and buyer, then these costs increase. And if the stock must be returned to the farm for some reason, then charges really begin to escalate. Certainly, it must be less expensive to bring the buyer to the animals than the other way around. That observation is correct, and so-called direct sales at the farm dominate in the industry (Figure 13-4).

In 1987, some 80 percent of slaughter stock sales were made directly

FIGURE 13-4. The Packer Buyer on the Farm [Note the mobile telephone connection with headquarters.] (Source: IBP)

(including buying stations). This was quite constant for the three species of slaughter stocks, but not for feeders (USDA, P&SA, "Résumé 1987," Table 3). Direct sales typically involve a buyer visiting the farm to view and bid on a lot. Often the fleet of buyers is highly organized and connected by mobile telephone to the head buyer who coordinates purchases and prices. For feeders, the buyer is more likely to be an order buyer or a dealer assembling supplies for subsequent resale.

The sale itself consists of meeting at an appointed time, with the animals to be sold penned for easy inspection. Price and other sale terms are then negotiated, and a sale/no sale decision reached. (For a detailed discussion of this process see Ward 1988, Chap. 3). Of course, this simple description camouflages considerable negotiating complexity for the seller, as described under the comments and variations sections below. But before proceeding to the particulars, it is helpful to examine why numerous sales are *not* made directly.

Direct sales are not feasible for the smaller producer (where small is defined as less-than-truckload lots). Costs of buying and shipping are often prohibitive for small lots. Buying varied lots that require individual inspection and possible dispersal for different uses is also not feasible. Finally, the buyer seeks an area of moderately dense production so that his/her time can be used effectively. Direct sales, then, are reserved for the bulk of the industry that is focused on producing uniform animals in large numbers for specific uses. That leaves fringe producers, fringe producing regions, and any animals requiring special attention dependent on other market outlets.

Livestock Types/Grades Accepted

Direct purchases are typically made by packers who specialize in very particular classes and grades. Thus, the needs are quite narrowly defined and homogeneity is mandatory.

Cost

With the buyer appearing at the farm and the buyer arranging for shipping, seller direct costs are virtually nonexistent (Johnson 1972, p. 19). However, the buyer looks for some seller contribution to costs such as a pencil shrink reduction or a "cutback" (see Chapter 7). The challenge to the seller is to understand these indirect charges and to minimize them during negotiations. A typical farmgate slaughter steer sale will include a four-percent pencil shrink (USDA, ERS 1986, Table 14).

Convenience

Direct sales can be described as highly convenient for the seller. A properly arranged sale, though, does require planning to have all animals sorted and penned properly and filled according to the expectations of the buyer.

Cost of Cancelling

Cancelling a sale means simply sending the buyer packing and returning the stock to the feed lot or range. Thus, they are minimal.

Price Uncertainty

Negotiating a fixed price prior to the sale virtually removes any price uncertainty. But many sales are for delivery at some future time because packer buyers, in particular, attempt to schedule supplies in advance of needs. In such cases, it is common to agree not on a fixed price but on a means of determining the future price. These future, or formula, price agreements are typically keyed to a market quote or to a futures price (see Chapter 7). As prices can move considerably, even in the very short-term, formula agreements involve an unknown price risk. The accurate projection of very short-term prices is nearly impossible, but one would want to avoid key periods of particularly high price variability potential. These times can include weeks of major crop reports and production expectations (see Chapter 10).

Duration of Relationship

The reality of the U.S. livestock market in the 1990s is that the number of large buyers within a reasonable distance of a farm is strictly limited. Studies have placed the average number of packer buyers competing for slaughter cattle at two to four, while lamb slaughtering plants buy from 10 to 40 percent of the area supply (Ward 1979 p. 16; USDA, P&SA 1987, p. 49). A typical supply area is a 100-mile radius (Hayenga et al. 1986). With these small numbers, the same buyer will be used time and again. As a result, relationships become long term and should be treated that way.

Payment Risk

The same comments made previously apply here; notably, that the P&SA requires packers and buyers to pay for purchases within a specified period, and to carry a bond to cover outstanding purchases (see Chapter

11). However, those bonds are typically insufficient to cover debts of a financially troubled firm, so the seller should not depend on these regulations to assure payment. Rather, caution is advised.

Equity Requirements

None.

Comments/Suggestions

The direct sale sets the stage for a nearly classic confrontation between the occasional seller and the professional buyer. But the sale need not be confrontational; the buyer depends on the seller just as the seller depends on the buyer. Both are looking for a profitable exchange and, in the long term, a mutually profitable arrangement is of benefit to both. The quest for short-term advantage can nonetheless upset this rosy scenario.

In the short term, the seller is at a competitive disadvantage compared to the better-informed buyer. A successful transaction, therefore, requires the seller to overcome his/her disadvantage to the degree possible. The first step in the process is to reduce the complexity of the price. If, for example, the buyer demands an estimated discount for mud carried by your animals, see that they are clean so the issue does not arise. Other components of net price involve pencil shrink, weighing point, and other factors. With a sharp pencil and calculator in hand (Chapter 7), it is possible to reduce these factors to a common base (e.g., farmgate price) for comparison. Inexpensive portable computers are becoming available which can further simplify this procedure. The key is to know what price is being negotiated.

Beyond that, it is imperative for the seller to be as knowledgeable as possible about market conditions. The first component is knowing prevailing prices, the second is knowing the yield potential of your animals as feeder or slaughter stock. If a good grasp of both of these factors is problematic, selling direct becomes a chancy activity.

Variations – Forward Selling

As noted, it is common for packer buyers in particular to seek commitments for future shipment rather than immediate delivery. This allows planning plant operations with a minimum of penning space. However, given the volatility of livestock prices, future commitments involve price risk which is shifted back to the seller through some future pricing system. Typically, these are based on "Yellow Sheet," futures, or other reporting system price quote for some future date (see Chapter 7). Some sellers may be unwilling, or unable, to accept that uncertainty and should seek alternative arrangements or markets.

In other cases, the forward sales arrangement is longer: up to 30 days. Pre-selling helps scheduling as well as assuring that the producer will have an outlet during the relatively brief period when the stock reaches optimal market readiness. Payment at the time of the sale also shifts price risk off to the buyer although, as noted, other techniques are frequently used to determine the price at the time of the sale so price risk effects are neutralized. Forward sellers can expect a formal contract agreement that specifies such things as grades, weights and quantities, and includes a penalty for noncompliance. Forward sales are limited (annually, some 2.5 percent of feeder calf sales nationally), but there is some evidence of an increasing trend over time (Petritz, Erickson, and Armstrong 1981 pp. 93-95). Figures for slaughter sales contracts indicate 77 percent of cattle feeders responding to a 1988 survey had forward contracted at some time while in 1989, 12 and 23 percent of cattle and hog feeders had forward contracted, respectively, (Smith 1989, Table 27). Contracting was found desirable because it reduced the need to hedge (see Chapter 14) (*Beef Today*, September 1988).

Grade and Yield Selling

Selling on inspection involves two complexities. First, an experienced buyer must be used, and, second, price is based on estimated carcass yields. This cost and uncertainty combine to make stock less valuable to the packer so that, on average, prices will be lower than in their absence. Conceptually at least, it is possible to raise prices by eliminating these factors. And they may be eliminated through pricing the carcass and not the live animal — a process known as grade and yield selling. In 1987, 30 percent of all cattle were sold this way, compared to two-thirds of sheep and lambs and 14 percent of hogs (USDA, P&SA, Résumé 1987, Table 11). Grade and yield sales have increased steadily, with less than 15 percent of cattle, seven percent of sheep, and but three percent of hogs priced this way in 1967 (USDA, P&SA, Résumé 1977, Table 12). To some extent, this trend is related to the increasing size of feeding operations. There is evidence that larger feeders use grade and yield sales far more than do smaller operations (Rhodes, Gonzales, and Grimes 1981; Meyer and Lang 1980). Clearly, this procedure is not employed for feeders.

Typically, a packer will establish a daily live weight price for a preferred weight range, say 210-240 pounds for hogs. This is converted into an approximate carcass weight price by assuming a yield percentage. Thus, if the live price is $50/cwt and the yield percentage is 73 percent, the carcass value is $68.49/cwt. This is the base price. Actual prices are established from a schedule of premiums and discounts around this base. Two schedules are used, one for yield and one for quality (grade).

Considering the quality (grade) schedule first, packers typically select a relatively low quality as the base so that most sales receive some premium. The premium schedule is particular to each firm, and plants within firms, with even the conditions for establishing the quality typically set by the firm. As a general matter, the grade schedule is based on fattiness and muscling. The lack of a standardized system clearly makes it very difficult for a seller to estimate the final price, and to compare effective prices across firms.

Yield pricing is keyed to weights differing from the base, or ideal live weight for the packer, say 1000-1100 for steers or 210-240 pounds for barrows. Weights above, or below, those levels are discounted according to a schedule established by the packer. An example of a typical settlement report for a lot of eight hogs is shown in Table 13-1.

Moving from left to right across the table, the live weight is converted to the carcass weight by dividing by the standard yield. This gives the base carcass price which, multiplied by the actual carcass weights, gives the base carcass values. The standard yield is merely a method of translating the live price into a carcass, but it is the carcass price that is important since the producer is paid based on the carcass (not the live) weight. This also means that no specific premium for a yield above the standard is paid. The "premium" is reflected in the payment at the higher carcass weight.

The final adjustment to the price received is made through the grade premium. The premium in each weight class is added to the cwt carcass price and, when multiplied by the carcass weights, gives the total payment. In this example, the farmer receives $987.39 for the lot.

Below the main table is a summary of the price components. The lot value and grade gain, or premium, are computed as above. The yield gain is the difference in value for the animals sold compared to what it would have been had they yielded only the standard. However, note that this amount is included in the base carcass value and is not awarded as a distinct premium. That is, it is a hypothetical value. The "sort factor" is the value loss for weights that fall below the ideal level. Like the yield gain, this difference is reflected in the live price by weight range and does not enter directly in the computation of the lot value.

Despite their obvious advantages, grade and yield sales have several limitations that restrict their attractiveness for livestock producers. The individualized premium systems make it very difficult to compare prices among packers. Perhaps more significant, there is a high degree of trust

TABLE 13-1. Typical Grade and Yield Settlement Reports for Hogs (Source: USDA, P&SA 1984, p. 13)

Base live price 1/ . . . . $50.00   Average base live price 2/ . . . . . . $45.59   Total live weight 3/ 1895
Total head 4/ 8   Base carcass price 5/ . . . . $68.02   Tatoo number 1234

6/ Weight Range Live	Carcass	7/ Live price	8/ No. head	9/ Std. yield	10/ Base carc. price	11/ Carc. weight	12/ Base carc. value	13/ Grade number	14/ Grade prem. cwt	15/ Grade prem. total	16/ Total carc. price	17/ Total value
231-240	170-176	50.00	1	73.5	68.02	170	115.63	2	+1.00	1.70	69.02	117.33
241-250	177-184	49.75	1	74.0	67.22	181	121.66		+2.00	3.62	69.22	125.28
241-150	177-184	49.75	3	74.0	67.22	537	360.97	1	+1.00	5.37	68.22	366.34
251-260	185-192	49.25	1	74.0	66.55	189	125.77	2	+2.00	3.78	68.55	129.55
251-260	185-192	49.25	2	74.0	66.55	374	248.89	1 3	- 0 -	0	66.55	248.89
Totals/averages		49.59			67.05	1451	972.92			14.47	68.04	987.39

Total live weight 21/ 1895
Total hot carcass weight 22/ 1451

Overall lot summary:

Actual yield 18/ 76.57
Standard yield 19/ 73.94
Yield difference 20/ 2.63

	Total	Per live cwt
Lot value 23/	$987.39	$52.10
Yield gain 24/	.$33.16	$1.75
Grade gain 25/	.$14.47	$0.76
Sort factor 26/	.$- 8.13	$- 0.43

303

required since a packer employee may do the grading and the packer is responsible for keeping track of the carcasses and seeing that each seller is paid for his/her own animals. A system must be established for the inevitable cases when carcass identification is lost (although recent regulations for permanent hog identification could minimize this). The producer accepts more risk in a grade and yield sale because sharp discounts apply to bruised meat (see Chapter 9) and condemned carcasses. Compounding these problems is the general lack of information, and misinformation, among producers about grade and yield selling (USDA, P&SA 1984, pp. 4-5). Grade and yield selling, then, provides opportunities, but only for those who are willing to spend the time necessary to make a detailed comparison of prices from area packers.

Contracting and Joint Ventures

The preceding descriptions applied to relatively straightforward exchanges. Although the settings and facilities differ, the systems share the common characteristic of transferring full title at, or near, the sale date. Such clear ownership arrangements, while they dominate the sector, are not appropriate in all cases. Here, several alternatives are described, notably, contract production/custom feeding and joint ventures/cooperatives. Each serves a particular need but shares the characteristic of nontraditional ownership patterns. These arrangements typically allow for risk shifting or improved market access.

Contract Production/Custom Feeding

Under these systems, the animals may belong to the buyer, with the farmer providing specified services on a fee basis. With custom feeding, for example, the buyer pays a fee on a per pound of gain or similar basis. Pricing could be keyed to local markets, or to futures quotes, with specific schedules established for pencil shrink and other quality factors (see Ward 1979, pp. 13-15). Alternatively, the owner and feeder may divide the slaughter value on a shares basis. An example of the contract production system would be a farmer maintaining and farrowing sows for the owner. Custom feeding and production is, as yet, relatively uncommon. Custom feeding of slaughter cattle in 1987 was only 4.9 percent of the total, up 30 percent in one year, but still below the high point of 7.9 percent in 1967. For hogs and sheep, the numbers are less than one percent and 11.8 percent, respectively (USDA, P&SA 1987, Tables 12, 13, 14, and 17). There is some ambiguity in the interpretation of these numbers since they refer to feedings "by and for meat packers," but it is safe to assume the great majority is *for* (not by) packers. Data on contract production are more scarce, but VanArsdall and Nelson estimate 12 percent of feeder pigs in the Southeast were sold that way in 1980, but only one percent in the North Central region (1984, Table 58). For sheep, packers are credited with owning or controlling 27 percent of slaughter (USDA, P&SA 1987, p. 14). Thus, despite the data limitations and definitional problems, it appears that custom feeding of one form or another is a notable component of the industry in recent years.

Contract production is, in fact, more involved than implied above. There are several forms of contracts. These are generally referred to as Marketing Services Contracts, Market-Specification Contracts, Resource-Providing Contracts, and Production-Management Contracts (Holder 1976, pp. 92-95). Market-Specification Contracts refer to an ongoing exchange agreement between seller and buyer, while the Resource-Providing Contracts are those where the buyer provides some substantial input such as the breed stock or feeder animal. Under a Production-Management Contract, the buyer also specifies some production practices such as the ratio. Unfortunately, the available data do not allow quantification of the degree of use for the several forms of contracts, but the Market-Specification and Resource-Providing ones can be expected to dominate.

Resource-Providing Contract production and custom feeding remove many of the risks of livestock enterprises from the farmer. Risk is associated with fluctuating prices, market access, and death loss. If the farmer does not own the animals, most of these factors do not apply. At the same time, the producer forgoes many of the discretionary opportunities of the independent entrepreneur as well as the associated profit opportunities. Buyers benefit by exercising greater control over the timing and quality of purchases, as well as possible reduction in procurement costs. One study estimated the benefits of contracting to the hog packer to be $2.55 per animal, and an additional $1.85 if plant capacity utilization could be maintained at 100 percent (Snyder and Cardler 1973). However, this is done at the expense of increasing exposure to price and other risks, as well as some loss in flexibility. The large buyer is, in some ways, better equipped to accommodate risk (use of futures contracts, for example, is more economical for larger enterprises) but the optimal choice may be to leave it with the producer.

Market-Specification Contracts are a middle ground. They delineate what, when, and where a sale will be made as well as other aspects of the transaction (Figure 13-5). Producers are assured an outlet and at least the means by which price will be determined. Buyers are given greater control

FIGURE 13-5. Sample Slaughter Cattle Sales Contract (Source: Ward 1977, App. 1)

AGREEMENT made this _____ day of _____, 19 ___, between _____
of _____ (hereafter referred to as Seller) and _____
of _____ (hereafter referred to as Buyer), witnesseth:

1. QUANTITY: Seller hereby agrees to sell and deliver and Buyer agrees
   to buy and receive _____ pounds of live cattle (the equivalent of _____
   units of 40,000 lbs. each), based on gross delivered weight at _____
   less ___ % shrink, or the equivalent thereof, producing a net delivered
   weight.

2. QUANTITY VARIATION: In the event of variation in quantity delivered,
   Seller has the option of adding to or subtracting from the number of
   delivered in order to bring the net delivered weight to 40,000 lbs.
   per unit, or Seller may instruct Buyer to adjust the amount of total
   payment to the extent that the variation in delivered pounds will be
   multiplied by the price differential between the contract price and the
   Buyer's prevailing prices for these weights and grades in effect at
   time of delivery. However, no variation beyond 2,000 lbs. per unit
   will be permitted.

3. DESCRIPTION AND PRICE: Cattle shall be described and priced as follows:

   ⬨ Steers weighing between 1,000 and 1,200 lbs. live weight at a price
   of $ _____ per live cwt. When the live price is divided by the equivalent
   dressed price is $ _____ per cwt. for dressed beef carcasses weighing
   612 to 816 lbs. (hot weight).

☐ Heifers weighing between 900 and 1,050 lbs. live weight at a price of $ _____ per live cwt. When the live price is divided by the par dressing percent of 62.75%, based on hot carcass weight, the equivalent dressed price is $ _____ per cwt. for dressed beef carcasses weighing 510 to 714 lbs. (hot weight).

4. GRADE REQUIREMENTS: All cattle shall be U.S.D.A. quality graded and shall consist of 80% Choice, balance Top Good.

All cattle shall be U.S.D.A. yield graded and shall consist of 80% Yield Grade 3, balance yield grade 4. Additional Yield Grade 3 carcasses shall be priced in accordance with the par dressed price, and additional Yield Grade 4 carcasses shall be priced at $2.00 per dressed cwt. under the Yield Grade 3 price. Yield Grade 1 and 2 carcasses shall be priced at $1.00 per dressed cwt. over the Yield Grade 3 price. Yield Grade 5 carcasses shall be priced at $5.00 per dressed cwt. under the Yield Grade 3 price.

All carcasses shall be free of yellow fat, excessive bruise or grub damage, or other abnormalities. In the event that more than 10% of

FIGURE 13-5 (continued)

the cattle produce carcasses damaged by grubs or bruises, all damaged carcasses above and beyond the allowable 10% shall be reduced in price by $2.00 per dressed cwt.

Cattle with a higher dressing percent or higher grade than herein provided, or cattle not meeting dressing percent, weight, or grade specifications will be priced in accordance with Buyer's prevailing price differentials for these weights, and grades in effect at the time of delivery.

5. DELIVERY DATE AND LOCATION: Said cattle are, or will be, located at _____ in _____ County, State of _____. Said cattle are to be delivered by Seller at _____ on the day specified by Buyer during the week commencing _____, 19___, in good and merchantable condition and suitable for immediate slaughter to produce meat for human consumption. Title to said cattle shall pass to Buyer upon delivery. By mutual agreement of Buyer and Seller, the delivery date may be changed to the week prior to or the week following the date specified herein. Any further variation in delivery date shall be subject to the provisions of Section 9 below.

6. CREDIT REQUIREMENTS: Seller agrees to furnish Buyer necessary credit information and documentation, and upon approval by Buyer's Credit Department, Buyer agrees to pay Seller $1,000 per 40,000 lbs. unit as part payment for said cattle. In the event that Buyer's Credit Department fails to approve this contract within 20 days from the date of this contract, Seller or Buyer shall have the right, at its option, to cancel this contract upon written notice to the other party, provided that this notice is mailed or delivered within 30 days of the date of this contract.

7. FINAL PAYMENT: Buyer agrees to pay Seller the balance of the purchase price for said cattle before the close of the next business day following delivery and determination of the amount of the purchase price.

8. LIENS, SECURITY INTEREST: Said cattle are, on the date hereof, subject to a lien, security interest, or chattel mortgage, in the amount of $ _____ in favor of _____, Seller represents and warrants that he will obtain the release of all liens, security interest, and chattel mortgages on said cattle prior to final payment therefor and will submit proof thereof to Buyer. In the event Seller fails to submit proof of such release, final payment for said cattle will be made jointly to Seller and lien holder.

FIGURE 13-5 (continued)

9. BREACH OF CONTRACT: Upon Seller's failure for any reason whatever, to deliver to Buyer all cattle purchased hereunder as herein required, Seller shall promptly refund to Buyer all money advanced on such undelivered cattle and shall also be liable to Buyer in the amount by which the prevailing market price on the date such cattle should have been delivered exceeds the contract price for such undelivered cattle. Seller shall reimburse Buyer for its reasonable expenses incurred in the collection of damages herein provided.

10. ENTIRE AGREEMENT: This contract contains the entire agreement between Buyer and Seller and cannot be varied orally. It shall be binding on the respective heirs, successors and assigns of the parties.

IN WITNESS WHEREOF, the parties have executed this agreement on the date first above written.

_____

By _____
            Seller

_____

By _____
            Buyer

over purchases without the need to share risks or increase capital requirements. Thus, this form of contract holds potential for growth in the future, especially if price uncertainty continues (USDA, World Ag. Outlook Board 1984, p. 43). There is no analytical foundation for the oft-held producer concern that contracting depresses livestock prices (USDA, P&SA 1987, pp. 51-53).

Joint Ventures and Cooperatives

Joint ventures refer to instances when sellers and buyers have some shared equity interest. Simply stated, a joint venture is a partnership. A Resource-Providing Contract is actually a form of joint venture if the buyer owns the animals being fed or bred, but generally the investment is more permanent (such as joint ownership of production facilities or packing operations). These joint operations may be "forward" or "backyard," where the direction distinguishes between the seller investing in the buyer, or vice versa.

Cooperatives can typically be characterized as special forms of joint ventures where the investment is made on a group, rather than an individual, basis. Since most cooperatives are composed of producer groups, the investment is forward into marketing activities. Cooperatives are involved in a range of marketing activities, from price negotiation to operating public (auction and terminal) markets and including slaughtering and processing. In 1975, six cooperative packers slaughtered less than one percent of cattle and 2.5 percent of hogs. The total share of cooperative firms at all levels of livestock marketing was 13 percent (Torgerson 1978, App. Tables 26 and 27). Subsequently, cooperative involvement with these activities gyrated, but with a definite downward trend (Nelson 1985). For example, in 1987, Land O' Lakes (a co-op) sold its major Spencer Beef division to Excel, a private firm. In 1975, cooperatives handled about 15 percent of all livestock at some point in their production/ marketing/processing cycle (Torgerson 1978, App. Table 27).

There is nothing inherently different about the way a cooperative operates compared to proprietary companies. The cooperative arrangement, however, necessitates a capital investment by the member owners (an average of $39, in 1978, when packing operations are excluded). For cooperative packers, members had an average investment, in 1978, of $1,400 (Torgerson 1978, p. 19). Federal law limits the interest producers can be paid on these investments.

The low margins prevailing in the sector (see Chapter 15) and the limits placed on returns to member equity have made cooperative livestock marketing enterprises highly chancy. As a further impediment, producers often enter marketing as an eleventh-hour effort to preserve a market when a private firm withdraws (see Hogeland 1982, p. 1). In such cases, producers should ask if they can operate more efficiently than the private company, or are they simply willing to accept a lower return on capital. In cooperative meat packing, the latter has historically been the case, with cooperative packers receiving returns substantially below private firms (Torgerson 1978, pp. 19-20). The operation of marketing facilities has been more successful.

Several other options exist for producers organized as a cooperative. They may choose to negotiate a higher price, using the threat of withholding a substantial supply as leverage. The inability to maintain that control over a broad group of producers has doomed most of those efforts (Torgerson 1978, pp. 38-41). Joint ventures with packers, where producer and packer share the carcass value, have also failed as a rule, due in large part to the inability to document costs (Holder and Hogeland 1982, pp. 16-17). Thus, care is advised before initiating any of these efforts. At a minimum, members are advised to understand the concept of a cooperative and be willing to commit a certain number of head to the cooperative activity (see Hogeland 1982, pp. 46-49).

Sales to Consumers

As a final approach to be treated here, producers may wish to consider direct sales to consumers. Such sales are typically referred to as "freezer beef" sales, or some variant thereof. Data on volumes are difficult to come by, but a first approximation is the amount of non-federally inspected slaughter. In 1987, that was four percent of cattle and sheep and three percent of hogs (USDA, ERS 1989, Table 64). That figure ignores both the amount of federal slaughter going to this use as well as the home consumption of some "farm slaughter," but as an estimate it probably indicates the relative importance.

Producers are attracted to this market because it allows them to capture much of the margin otherwise going to intermediaries and, possibly, packers. Of course, the producer also "captures" much of the work as the producer must identify customers, arrange for slaughter, deliver the animal (and possibly the meat), and secure payment. This can be profitable, but numbers are necessarily limited since beef sales seem to peak at about 50 head (100 customers) per year (Rasmussin, Lesser, and Anderson 1982). These are clearly part-time producers; many are in areas where there are no concentrations of livestock to facilitate the easy marketing of large volumes. Theirs is a small, high margin business.

Based on my own work, the principal problems encountered by this group are: (1) the identification of potential new consumers once the supply of friends and coworkers is exhausted, and (2) receiving payment. According to the best available data, consumers are relatively high income and can be sought in appropriate "high rent" areas. Payment could be made through charge cards that would both ensure payment and ease the burden of payment before consumption. Producers have sold by the animal or by the side, thereby avoiding the problems associated with balancing sales of popular and unpopular cuts (Lesser 1979).

Electronic Marketing

Electronic Markets are not presently markets in a strict sense since their availability is geographically very limited. By mid-1980, there were less than one dozen of these markets in the U.S. and about four in Canada. Estimated numbers sold are in the range of one million feeder pigs and 100,000 lambs annually, minuscule in terms of the size of the industry (U.S. Gen. Act. Office 1984, pp. 1-2; Hogeland 1987, Table 1). Yet this approach provides a new means of exchange within a system that was largely developed centuries, if not millennia, ago. It therefore justifies an evaluation of its role and prospects.

History of Electronic Trading

The Canadians were the principal innovators in this area, developing a teletype auction for slaughter hogs in 1961. Initial work was done by the Ontario hog producers marketing board in 1961, with subsequent adoption by the boards in Alberta, Manitoba, and the Maritime Provinces. Among the 50 states, Virginia took the lead in the early 1960s with a telephone auction for slaughter hogs which soon expanded to slaughter sheep. In 1965, a Missouri livestock coop began trading feeder pigs electronically, video files were added in the mid 1970s, beginning in Montana. By the 1980s, considerable interest had arisen in this area as both the result and cause of a multi-million-dollar U.S. Department of Agriculture budget to develop and test prototypes for livestock and other commodities. Results of these projects were disseminated in several conferences and publications (Virginia Tech 1983; Sporleader 1980).

Forms of Electronic Marketing

The several forms of marketing through electronic means may be grouped as teleauctions, teletype systems, and computer-based systems. These are listed in order of increasing sophistication of the electronic systems used. All share the common characteristics of remote access and descriptive selling with post-sale delivery. That is, the livestock sold generally remain on the farm/feed lot, are sold based on description (sometimes aided by video pictures), and are shipped directly from the farm, bypassing the physical collection and dispersion (literature review in Gunter, Lesser, and McLaughlin 1986, pp. 9-32).

With *telephone clearing houses*, a market agent collects and matches bids and offers from trades via the telephone. Only a limited amount of information is exchanged, and there is no direct interaction between buyers and sellers. Moreover, it can take 15 minutes for the clerk to poll all possible buyers, at which point many have lost interest. Slightly more complex are *teleauctions* under which products are sold by an auctioneer connected to buyers through a conference loop telephone call. This makes the system readily available and relatively low cost, although the number of traders who can participate simultaneously is limited by the capacity of the conferencing interloop. The sale may be facilitated by distributing descriptions of lots and sale terms. Once underway, only limited information is exchanged over the system. The buyer is restricted to an affirmative response to the auctioneer. Since all information is exchanged orally, errors and misunderstandings are always possible.

Teletype networks have been used successfully for over 20 years, beginning in Canada. Buyers linked through teletype terminals can receive printouts of the offerings just prior to the sale. The actual auction is done in the Dutch (descending) mode, with the buyer hitting a key to register acceptance of the bid price (Figure 13-6). The system maintains records that enhance the information exchange process, but only at the cost of maintaining a terminal in each potential buyer's office. Presently, the teletype machine is technologically outdated since multi-function microcomputers are available at lower cost.

Computerized electronic trading systems link buyers and sellers through a network of computer terminals. This allows both a centralized exchange along with direct contacts between seller and buyer for the arrangement of transportation, etc. Costs of the system depend, in part, on how the central computer is paid for (whether owned, rented, or secured on a time-sharing basis). Time sharing has the benefit of providing back-up capacity, an essential feature since mainframe computers have a reasonable amount of "down" time that would paralyze the marketing system. Variable costs have been estimated at ten cents per head (billed to the seller), although the figure is variable according to lot sizes and total animal numbers (Rogers and Purcell 1982). However, the systems require participants to have computer terminals (or microcomputers with a tele-

FIGURE 13-6. Trading Livestock Over a Teletype [These systems are easy to operate, but use antiquated technology by standards of the 1990s.] (Source: Johnson 1972, p. 63)

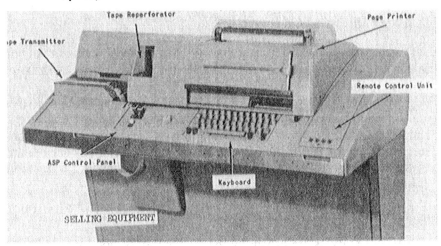

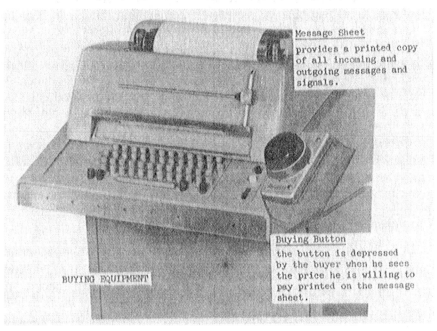

phone linkage), which is impractical for small, occasional sellers unless provided at some centralized location like a county extension office or coop.

Increased Marketing Efficiency

Conceptually, electronic systems provide multiple benefits to both sellers and buyers (see, e.g., Bell et al. 1983; Henderson 1982; Ward 1982). By limiting the physical handling of animals, especially by omitting the intermediate collection at market, both trucking costs and stress are reduced. Younger, more disease-prone animals can be expected to benefit the most from lack of intermingling with other, possibly disease-carrying, animals and facilities. In one study, this factor has been ranked as the most significant benefit by both sellers and buyers (Ward 1982, Tables 2 and 4).

Improved Information

Efficient markets are driven by the availability of good information at low cost to all participants; the perfect competition model, in fact, has "perfect information" as one of its underlying requirements/assumptions. Yet, the reality is that information is not freely available in an age of "private treaty" sales. Further, as frequent market participants, buyers generally have better access than the occasional seller, the producer. An electronic system that generates volume and price information enhances the general efficiency of the system and helps equate the position of sellers and buyers. More specifically, improved information can, for example, direct livestock to higher valued markets (thereby reducing transshipping) and other efficiencies. A final component of this group of benefits is improvement in the accuracy of information. At least, this is true of the conceptual level; demonstrating effects in the field is a different matter.

Competition

A reality for many producers is the limited number of buyers willing and available to bid on any lot. This problem is compounded in low-density producing areas for smaller producers and among species produced in lower volume. Sheep present a case in point, with individual plants purchasing virtually the entire production in the states of Washington, New Mexico, and Texas (USDA, P&SA 1987, p. iii). Electronic marketing can reduce the cost of participating in a market by removing the need to be present. Thus, these markets can enhance competition and strengthen prices. Confirmation exists from several sources but does not necessarily apply to all classes of all markets. Ward (1982, pp. 13-14) found video auction prices for feeder cattle averaged $.83/cwt higher than prices at the Oklahoma City market for the same week. Bell et al. (1983, p. 11) report that "virtually all" electronic livestock markets have shown appreciable price increases, from $1/cwt for slaughter hogs in Ohio to $3/cwt for market lambs in Virginia. However, the attribution to competition, efficiency, and/or other sources could not be made. Rhodus, Baldwin, and Henderson (1989) attribute the $.94-$.99/cwt increase in slaughter hog prices during the experimental HAMS program (Hog Accelerated Marketing System, 11/80-6/81) to enhanced pricing efficiency.

The competitive balance changes in another way. Animals shipped to market, as noted above, are expensive to withhold for return or subsequent sale. Thus, producers tend to sell their animals even if they fetch considerably less than the expected price. With the stock still on the farm, as would be the case for electronic selling, it is less costly to declare a "no sale." The producer thus becomes a more competitive "buyer" in the market, moving prices up.

Despite these factors, there are strong disadvantages. If not, electronic livestock marketing would be far more prevalent today than it is. Certainly, for the buyer, the accuracy of the description becomes paramount in the absence of visual inspection.

Descriptions can fail from either the ability of the grader or, more likely, the validity of described characteristics to represent the animal. Additionally, there is considerable latitude within a grade that cannot be expressed in descriptive terms (see Chapter 12). The use of video can minimize this set of problems, but technical limitations persist (Ward 1982, Table 5).

Performance Concerns

Exchanges, particularly on a sight-unseen basis, require considerable trust, something potentially lacking when there is no direct contact between sellers and buyers. Some marketing agencies have found it necessary to guarantee seller performance before buyers can be attracted. Even then, selling agencies with good records find entry into electronic exchanges the most successful (Russell and Purcell 1984, p. 15).

Cost Factors

While computerized exchange may, in the long term, be far less costly than conventional methods, it does involve substantial overhead on the market organization and individual participant levels. These costs can only be brought down to reasonable levels with high volume. Yet start-up volumes will be low, meaning a period of high costs must be endured before long-term benefits become evident. One of the intents of USDA funding was to aid this transition period, but the problem may simply have been moved forward in time until the subsidy was removed. This cost problem plagued the USDA-sponsored HAMS project in Ohio. The system did not attract large producers who were able to sell direct to packers, so that lot sizes, and hence unit costs, remained high. Actual costs were in the $2.60-$3.10/head range, about twice the cost of country auctions (Baldwin and Henderson 1981).

Institutional Issues

Electronic marketing will upset entrenched trading patterns, something those who are potentially disadvantaged will resist. Auction market operators and order buyers have expressed concern most vocally, and could apply pressure to packer buyers. The packers themselves may be hesitant to participate, especially if it would involve some loss of competitive advantage.

Other, more prosaic, issues arise as buyers and sellers attempt to determine how to sell/buy livestock. Timing of commitment to sell and the sale and delivery dates are significant in rapidly fluctuating markets. The length of time it takes to sell and buy become further constraints. Thus, careful attention to detail is an essential condition to successful electronic markets, as with any marketing activity.

With hindsight, it is clear that electronic markets will not replace all other marketing systems nor, in all likelihood, will they become the dominant method. In particular, direct purchasing of truckload lots is such an efficient system that it will be difficult to remove it from the number one position. Electronic marketing does, though, have a major potential role at the fringes of these systems, where fringe may be defined as lower-density producing regions, smaller producers, limited numbers of buyers, etc. There, electronic trading would compete with auction markets and other rather high-cost systems. But even in those cases, significant penetration will depend on hard work and commitment by producers and producer groups.

SYNOPSIS

Income from livestock sales can be maximized by instituting a marketing program or strategy as compared to sequential sales. Establishing a marketing strategy entails identifying specific goals (and limitations) of the livestock enterprise. Once the strategy is set, it is necessary to introduce a plan of action for the realization of that strategy. One aspect of the strategy/plan is the awareness of major marketing alternative's characteristics, that is the subject of this chapter.

Six principal alternatives exist, each exhibits a different mix of costs, convenience, and price risk, among other facets. The six are:

- Public Markets
- Buying Stations
- Direct (Private Treaty) Sales
- Contracting and Joint Ventures
- Direct Consumer Sales
- Electronic Marketing

Public (auction and terminal) markets are convenient but costly, a major factor in their long-term decline to about 10-20 percent of sales volume. Market representatives can assist newer and smaller sellers in understanding of market conditions. Studies have indicated that pooled (commingled) animals in uniform, truckload-sized lots bring higher prices. Buying stations are no longer a major outlet but are important in some regions. They are being abandoned by packers largely due to high cost.

The major form of exchange is direct or private treaty sales, which account for some 80 percent of sales by number (less for feeder stock). These sales are relatively inexpensive but do require some preparation by the seller in sorting and penning as well as keeping abreast of prices. A small portion of sales is done for future delivery — up to 30 days. Larger future sales contracts can involve considerable price risk for producers. About one-third of all direct sales are made on a grade and yield basis where the value is based on a carcass evaluation. This system shifts some risk from packers to producers but can be beneficial for producers with uniform lots.

Custom feeding (where the packer or other party owns the animals and pays per pound of grain or other fee) shifts most risks from producers to packers or others. Due to that factor, it has remained at about ten percent of total sales. Cooperative shares in marketing have remained steady, while there is a continued downward trend in slaughter.

Direct sales to consumers (e.g., freezer beef) is potentially profitable for smaller, well-located producers since the entire producer-retail margin can be retained. Considerable time, however, must be invested. Finally, electronic marketing is not a market per se but rather a means of price discovery. In these markets, animals are sold on a description basis or through video images. Conceptually, they should be more efficient and studies have shown them to be so, but start-up costs and "habit" have limited them to a very small percentage of sales.

Study Questions

1. What is a "marketing strategy" and how does it fit into a livestock production/marketing plan?
2. If you live equidistant between a livestock auction and a terminal market that had similar fee structures, how would you decide which of the two to use?
3. How should you prepare for the packer buyer who will be calling at your feedlot later in the week?
4. Under what conditions are grade and yield sales particularly attractive?
5. Identify and define five distinct forms of risk that are involved in any livestock marketing activity.
6. Now suppose you entered into a Market-Specification Contract with a feeder or packer. Which of the risks you identified in No. 5 would be affected. How?
7. Imagine that you are on the Electronic Marketing Committee for your local livestock producers' coop. Outline the pitch you would give to area packers on why they should use your system.
8. For your livestock operation describe: (a) the marketing alternatives available, (b) the relevant management objectives and restraints, and (c) a strategy that maximizes your expected profits within those restraints.

REFERENCES

Baldwin, E. D. and D. R. Henderson. "HAMS: What Happened and Why." Ohio State Univ., Coop. Ext. Service, No. 640, December 1981.

Bell, J. B. et al. "Electronic Marketing: What, Why, How." Virginia Coop. Ext., Pub. 448-004, November 1983.

Chambliss, R. L. and J. B. Bell. "Selling Feeder Cattle in Commingled Lots: A Pilot Study of Pooling in Livestock Auction Markets." Coop. Ext. Service, Virginia Polytechnic and State Univ., Virginia Ag. Econ., No. 257, April 1974.

Gunter, F. W., W. H. Lesser, and E. W. McLaughlin. "Electronic Trading of

Fresh Fruits and Vegetables to New York State Institutions: A Feasibility Analysis." Cornell Univ. Dept. Agr. Econ., A.E. Res. 86-20, August 1986.

Haas, J. T. "Veal Calf Pooling." USDA, Farmer Coop. Service, Mkt. Res. Rpt. 615, August 1963.

Hayenga, M., R. Deiter, and C. Montoya. "Price Impacts Associated with the Closing of Hog Slaughtering Plants," *North Central J. Agr. Econ.* 8(1986): 237-42.

Henderson, D. R. "Electronic Markets for Agricultural Commodities: Potentials and Pitfalls." Univ. Wisconsin, NC-117 WP-62, April 1982.

Hogeland, J. A. "The Future Role of Livestock Cooperatives." USDA, Ag. Coop. Service, Res. Rpt. No. 61, May 1987.

Hogeland, J. A. "Organizing Meat Packing Cooperatives: Recent Producer Attempts." USDA, Ag. Coop. Service, ACS Res. Rpt. No. 1, February 1982.

Holder, D. L. "Producer-First Handler Exchange Mechanisms for Livestock with Special Emphasis on Hogs." In B. W. Marion (ed.), *Coordination and Exchange in Agricultural Subsectors.* Univ. Wisconsin, NC-117 Monograph 2, January 1976.

Holder, D. L. and J. A. Hogeland. "Cooperative Lamb Slaughtering in the Northeast." USDA, Ag. Coop. Service, ACS Res. Rpt. No. 14, February 1982.

Johnson, R. D. "An Economic Evaluation of Alternative Marketing Methods for Fed Cattle." Univ. Nebraska, Ag. Exp. Sta., SB520, June 1972.

Lesser, W. "Marketing Freezer Beef in New York." Cornell Univ. Dept. Agr. Econ., A. E. Res. 79-12, May 1979.

Meyer, A. L. and M. G. Lang. *Carcass-Based Marketing of Cattle and Hogs.* Purdue Univ. Ag. Exp. Sta., Bull. 300, November 1980.

Nelson, K. E. "Issues and Developments in the U.S. Meat Packing Industry." USDA, Econ. Res. Service, Nat. Economics Div., 1985.

Petritz, D. C., S. P. Erickson, and J. H. Armstrong. *Cattle and Beef Industry in the United States.* Purdue Univ., Coop. Ext. Service. CES Paper 93, 1981.

Rasmussin, K., W. Lesser, and B. Anderson. "Marketing Alternatives for New York Fed Beef Producers." Cornell Univ., Dept. Agr. Econ., A. E. Res. 82-3, January 1982.

Rhodes, V. J., J. Gonzales, and G. Grimes. "Farm Study: How the Large Units Sell their Hogs." *Hog Farm Mgmt.* Park Producers Planner, December 1981.

Rhodus, W. T., E. D. Baldwin, and D. R. Henderson. "Pricing Accuracy and Efficiency in a Pilot Electronic Hog Market," *Am. J. Agr. Econ.* 71(1989): 874-82.

Rogers, P. and W. Purcell. "Electronic Marketing: Wave of the Future." Virginia Coop. Ext. Service, No. 334, July-August 1982.

Russell, J. R. and W. D. Purcell. "Participant Evaluation of Computerized Auctions for Slaughter Livestock – the Experience with Electronic Marketing Association, Inc.," *North Central J. Agr. Econ.* 6(1984):8-16.

Schroeder, T. C., J. M. Jones, and D. A. Nichols. "Analysis of Feeder Pig Auction Price Differentials." *North Central J. Agr. Econ.*, 11(1989):253-63.

Smith, J. N. and H. D. Smith. "Per-Lot Versus Single-Head Selling of Calves at Maryland Auction Markets." Univ. Maryland, Misc. Pub. No. 391, 1960.

Smith, R. D. "National Assessment of Producer Marketing Alternatives: Producers and Attitudes." Texas Ag. Exp. Service, April 1989.

Snyder, J. C. and W. Cardler. "A Narrative Analysis of the Value of Quality, Regularity and Value in Hog Marketings." Purdue Univ., Dept. Agr. Econ., SB12, August 1973.

Sporleader, T. L. (ed.). "Proceedings: National Symposium on Electronic Marketing of Agricultural Commodities." Texas A&M Univ., MP-1463, March 1980.

Stoddard, E. O. "An Economic Analysis of Cost of Service and Value of Service Tariffs in the Livestock Auction Industry." Univ. of Maryland, Dept. Agr. Econ., unpublished Ph.D. dissertation, 1975.

Sullivan, G. M. and D. A. Linton. "Economic Evaluation of an Alternative Marketing System for Feeder Cattle in Alabama." Alabama Ag. Exp. Sta., Circular 253, June 1981.

Torgerson, R. E. *The Future Role of Cooperatives in the Red Meats Industry.* USDA, Economics, Statistics and Cooperative Service, Mkt. Res. Rpt. 1089, April 1978.

USDA, Economic Research Service. "Livestock and Meat Statistics, 1983." Stat. Bull. No. 715, December 1984.

USDA, Economic Research Service. "Livestock and Meat Statistics, 1984-88." Stat. Bull. No. 784, September 1989.

USDA, Economic Research Service. "Livestock and Poultry: Situation and Outlook Report." LPS-21, August 1986.

USDA, Packers and Stockyards Admin. *Grade and Yield Marketing of Hogs.* P&SA 2-84, 1984, p. 63.

USDA, Packers and Stockyards Admin. "Packers and Stockyards' Statistical Report: 1985 Reporting Year." P&SA 86-2, 1986.

USDA, Packers and Stockyards Admin. "Slaughter Lamb Marketing: A Study of the Lamb Industry." January 1987.

USDA, Packers and Stockyards Admin. "Statistical Résumé." Various years.

USDA, World Ag. Outlook Board. "Agriculture in the Future: An Outlook for the 1980s and Beyond." Ag. Info. Bull. No. 484, December 1984.

U.S. General Accounting Office. "Electronic Marketing of Agricultural Commodities: An Evolutionary Trend." GAO/RCED-84-97, March 8, 1984.

Van Arsdall, R. N. "U.S. Hog Industry." USDA, Econ. Res. Service, A. E. Rpt. 511, June 1984.

Van Arsdall, R. N. and K. E. Nelson. "Characteristics of Farmer Cattle Feeding." USDA, Econ. Res. Service, A. E. Rpt. 503, August 1983.

Virginia Tech, Cooperative Extension Service. "Proceedings: Electronic Marketing Conferences, Atlanta, Chicago, Salt Lake City and Oklahoma City." Pub. 448-003, January 1983.

Wallace, L. T. *Agriculture's Futures.* New York: Springer-Verlag, Inc., 1987.

Ward, C. E. "Market Participation and Cooperative Price Evaluation of Marketing Feeder Cattle by Video Auction." Oklahoma State Univ., Dept. Agr. Econ., Paper A. E. 82108, October 1982. Also, OSU Extension Facts No. 463.

Ward, C. E. *Meatpacking Competition and Pricing.* Blacksburg, VA: Research Institute on Livestock Pricing, 1988.

Ward, C. E. "Slaughter-Cattle Pricing and Procurement Practices of Meatpackers." USDA, Econ. Stat. and Coop. Service, Ag. Info. Bull. 432, December 1979.

Ward, C. E. "Vertical Coordination of Cattle Feeding and Slaughtering in the Cattle and Beef Subsector." Univ. Wisconsin, NC-117 WP-14, December 1977.

Chapter 14

Controlling Price Risk Through Futures Markets

The difficulties of growing out stock are so great it is easy to overlook what is frequently the greatest threat to the economic well-being of livestock enterprises: price variability. Price fluctuations are large enough that what is a profitable level when stock is put on feed can, all too often, turn into a loss when those animals are market ready (Figure 14-1). Therefore, producers, at their peril, must consider means of controlling/minimizing price risks. There are several ways this may be done, including forward selling (Chapter 13). Risks are compounded when the instability of feed prices are taken into consideration. This chapter describes one major formalized means of limiting price risk: futures markets and their stepchild, options.

While futures are a vehicle for price discovery, for the most part they are *not* markets in the sense of physical exchange. Only some one percent of contracts actually lead to shipments of animals. Delivery options do exist but they are complex and costly. That option exists to eliminate arbitrage opportunities[1] that might arise if futures prices and cash prices did not converge at expiration. When the complexities of physical delivery led to fewer traders, a cash settlement alternative was added by the exchanges in order to further promote convergence.

A futures contract is a legal commitment for some specific future action. For livestock, the commitment would be for the sale or purchase of a specified quantity and quality of animals at some specified future date for the agreed price. What troubles many livestock producers — a practical group in general — is the possibility of committing something for sale that they may not actually have. Thus, a sell contract (short position) can be

---

1. Economists like to use the term arbitrage, which refers to the purchase and immediate resale of securities and related instruments in order to profit from a price discrepancy.

FIGURE 14-1. Monthly Prices for Slaughter Steers and Hogs, 1989-March 1990 (Source: USDA, ERS 1990)

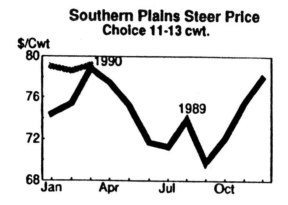

sold and payment received, even by someone who neither has nor intends to acquire livestock. The condition is, of course, that the sell must be offset by the cash purchase of a buy (long position) at some later date. Once an individual owns a legal commitment to both buy and sell the same product at the same point in time, the two offset each other and no net responsibility exists. Thus, one is dealing not in commodities, but rather in contracts.

Despite the abstraction, the use of futures can be helpful, if not essential, in one aspect of livestock enterprises. That aspect is controlling certain kinds of risk. In this section, the major uses of futures and options for livestock marketing are described along with typical strategies. Space constraints allow but a broad overview of the numerous issues involved in the effective use of futures and related options markets. Anyone pursuing actual use of these markets is advised to explore additional sources of information. On a semantic note, it is useful to recognize these are *futures* not *future* markets. For all of their obvious skill, futures traders lack any privileged knowledge of the future.

USES OF FUTURES FOR LIVESTOCK MARKETING

Risk

Understanding how futures markets can help the livestock sector requires an understanding of risk. Risk may be defined as the chance of an unfavorable outcome. Thus, if one is lured into playing Russian roulette for $1,000 a turn, the desired result is +$1,000. However, one also stands a one in six chance of a mortal consequence. That is risk—the probability of an unfavorable outcome. For the livestock producer, there is also the chance that things may turn out *better* than expected. Livestock prices may rise and/or feed costs fall. Thus, for livestock producers, one should more correctly refer to the outcome as "uncertain" (uncertainty of a favorable or unfavorable outcome).

We shall, nevertheless, retain the term "risk" in this discussion. This is because the upside and downside do not hold the same consequences. Most of us can accommodate an unexpected increase in income whereas a decline, especially if it can force bankruptcy, is a different matter. Thus, we shall be concentrating here on means of reducing the possibility, or amount, of potential loss. Regrettably, one also often reduces the potential for additional profits. It is often the restriction of profit potential which limits interest in futures markets.

When a livestock producer adds a brood animal to the herd or places an animal on feed, it is done with an expectation of profit. However, several events could diminish that expected return, possibly turning it into a loss. Prices could decline, feed costs increase, production (calving rate or gain rate) diminish, death occur, or some combination of the above. Such un-anticipatable results could be caused by weather, economic shifts at home and abroad, and by policy changes, among other factors. These, then, are the risks that are part of ownership. Futures markets are useful only in minimizing the profit risks (the purchase, sale, and feed price spreads). Anyone who speaks of "locking in" a profit through careful use of the futures market refers only to the per-pound margin; production and other risks are ignored. And futures are only partially useful in achieving risk-free profits, as we shall see. But within their limitations, futures markets can provide tremendous benefits that should and will be used by more livestock producers.

Concept of a Hedge

The risk-reducing use of futures markets is referred to as *hedging*. Hedging, in its most basic form, may be understood quite easily. Let us say a cattle feeder purchases 80 feeder calves weighing 600 pounds and expects to feed them out to 1,000 pounds in about 25 weeks. In 25 weeks, with a small degree of latitude, he/she will be selling 80,000 pounds of fed beef. At 80 cents a pound, the projected selling price, the anticipated gross revenue is $64,000. A five-cent-per-pound price decline would reduce gross returns to $60,000 (down by 6.25 percent, or $50 a head). This amount exceeds the long-term net profit to feeders and likely represents the difference between a profit and a loss. In an alternate scenario, prices could rise but, as noted, this presents no serious problem.

As a means of avoiding the risk associated with the price fall, the producer may, on the day the feeders are bought, sell two fat cattle futures contracts (the equivalent to 80,000 pounds at the prevailing futures market price of 80 cents). That is, he/she goes short in the market by taking an equal but opposite position in futures from the cash market. The producer's futures account is credited with the $64,000, equal with the projected cash sale.[2] However, when the time comes to market the cattle, cash and futures prices have fallen to 75 cents, and the cattle fetch $60,000 in the cash market. The obligation to deliver on the futures market is offset by buying a futures contract for $60,000. Net returns then are:

$$\$64,000 \text{ (futures value)} + \$60,000 \text{ (cash price)}$$
$$- \$60,000 \text{ (futures cost)} = \$64,000,$$

which meets the revenue goals set at the time of initiating feeding. Thus, the producer has circumvented the risk of a price fall.

Note, however, that he/she has also "avoided" the benefits of a price rise.[3] Suppose the live and futures price at the time of sale were 85 cents, or total revenues of $68,000. Total returns would then be:

$$\$64,000 \text{ (futures value)} + \$68,000 \text{ (cash value)}$$
$$- \$68,000 \text{ (futures cost)} = \$64,000,$$

or the same as under a price decline. Thus protecting him/herself from downside risk has also necessitated forfeiting potential profits. The exchange may be well worth it, but there are other approaches, including partial and selective hedges and options. These are described in a later section of this chapter.

This example is simplified so as to ignore costs of market participation and the value of capital. However, it does indicate the way in which futures markets can be of use to producers. One may, nonetheless, inquire where the money comes from when the sale price is below the purchase price. For every gainer there must be a loser. Just who are those people?

---

2. In practice, futures are traded on a margin or percentage of the contract value. For simplicity that factor is excluded here, but see below.

3. In practice, futures and cash prices may not be equal due to transportation and storage costs. The differential, called basis, fluctuates over the life of the contract. Thus, producers have shifted from price risk to basis risk. See later section.

There are two groups who take the opposite position of producers. A principal one is users, especially packers. For meat users, an increase in the live price represents a potential loss. To protect themselves, packers and others would wish to employ futures also, by *buying* live animal contracts in advance. When it came time to buy the livestock in cash markets, the futures contracts would be sold. Using the same figures as above, and assuming a price decrease of 5 cents/lb:

$$-\$64,000 \text{ (futures cost)} - \$60,000 \text{ (cash price)}$$
$$+ \$60,000 \text{ (futures value)} = -\$64,000.$$

Note that the packer's $4,000 decrease in cash buying cost was offset by a $4,000 loss in the futures position by hedging. (The final value will always be negative since procurement is an expense.) But, if the price rose to 85 cents/cwt, then

$$-\$64,000 \text{ (futures cost)} - \$68,000 \text{ (cash price)}$$
$$+ \$68,000 \text{ (futures value)} = -\$64,000$$

so that the packer is protected from the cost effect of a live animal price rise. Thus, producers (sellers) and packers (buyers) take opposite positions in the market and, to a large degree, they exchange futures "profits" and "losses."

The markets, however, do not work so perfectly that there is necessarily a hedger on the opposite side of the market just when you wish to buy or sell a contract. Markets in which hedgers are the primary players are called "illiquid." Illiquid markets can result in large bid-ask spreads. It is the speculator who frequently provides the necessary liquidity in futures markets. One class of speculators (known as spreaders or scalpers) does not retain a contract for long, certainly not overnight. Rather, profits are made on tiny, up and down, price movement and multiple trades. Yet these speculators provide an important function by giving improved liquidity to the market, the opportunity to buy and sell when you choose. A second class of speculators (day traders and position traders) buys on fundamentals and holds contracts for longer periods based on their expectations about price movements. These speculators, who can make large profits (and losses), are often viewed with suspicion by producer hedgers who see them as influencing the contract price and thereby causing loses for the hardworking rancher and feeder. However, such a view ignores the need for speculators to absorb the risk hedgers' wish to transfer.

One such case arose in early 1986 when several cattlemen's associations voted to terminate cattle futures contracts. This was a period of volatile (actually, declining) live and futures prices that coincided with a new federal policy encouraging dairy farmers to have their cows slaughtered. Preliminary investigations by the Commodity Futures Trading Commission (CFTC), the federal agency responsible for overseeing and regulating futures markets, revealed no malfunction of the market or manipulation by speculators (Phillips 1986). The issue, nonetheless, remains unresolved. The CFTC does maintain certain restrictions, such as a limitation on the number of contracts owned by an entity, as a means of preventing market manipulation. (For an overview of the activities of the CFTC, see Tendick 1975.) Futures markets remain controversial, in part due to producer's lack of insight into their operations.

OPERATION OF FUTURES MARKETS

Futures contracts are commodities that must be traded much like any other product. Indeed, the Chicago Mercantile Exchange, where the live animal futures contracts are traded, has its origins in the Chicago Produce Exchange, which emphasized butter and egg trading. Many of the practices of these markets date to that time and activity. Futures trading is still described as occurring in a "public market." That means the contract prices are public information, and the traders are required to call out the prices as trades are made. It is those prices that are recorded by market personnel and disseminated worldwide by wire and other services (Figure 14-2).

The spread of price information plays an important role in the efficient operation of these markets. Yet, it creates havoc on the trading floor; there is so much noise that traders cannot rely on their ears for accurate information (Figure 14-3). An elaborate series of hand signals have evolved that convey information rapidly and, one hopes, accurately (Figure 14-4). The din and the gesturing are characteristic of these markets, but they both serve important purposes.

Technology is nevertheless creeping into these markets in the form of electronic trading. This is almost a necessary transition as trading becomes more international and, with the inherent time differences, more round-the-clock. To date, electronic futures trading has been limited to after hours transactions, it can be expected to change the trading floor itself over the decade.

While the futures markets retain some of the characteristics of public markets, it does not mean one can personally enter these markets and execute a trade. Trading is limited to members, with the number of memberships limited by the administering body of each exchange. Member-

FIGURE 14-2. Newspaper Report of Futures Prices

Livestock

CATTLE, Live beef (CME)
40,000 lb.; ¢ per lb.

80.45	68.30	Oct	80.22	79.65	79.92		14,393
78.12	71.00	Dec	77.45	76.62	76.85	−.15	26,411
77.80	72.50	Feb	75.62	74.97	75.35	+.30	13,042
78.05	74.15	Apr	76.35	75.62	76.02	+.30	7,779
75.45	72.52	Jun	73.75	73.10	73.45	+.30	4,534
73.80	71.05	Aug	72.15	71.65	71.80	+.70	1,634
72.60	71.02	Oct	72.25	71.90	71.95	+.70	703

Est. sales 17,697. Tue.'s sales 14,029.
Tue.'s open int 68,526, up 1,249.

CATTLE, Feeder (CME)
44,000 lb; ¢ per lb.

88.47	78.20	Oct	88.25	87.42	87.75	−.07	3,329
88.10	79.00	Nov	87.50	86.45	86.87	+.10	3,438
87.15	79.75	Jan	86.35	85.52	86.07	+.25	2,216
85.60	80.90	Mar	84.10	83.30	83.47	+.10	1,176
84.85	81.20	Apr	83.15	82.60	83.20	+.60	222
83.50	80.20	May	82.85	81.90	82.05	+.28	344

Est. sales 1,472. Tue.'s sales 1,517.
Tue.'s open int 10,725, up 70.

HOGS, Live (CME) 30,000 lb. ¢ per lb.

57.32	40.80	Oct	57.07	55.90	56.60	+.40	5,041
55.60	44.25	Dec	55.95	54.35	55.67	+1.00	13,114
52.95	45.25	Feb	53.05	51.92	52.77	+.55	5,346
49.05	43.60	Apr	49.25	48.35	49.15	+.58	1,828
51.95	47.70	Jun	52.80	51.50	52.35	+.68	691
52.15	48.30	Jul	52.50	51.95	52.50	+.30	201
50.50	46.90	Aug	50.60	50.00	50.75	+.35	58
46.30	42.90	Oct	46.05	46.05	46.10		32

Est. sales 11,072. Tue.'s sales 11,482.
Tue.'s open int 26,311, up 580.

PORK BELLIES (CME)
40,000 lb; ¢ per lb.

63.65	48.07	Feb	64.20	63.70	64.20	+2.00	4,245
62.95	49.20	Mar	64.07	63.57	64.07	+2.00	743
63.00	49.50	May	64.67	63.40	64.67	+2.00	401
62.52	50.75	Jul	64.52	63.20	64.45	+1.93	95

Est. sales 2,190. Tue.'s sales 354.
Tue.'s open int 5,497, off 41.

ships are sold conditionally on approval of the governing bodies. Members must document financial wherewithal to cover a specified number of contracts, and be of good character and reputation. This all helps to assure integrity.

For nonmembers — that is, most of us — it is necessary to find an agent, or broker, to execute a trade for our account. In practice, this is similar to buying and selling on the stock exchanges (except that futures can be bought on margin, which is not allowed for stocks). In recent years, the major stock brokerage firms have provided futures trading services also. One can also work through many banks and specialized commodity trading firms.

All relevant domestic contracts (that is to say, live animals, meat products, and major feed grains) are traded in Chicago. The Chicago Mercantile Exchange handles the live red meat animals and meat products. Major

FIGURE 14-3. Action on the Trading Floor on a Busy Day (Source: Chicago Mercantile Exchange)

feed grain contracts are traded at the Chicago Board of Trade, while the Mid-American Commodity Exchange (also in Chicago) trades half-size contracts from the other two. (The contracts are described below.)

Trading Activities

Once the buy/sell order has been recorded with a brokerage firm, it is telephoned to the trading floor (Figure 14-5). The order is logged in, time stamped, and passed on to a "runner" for transferal to the trader in the "pits." The trading areas are known as pits but are actually raised rings. Each ring, or part of ring, is for a specific commodity and the different contract months are traded in part of that pit (Figure 14-6). This facilitates the runner finding a trader for his/her firm, although colorful smocks are also used. The order is passed on to the trader who executes it, noting the buyer or seller on the opposite side of the market, and tosses it outside the ring. From there it is retrieved and recorded, and the customer is notified

FIGURE 14-4. Hand Signals Used in the Trading "Pits" (Source: Chicago Board of Trade)

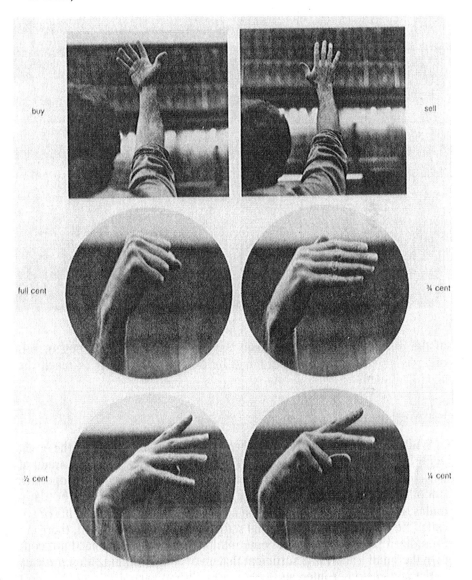

FIGURE 14-5. Scheme of the Chicago Mercantile Exchange Trading Floor (Source: Chicago Mercantile Exchange)

of the price. Orders may be made "at market" if a simple buy or sell order is given or a "stop" order to be executed when prices reach the specified level.

Contracts

While this process may seem haphazard, it is effective in the great majority of transactions. Buyers do indeed know to whom they trade at what price, and vice versa. To make certain each exchange operates smoothly, an independent "clearing corporation" checks all the daily trades, flagging those that are not in agreement, and settles accounts or so-called "out trades." If an informal settlement cannot be reached, there are formalized procedures which come into play. These details need not concern the small trader; it is sufficient that an oversight organization protects trader and customer interests.

Trading is, however, not quite as unfettered as that. Certain regulations limit price movements. Each contract has a minimum price move, the "tick" value. Similarly, each contract has a maximum allowable daily price movement. If price moves up or down by that amount in a day,

FIGURE 14-6. The Cattle Pit at the Chicago Merc (Source: Chicago Mercantile Exchange)

trading is suspended until the following day, and so on. Successive limit moves for three consecutive days can cause the exchange to raise the limit until the period of rapid movement passes. The limit fluctuation for hogs, for example, is 1 1/2 cents a pound, or $450 a contract (Table 14-1). These limits help to stabilize markets that might otherwise run away based on a single piece of good or bad news. Thus, holders of contracts are protected to some degree. However, a succession of daily limit moves early in the trading day can force one to hold contracts through periods of substantial price change. Users must be prepared for such possibilities even though they are rare (see below under margins and calls).

Contract Requirements

To be traded as commodities, that is, without individual evaluation, a contract must be highly specific. Futures contracts are so, detailing the product traded, the pricing terms, and delivery requirements. The product refers to a specific quantity of a certain product grade (e.g., "40,000 pounds of choice grade live steers"). Since a steer delivered in July is not the same as one delivered in January, contracts must specify the deliv-

TABLE 14-1. Futures Contract Specifications for Livestock, Meat, and Feed Grain Contracts (Source: Chicago Board of Trade 1989)

Commodity	Exchange	Size	Months	Minimum Price Tick (¢)	Minimum Daily Limit Move (¢)
live cattle	CME	40,000 lbs	Feb, Apr, Jun, Aug, Sep, Oct, Dec	.025/cwt	1.5/lb
	MCE	20,000 lbs		.025/cwt	1.5/lb
feeder cattle	CME	44,000 lbs	Jan, Mar, Apr, May, Aug, Sep, Oct, Nov	.025/cwt	1.5/lb
live hogs	CME	30,000 lbs	Feb, Apr, Jun, Jul, Aug, Oct, Dec	.025/cwt	1.5/lb
	MCE	15,000 lbs		.025/cwt	1.5/lb
pork bellies	CME	40,000 lbs	Feb, Mar, May, Jul, Aug	.025/cwt	2/lb
wheat	CBT	5,000 bu	Mar, May, Jul, Sep, Dec	.25/bu	20/bu
	KCBT	5,000 bu		.25/bu	25/bu
	MCE	1,000 bu		.125/bu	20/bu
	MGE	5,000 bu		.25/bu	20/bu

Commodity	Exchange	Contract size	Delivery months		
corn	CBT	5,000 bu	Mar, May, Jul, Sep, Dec	25/bu	10/bu
	MCE	1,000 bu		125/bu	10/bu
oats	CBT	5,000 bu	Mar, May, Jul, Sep, Dec	.25/bu	10/bu
	MCE	1,000 bu		.125/bu	10/bu
soybeans	CBT	5,000 bu	Jan, Mar, May, Jul, Aug, Sep, Nov	.25/bu	30/bu
	MCE	1,000 bu		.125/bu	30/bu
soybean meal	CBT	100 short tns	Jan, Mar, May Jul, Aug, Sep Oct, Dec	10/ton	1,000/ton
	MCE	20 short tns		10/ton	1,000/ton
soybean oil	CBT	60,000 lbs	Jan, Mar, May Jul, Aug, Sep Oct, Dec	.01/cwt	1/lb

CME: Chicago Mercantile Exchange
CBT: Chicago Board of Trade
MCE: MidAmerica Commodity Exchange

NYME: New York Mercantile Exchange
KCBT: Kansas City Board of Trade
MGE: Minneapolis Grain Exchange

ery—or maturity—date. For live hogs, there are separate contracts for seven months: February, April, June, July, August, October, and December. The maximum daily price movement and minimum price change are also specified. These contract terms are detailed in Table 14-1.

Finally, delivery must be tightly regulated. Thus, pre-notification is required and delivery is limited to business days Monday through Thursday. For slaughter steers, an official livestock yard receipt specifying quantity, quality grades, yield grades, and estimated hot yields registers the delivery. All cattle must average between 1,050 and 1,200 pounds, with an individual variation of +/− 100 pounds and a ban on weights above 1,300 pounds and below 950 pounds. Discounts apply to broader weight deviations and more than eight head with yield grades of four. Basic requirements are listed in Table 14-2, but specific details should be checked with the exchange in the unlikely event of delivery.

Margins and Calls

Margins on futures contracts—the proportion of cash which must be advanced at the time of purchase—vary, but are typically within the 5 to 15 percent range. This means that, for example, only 10 percent, or $1,500, need be put up for a $15,000 live hog contract (30,000 pounds @.50/lb). The minimum margin is established by the commodity market, but most commission firms require something more than the minimum. Since they hold the physical product, hedgers typically have lower margin requirements than speculators. Compared to many other investments, such as stocks, this is a small cash requirement and futures are therefore referred to as highly leveraged. Leveraging is attractive because it makes it possible to enter the market with limited cash resources compared to the value of the contract. For livestock producers with limited cash flow, this is highly desirable. In addition, hedgers may find banks who are willing to finance their hedging activities since the goal is risk management. At the same time, speculators are attracted because a small amount of capital can be rapidly transformed into a large profit.

Consider the live hog contract described in the preceding paragraph. A 1.5-cent price rise (the daily limit move) would yield a return to equity of 30 percent (.015/lb. × 30,000 lbs. = $450/$1,500 margin = 30 percent). This turns a small fortune into a large one. Regrettably, futures speculation is notorious for turning large fortunes into small ones (just consider the fate of the billionaire Hunt brothers). The same leverage effect operates for price *declines*, so that a decrease of five cents would wipe out the entire investment in the contract.

Producers do not (or should not) speculate, but leveraging affects them

TABLE 14-2. Basic Futures Delivery Requirements, Selected Commodities (Source: Chicago Board of Trade 1989)

COMMODITY	DELIVERY GRADE	DELIVERY POINTS
Live feeder cattle	600-800 lbs of feeder steers	cash -- based on cattle FAX quotes
Live beef cattle	USDA choice steers, yield grades 1, 2, 3, and 4, average weight 1,050-1,200 lbs.	Peoria, IL, Joliet, IL, Omaha, NE, Sioux City, IA, Guymon, OK, Greeley, CO
Live hogs	USDA grades 1, 2, and 3, barrows and gilts, average weight 210-240 lbs.	Peoria, IL, Omaha, NE, E. St. Louis, IL, Sioux City, IA, St. Paul, MN, Kansas City, MO, St. Joseph, MO
Frozen pork bellies	Green, squarecut, clear seedless bellies, 12 to 14 lbs. or 14 to 16 lbs.	Approved warehouses

Note: A "certificate delivery system" was instituted for live cattle in December 1983. See Hudson, Hieronymus and Koontz 1988.

also. A price decline of two cents on a live cattle contract at 80 cents would reduce the contract value by: $\$.80 \times 40{,}000$ lbs. $- \$.78 \times 40{,}000$ lbs. $= \$800$, or 2.5%. But the hedger put up a ten-percent margin, or $.10 \times \$32{,}000 = \$3{,}200$. The 2.5-percent decline in the contract value is leveraged into a $\$800/\$3{,}200 = 25\%$ decline in his/her margin deposit. More importantly, the ten-percent margin requirement remains in effect, but the margin is now only: $(\$3{,}200 - \$800)/\$31{,}200 = 7.7\%$. A "margin call" for a variation margin of $.10 \times \$31{,}200 - \$2{,}400 = \$720$ will be made immediately, payable on short notice.[4] Failure to pay means the contracts are sold, with the producer still responsible for any outstanding amount. Of course, if a price movement raises the original margin deposit to more than ten percent (in this example), the surplus amount may be withdrawn.

The $720 from the above example may not seem large, but the amount could be greater, and could occur repeatedly. What is more, this is a per-contract amount. Thus the hedger *must* arrange for a source of funds to meet possible margin calls. Lending institutions and brokerage houses can assist in estimating and establishing such accounts.

HEDGING STRATEGIES

Routine Hedge

A routine hedge, which is a complete hedge applied throughout the production period as a regular matter, is the easiest to administer. Essentially, no planning is required in this case; the process is reduced to a mechanical one. Theoretically, routine hedging will give the same long-term average profit as will no hedging, but without the risk of adverse price movements. This projection ignores costs associated with market activities, and says nothing about the short-run comparison between a fully hedged and fully unhedged position. In particular, a routine hedge requires the producer to give up any and all windfall profits, often the sustaining force in the sector. Thus, truly routine hedging is not very popular.

Even one seeking a fully hedged position will find several intervening risks that necessarily raise the risk of even that cautious position. One major factor is the lumpiness of the contracts. A feeder cattle producer,

---

4. This is a bit of a simplification, because, in practice, there can be a difference between the initial margin and the maintenance margin. See Marshall 1989, p. 27.

for example, with an anticipated 60,000 pounds of calves at market weights will have a difficult time accommodating the lumpiness of the 44,000 pound contracts. Purchasing two contracts would mean speculating on 28,000 pounds, whereas one contract would mean being unprotected for 16,000 pounds. The Mid-American Exchange with its smaller contract sizes (Table 14-1) would help, but not entirely eliminate, the problem. The risk of projecting market weights is a related matter. Even if one uses normal gain rates to project market weight, there is an unknown possibility (risk) that actual gain will fall short due to weather, death loss, or other factors. Thus, it is generally not advisable to hedge the entire projected quantity, but rather to make an allowance for production risk. Historical data are a good basis from which to project production risk. Finally, there is traditionally a difference between the cash and the futures price. Variations in that difference, known as basis risk, mean that even a fully hedged position does not remove all price risk from the producer. Basis risk is discussed later in the chapter.

Selective Hedging

Selective hedging is a more complex undertaking since it requires ongoing evaluations of when to place or lift a hedge. With volatile commodity markets, selective hedging necessitates frequent evaluation of the performance of the relevant futures market(s). If regular reviews of the market are not feasible, either personally or through the assistance of service organizations like advisory services and cooperatives, then selective hedging should probably not be undertaken, at least during periods of volatile prices.

The second requirement for a successful selective hedging operation is an appropriate procedure, or *strategy*. In the absence of any systematic decision rules, it is all too easy to act on hunch or rumor. That approach can quickly become tantamount to speculation, the opposite of what hedging is attempting to accomplish. One obvious decision rule is not to hedge against declining prices if prices are far more likely to remain stable, or rise, than they are to fall further. Such might be the case for feed grains that are at the support price. With the federal government legally committed to buy at that price, further declines are highly unlikely, and a hedge unnecessary. (See any farm policy book for a description of commodity price support programs or Glaser 1986.)

With red meats, there are no such federal price supports so that the projection of a "price floor" is less certain. However, if the number of cattle reported on feed, or the number of sows farrowing, suggest a notable supply downturn in the future, then prices can be expected to stabilize or advance over the period when those animals will come to market. (See Chapter 4 for a discussion of data sources.) However, such a projection refers only to supply; demand could shift over the period (as has happened unexpectedly in recent years) and lead to lower-than-anticipated prices. Thus "price floor" projections for livestock are problematic and perhaps an insufficient base on which to make selective hedging decisions.

As an alternative, many traders use certain "rules-of-thumb" about patterns of futures price movements as decision criteria. The most common of these rules involves a decision about when (and hence whether) to place a hedge, but, once placed, the hedge is held until the stock is sold. More sophisticated strategies involve decisions placing and removing hedges, while the most intricate include patterns of multiple placement and removal of hedges. In many cases, it has been possible to use analytical techniques to determine the effect of these strategies on profits. Those studies are helpful in determining which approach to use. Nonetheless, it is important to recognize the analyses are based on *historical* data that will not necessarily apply to future price patterns. Simply stated, what worked in the past will *not necessarily* work in the future, so that not even the best conceived and successful hedging strategy should be followed blindly into the indefinite future.

Single Placement Strategies

The concept of a single placement strategy is to defer placing the hedge until some condition (or conditions) are met, and then to hold the hedge until the sale (purchase). This is known as a pure hedge. Typically, the condition imposed is that the hedge will be profitable. Two measures of profitability are often used: full production costs and variable production costs. No hedge would be placed unless the localized futures price exceeds the estimated production cost. Assume that, today, the closing futures price projected for your area is $49.75/cwt. Then, if your full production costs are $51.25, you would not hedge today, while production costs of, say, $49.50 would lead to a hedge decision under this decision rule. Of course, as variable costs are less than full costs, the use of the variable cost criterion will lead to more frequent hedging decisions. Worksheets such as those shown in Figure 14-7 help to compute production costs and make decisions on when to place a hedge.

Under conditions when the localized price could be predicted correctly, these strategies limit the number of times producers "lock in" an unprofitable price, and thus can increase the profit potential of hedging over routine hedging. Raising the requirements for placing a hedge, that is, moving from variable costs to total costs or using variable costs plus a dollar or more (feeder cattle), increases the number of days when no hedge is in effect. As a result, average profits increase, but at the expense of greater risk of loss (Miyat and McLemore 1982, pp. 6-7, Tables A-2 - A-9). As a reasonable balance, one might choose to place a hedge only when variable costs, plus a modest increment, are covered. But the producer will remain exposed to the full price risk during periods of low and declining prices.

A second limitation of this approach is couched in the term "localized" futures price. A localized price refers to projecting the cash price in your local area. That price will be the futures price plus or minus a time and location factor known together as the *basis* (Figure 14-8). That is, the basis is the difference between the cash and the futures price. It may be positive or negative depending on a number of factors including interest rates, seasonality, transportation costs, and projections of future supply and demand as well as local conditions like the number of animals on the market and the needs of feeders and packers. The net effects of these factors are difficult to predict, causing what is known as "basis risk." Due to basis risk, no hedge can be perfect in that a hedger can achieve a total risk-free position by using futures. Nonetheless, the use of futures can be justified by the case that the basis is more predictable than the cash price itself.

Hedge-Lifting Strategies

There is, of course, no reason why the hedge, once placed, must be maintained throughout the production period. A simple (but not recommended) rule for dropping a hedge is to let the position be sold out rather than to respond to a margin call. This strategy is especially enticing as it is automatic — a failure to do anything leads to the desired result — and removes the threat of pressure on cash flow caused by meeting margin calls. This approach does not consider objectives or outlook and, hence, can often be suboptimal.

For the hedger who has gone short (sold a position), a margin call occurs when the futures price increases enough to take the equity in the customer account below the maintenance margin level. This means the price of buying an offsetting contract is rising. Assuming the cash price is rising simultaneously, lifting a hedge during periods of rising prices can be beneficial. After all, the hedge is, for the seller, intended to protect against *falling* prices. Problems arise if the upturn is but a temporary precursor to a decline, in which case lifting the hedge at that point is problematic. No one can know with certainty what will be the case, but empirical simulations have suggested that abandoning the hedge on the *second* mar-

FIGURE 14-7. Example of Means for Computing Costs as an Input into Making Hedging Decisions (Source: CME 1977, pp. 2 and 5)

Hedging Worksheet (for one crop)

(All prices in dollars and cents per hundredweight)

Number of Head _____

Est. Avg. Marketing Weight _____ Lbs.

Est. Marketing Date _____

Hedging Potential
(Max. No. of Contracts) _____

Futures Delivery Month _____

Cost of Delivery Basis _____

Date	Today's Futures Price	Estimated Basis (Historical)	Localized Futures Price	Break-even Price	Estimated Profit (Loss)	Hedging Strategy
	A	(-) B	(=) C	(-) D	(=) E	F

Initiated Futures Hedge   Date _____   No. Contracts_____   Price $_____

Offset Futures Hedge   Date _____   No. Contracts_____   Price $_____

Profit (Loss) on Futures $_____

Sale (Cash) Price of Livestock $_____

Realized Sale Price $_____

Cost of Production $_____

Profit (Loss) per hundredweight $_____

FIGURE 14-7 (continued)

I. Variable Costs

A. Feeder Purchase _____ lbs. @ _____ /lb. =

B. Feed Cost
1. Corn _____ bu. @ _____ /bu. = _____
2. Silage _____ tons @ _____ /ton = _____
3. Supplement _____ lbs. @ _____ /lb. = _____
4. Other _____ @ _____ _____
5. Other _____ @ _____ _____

Total Feed Cost (Cost of Gain) . . . . _____

C. Other Variable Costs
1. Vet & Medicine _____
2. Death Loss (2% of Line A for calves, 1% for yearlings) _____
3. Marketing, commission charges, trucking expense[1] _____
4. Power, Fuel, Equipment Repair[2] _____
5. Interest, insurance, taxes on feed and cattle[3] _____
6. Hedging Costs
   Brokerage Commission _____
   Interest on Margin _____
7. Miscellaneous _____

Total Variable Costs . . . . _____

346

II. Fixed Costs

A. Buildings & Equipment[4]

B. Labor _____ hrs. @ _____ /hr.[5] = _____

Total Fixed Costs .... _____

Total of All Costs (Variable + Fixed) Per Head .... _____

III. Break-even Price (per hundredweight)

(Divide total costs per head by market weight per head.) .... _____ /cwt.

[1]Include commission cost of purchasing feeder animal and hauling to farm, and commission costs of marketing fed animal and hauling to market.

[2]Power, fuel, and equipment repair costs are estimated to be $4 to $6 per head.

[3]Equal to 10.0% (.10) times the following: purchase price of steer plus value of corn fed plus value of corn silage fed plus one-half of the value of supplement, salt, and mineral fed. This value is then multiplied by the fraction of the year that is the feeding period.

[4]Equal to 14% of the current investment in shelter, silage storage, and feed-handling and other equipment associated with the cattle-feeding enterprise. It is assumed that storage for corn is included in the price of corn.

[5]Estimated to be 4 to 6 hours per head typically, with long-fed cattle requiring the larger amount.

347

FIGURE 14-8. Example of a Basis (Source: CME 1977, p. 27)

Live Hogs

····· Futures
------ Peoria
—— Local

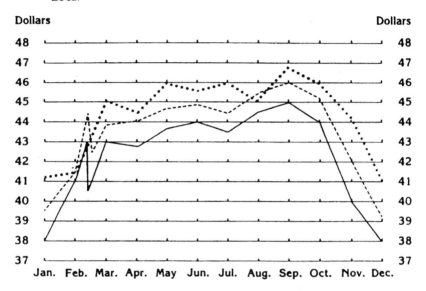

gin call leads to higher average profits, and especially lower profit variability, than does using the initial call as a decision point (see, e.g., Miyat and McLemore 1982, Appendix tables). Arguably, successive margin calls are better indicators of a rising market than is a single upturn. In any event, some limited number of margin calls should be considered a normal part of hedging and not necessarily a sign to liquidate a position.

For those holding a long position (purchase), the situation is similar. With this group, a price fall is negative since it indicates a lower price when the contract is sold. Hence, margin calls come at a time of falling prices. But because the hedge is not needed as protection against price declines, the option arises to exit the position when prices fall. The same caution noted above regarding the permanence of the price trend nonetheless also applies in these instances.

Selective Hedging Strategies

If, under certain price conditions, it can be profitable to delay placing a hedge, and to lift that hedge prior to the consummation of the sale/purchase, why limit oneself to a single set of transactions? Many hedgers indeed find that buying and selling futures contracts repeatedly can increase the average profit while only slightly increasing the variability of profits (the risk). Such multiple hedging strategies require discipline because numerous periods exist during which no hedge is in place. During unhedged periods, the producer is completely exposed to price risk.

Here we shall consider but two of the simplest strategies: those using *moving averages* and those that rely on so-called *point-and-figure* analysis. Both circumvent the problems identified above under single buy/sell strategies. That is, they rely on an analysis of the futures prices themselves (hence the name, technical analysis), and do not require an ongoing projection of localized prices. At the same time, they do involve projecting trends so that no separate judgement such as whether to rely on the first or second margin call is needed. Indeed, technical analysis is really the projection of trends. The recommendation for the holder of a short position is that a sale be made in a down trend and a purchase on an upward one. For "longs," the advice would be the opposite.

Selective hedging strategies can become extremely complex, more complex than space will allow us to explore here. Yet, they all are based on a simple concept, that of inertia. Rising prices are considered more likely to continue rising than not, while falling prices are anticipated to perpetuate the fall. Thus, these analyses give hedging signals for producers (sellers of products) on falling markets, and liquidation signals on rising markets. After all, producers do not need to be protected from the "risk" of higher prices, only lower ones. Selective hedging strategies, then, involve repeat buy/sell decisions for hedgers. The trick is to determine when the market is indeed rising, and when it is falling. This is the function of technical analysis.

The *moving average* approach involves an examination of the relative movement of shorter and longer term price trends. As a simple example, the short term could be today's price; the longer term, average prices for the past three or five days (Figure 14-9). When the short-term price rises above the longer term one, it signals an upward trend which is interpreted either as a signal to short sellers to exit their positions or as a signal to long hedgers to initiate a buy position. The shorter term price cutting the other from above indicates a declining price trend, giving the opposite hedging

FIGURE 14-9. Simple Two-Price Moving Average System, with Buy/Sell Signals (Source: Franzman and Lehenbauer 1979, p. 3)

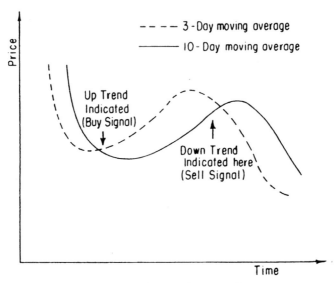

Illustration of crossing action of two moving averages.

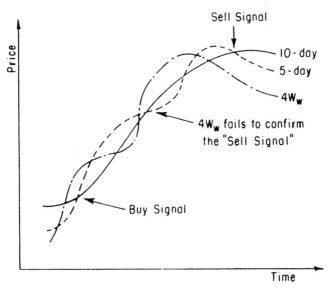

Illustration of buy and sell signals from three moving averages.

signals. There is no reason the number of price trend lines need be limited to two; frequently a third, yet longer term line, is used, this time to confirm any trend. Thus, in Figure 14-10, there are buy and sell signals as above, but the cross in the center of the illustration is not a sell signal because it is not confirmed by the longest term average which falls on the opposite side of the ten-day line. This point indicates a brief aberration and not a trend requiring an adjustment of the futures position.

The trick to a successful moving average system is the "optimal" selection of the proper averages. In general, the shorter the period of the averages (e.g., three vs. ten days), the less stable are the lines and the more buy/sell signals are generated. A second choice is that of treating each day's price equally (a simple average, or uniform weighing), or giving more recent data a higher importance by weighing it more heavily. For example, assume three recent prices of $62.98, $61.00, and $63.42. The simple average is:

$$(61.00 + 62.98 + 63.42)/3 = \$62.47$$

One weighted average would be:

$$(61.00 + 2 \times 62.98 + 3 \times 63.42)/6 = \$62.87$$

Note the weighted average, in this case, is higher as the price is rising, giving more importance to the more recent prices. Finally, the rule can include a "minimum penetration" requirement, meaning that the sell/buy signal is not given until the longer average line is crossed by some amount (for example, say three cents or eight cents). Clearly, the number of possible decision rules is huge.

It is possible analytically to determine the optimal weighing pattern. This is done by using past data and simulating what would have happened if each weighing practice were followed at that time. These analyses attempt to use long time periods to remove the effects of generally rising or falling markets, or other unusual occurrences. The outcomes are, nonetheless, dependent on the period used and do not necessarily apply to the future. Thus, it is best to use the most recent analysis available, and the reader is encouraged to seek the council of extension agents, commodity advisory services, or other specialists. However, the analyses tend to show that the decision systems using minimum penetration amounts and weighted averages are more profitable with less risk. These strategies tend to restrict the number of trades, but fewer trades alone are not necessarily the key to the optimal strategy (recognizing, of course, that transaction costs must be considered in the strategy) (Franzman and Shields 1981a;

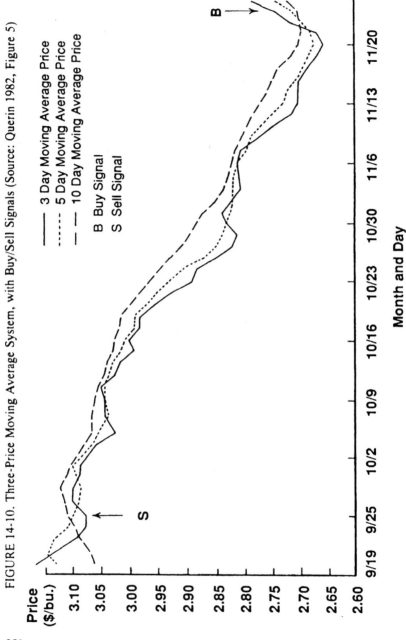

FIGURE 14-10. Three-Price Moving Average System, with Buy/Sell Signals (Source: Querin 1982, Figure 5)

352

Franzman and Lehenbauer 1979). However, many strategies improve profitability over pure hedging with only moderate increases in risk. Thus, while the optimal is ideal, benefits can be gained by other strategies as well. More significant than a lengthy search for the optimal, perhaps, is using an adequate one in a disciplined manner.

The *Point-and-Figure* approach is but another, albeit more complex, means of recognizing a trend. It may best be explained with the use of an example, limited here to the simplest "double top" and "double bottom" systems. Assume feeder cattle futures for ten weeks are as follows:

WEEK	HIGH	LOW
1	$ 90.30	$ 89.00
2	90.90	89.00
3	92.30	91.45
4	93.35	92.40
5	93.30	92.40
6	94.00	91.45
7	92.35	91.55
8	92.65	90.80
9	92.50	91.50
10	92.00	90.45

Now let us begin by placing zeroes in a grid, one zero for each 20-cent price movement. Thus for week one, an "X" is placed in all boxes up through $92. The following week, with a price rise to $90.90, three additional X's are placed vertically in the same column. This continues for two more successive weeks until X's are placed up to the $93.20 box, again in the same column (note that an X is not placed until the price change covers the entire range of the box).

The following week, however, showed a price decline. The amount was $93.35 minus the low for week five of $92.40, or $.95. This is greater than the reversal amount, computed as box size times a reversal constant (here 3) or $.20 × 3 = $.60. Thus, a "zero" is placed in the column to the immediate right of the X's, one box below the highest level achieved for the X's down to $92.40. The following week (#6) brought a new low and zeroes are plotted down to $91.40. With week seven came a reversal to a low of $91.55. The amount of the reversal is $92.35 − $91.45 = $.90, which again is greater than $.20 × 3 = $.60. Thus, a reversal has been signaled, leading to the placement of an X in the adjacent right-hand column up to $92.30, and down to one box above the lowest zero position. The succeeding week, number 8, brings another increase to $92.65, and the X's are extended up to that level. For the remainder of the ten weeks, prices again decline, leading to another column of zeroes (Figure 14-11). The final three columns are an example of a double bottom — a rise flanked by declines — which is a recommended sell signal.

As with moving averages, the hedger is seeking the optimal use of this strategy. Key decisions are the size of the "box," the unit of recording price graduations, and the rule for defining a change in price direction. Often, three times the box value is used (as in the example above). As with moving averages, the smaller the box value and the smaller the weighing factor, the more frequently buy/sell signals will be generated.

In general, hedging strategies based on technical analysis, of which but a few forms are reviewed here, can improve performance compared to

FIGURE 14-11. Sample Point-and-Figure Plot Indicating a Sell Signal (Box Size = 20¢, Reversal Number = 3) (Source: Franzman and Lehenbauer 1979)

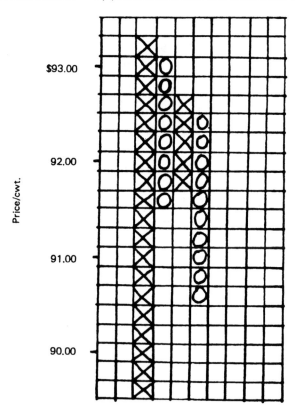

routine hedging.[5] That is, average profit increases with only a small rise in risk (variability of returns) compared to routine hedging. However, in an absolute sense, hedging strategies are not money making and should not be considered that way.

BASIS RISK

The basis, as noted, is the difference between the cash and futures price at a particular location. The basis becomes involved in hedging plans because the producer typically sells (or buys in the case of feeder stock) on the cash market. Thus, any unpredictable variation between the cash price and the futures price upsets the most careful risk-shifting strategy. This "unpredictable variation" is the basis risk, and it must be considered in any hedging plan. The existence of basis risk means that no hedge will be perfect in that it allows the producer to lock in an absolute margin prior to the actual sale. Nonetheless, futures brokers argue that it is easier to predict the basis (the relative price level) than it is to predict the absolute price level (the cash price). Thus, hedging becomes a process of moving from cash price risk to basis risk.

The basis is affected by a number of factors, both local and international. Factors that tend to vary the cash price relative to futures prices are: changing expectations of future supply (due to such factors as seasonality and low grain prices), adjusted imports or slaughter expectations, interest rates (low rates favor the delay of marketings), and transportation costs. Expansionary points in the livestock cycle mean gilts/heifers are being held for breeding stock, pushing current supplies down. The reversal of these conditions leads to the inverse effect on relative cash/futures prices.

Local conditions are also influential. If local markets are small compared to potential supply, and relatively isolated, there is a greater chance for the cash and futures prices to become "disengaged." When local supplies are heavy, buyers can run short of capacity, markets, or storage space so that prices face a downward pressure, and vice versa. This can happen at the same time national conditions are tending the opposite way. The more isolated an area, in terms of transportation charges to alternative markets, the greater the price divergence can be. Finally, if the product differs from those specified in a futures contract, yet another source of divergence exists. That is, if the product to be sold is heifers while the contract calls for steers, the cash/futures prices may vary more widely. Or if one attempts to hedge sheep (for which no futures contract exists) based

---

5. Numerous other forms exist, including trend line analysis, stochastics, Elliot wave theory, and the Relative Strength Index.

on the traditional relationship between slaughter sheep and hog prices, to take a single example, a wider cash/futures uncertainty should be anticipated.

Due to these several factors acting simultaneously, the futures price may be above, below, or even equal to the local cash price (see Figure 14-8), but a premium of the futures over the cash is the more common condition. The means of *predicting* the basis is predominantly the use of historical data, combined with judgement and data on future events. That is, if the 180-day futures spread for hogs is normally 10 cents, but the farrowing report (see Chapter 4) leads to the anticipation of increased supplies in your state only, expect a widening of that spread. (Be careful here, though, for a *general* supply increase will probably cause a decline in the futures price itself so that no further adjustment to your local condition may be needed.) A good historical cash price series, related as closely as possible to your actual marketing situation, is therefore the first and major step in controlling basis risk. Price, Reed, and O'Connor (1985) found that, while the live cattle basis was predictable, that for feeder calves was quite unstable. In either case, the basis varied by contract month and must be computed separately.

The entire risk may be shifted through a combination of futures and forward markets. Note that (for feeders), one can sell a futures contract (go short) as well as sell the animals for future delivery (when that option is available from local buyers). Now, the difference between the futures and the cash price is the basis, and this combination of sales fixes that amount so that the basis is hedged. In fact, future delivery prices are typically quoted as variations from the futures, such as "five cents below the March futures." Thus, forward selling is setting the margin; combining that with a futures contract firmly establishes the future cash price. This approach is, of course, workable only if delivery to the buyer is planned, for future sales (unlike futures contracts) are based on the expectation that delivery will be made. Further, the buyer will expect some return for accepting the risk of basis movements in the future. This return will come in the form of a lower price, so it will sometimes be more profitable to "self insure" (accept the basis risk) than to shift it all to others. But the opportunity is there.

HEDGING THE MARGIN

As shown, hedging livestock prices can stabilize the prices of these products while preventing paper losses during periods of declining prices. Paper losses refer to what would have happened had some course of action been followed or not followed. A person's paper stock market profits are, then, the profits that could be realized if the stocks were sold today. Actual profits when the stock is indeed sold could be quite different. The same distinction can be drawn with profits on livestock operations. Hedging can be instrumental in preventing losses *compared to* the nonhedged position. But *absolute* losses may nonetheless occur. This happens if feed prices rise unexpectedly, obliterating the margin. The remedy here is to hedge feed prices also. Hedging feed costs is appropriate even for the producer who grows much of his/her own corn and soybeans. In that instance, a rise in feed prices increases the value of the crop. Feeding it causes a loss in opportunity value (the profits which would have been earned had the feed been sold).

When examining hedging feeding costs, let us focus on cattle feeding because it presents the most involved system. Cattle feeders may hedge (a) feeder cattle, (b) fat cattle, and (c) feed costs, or a total of eight combinations from no hedging to hedging of all three components. Hedging all three truly "locks in" a margin since the major costs and sale prices are fixed ahead of time, and possibly prior to initiating the feeding operation. As an example, assume a margin per head of $30 can be expected. If the margin increased due to a fat cattle price rise, or a fall in costs, the number of head put on feed might be expanded. Alternatively, a thinner margin could lead to a commitment to fewer head, if that were consistent with cash flow requirements and effects on the distribution of fixed costs. At a negative margin, feeding should probably be delayed until the profitability outlook improves. In any case, it is important to recognize that the "locked in" margin is but a target. The basis risk can affect realized prices, as can death loss and other production risks. Finally, some inputs (forage is a major example) have no futures market.

The first step in margin hedging is the careful calculation of costs (see Figure 14-7 for a typical form). This exercise will help in determining whether to undertake production at all. Once the decision is made to go ahead, the decision to hedge each component must be made separately, using the criteria described above. Feed prices, however, must be treated somewhat differently as they are the result of federal programs that support prices in several ways. While the price support program is a complex system, the overriding practical effect concerning feed users is the price floor which is set annually. As prices approach the floor, the risk of further declines is negligible, while potential price increases are unaffected. Thus, total risk is asymmetrical, but the risk faced by the feeder is unaffected. However, prices typically stay near the floor when stocks are high, meaning that a significant price increase would follow only if a very large shift in supply or demand occurred. Such an event is unlikely, so hedging of feed is often done on a more selective basis than is livestock hedging, operating as it does in an unregulated market with no long-term stocks.

Again, the optimal hedging strategy for each component is most desirable, but is difficult for the individual to identify. One should not despair, though, because it has been shown that many multiple hedging strategies can increase profits substantially. In one study, multiple hedging of fat cattle, feeder cattle, and corn led to a 50-percent profit increase and a reduction in risk (by one measure) of over 40 percent (Franzman and Shields 1981b). Thus, efforts put into multiple hedging can yield attractive returns.

OPTIONS

Futures options provide an attractive, and relatively new, means of shifting some forms of price risk. Options are particularly attractive since they can allow price insurance *without* foregoing the benefits of a price rise. At the same time, these options have been permitted for only a few years (since October 1984, to be exact), following a 50-year ban that resulted from a series of scandals and illegal uses of options contracts. When the ban was lifted, strict regulatory control was implemented to avoid the abuses that had previously plagued options markets. Options are available for the following commodities of interest to livestock producers:

- Live Cattle (Chicago Mercantile Exc.) — 40,000 pounds
- Feeder Cattle (Chicago Mercantile Exc.) — 44,000 pounds
- Live Hogs (Chicago Mercantile Exc.) — 30,000 pounds
- Corn (Chicago Board of Trade) — 5,000 bushels
- Soybeans (Chicago Board of Trade) — 5,000 bushels
- Soybeans (Midwest Commodity Exc.) — 1,000 bushels

Commodity specifications for options are, in fact, identical to the related futures contract. This is the case because an option is the right, but not the obligation, to trade at a pre-specified price. The actual trade is the exchange of a futures contract, so that an option is actually an option on a futures contract.

Understanding futures options is most easily done by means of an example. Consider the case of a planned future sale of 30,000 pounds of hogs (one contract's worth). As a seller, one is concerned about the possibility of a price fall. Through options, price declines can be protected by buying a *put option*. A put gives the privilege of *selling* at a specified price known as the strike price. If price falls below that level, the option is utilized (exercised), and the strike price becomes the sale price. Should the price rise above the strike price, there is no benefit to exercising the option and the market price becomes the effective price. In these ways, futures options provide floor, but not ceiling, prices. With the additional benefit of no margin calls for buying options, they are attractive indeed.

For anything so potentially beneficial, one expects a cost, and so it is with options. The cost is the price or premium of the put. That cost, or total value, is separable (at least conceptually) into *time value* and *intrinsic value*. Intrinsic value is the difference between the futures price and the strike price. Thus, if the futures price were 53 cents per pound and the strike price were 55 cents, the intrinsic value would be two cents. Strike prices of 53 cents or less would have zero intrinsic values, respectively. In practice, options are available for various contract months, with a range of different strike prices corresponding to each contract. In late July 1990, for example, live cattle options had strike prices of 70 to 82 cents. (The near-term futures prices at that time were near 78 cents.)

The time value component of the premium reflects both interest rates and a risk premium. The total value is then paid for the privilege of holding the put. Should the option not be used, the total premium is the maximum cost that can be incurred.[6]

Similarly, livestock producers, as major purchasers of feed grains, may wish to protect against price *rises* for those inputs. This may be done under the options system by *purchasing a call option*. A call option is the right, but not the obligation, to *purchase* a futures contract at a designated price. Call options, too, have premiums that are determined just as with put options. However, because the risk of a price rise and price fall at any one time are not identical, the price of a put and call for an otherwise identical transaction are not equal. Thus, for a day at the tag end of July 1990, the 56-cent strike price for an October live hog option at the Chicago Mercantile Exchange was 1.00 cent per pound for a call and 2.90 cents for a put. Options costs would then be $300 and $870 for the call and put, respectively. The market makers were clearly betting that prices were heading down.

Consider, for a moment, the other side of the market. If *buying* a put option limits your cost to the total value paid but allows you potentially

---

6. As of this writing, the full price of an option must be paid at the time of its purchase. In 1989, a proposal was made to allow margining of options: the up-front payment of a portion (say 10 percent) of the total cost. Its eventual acceptance or rejection is not clear at this time. (See *Federal Register*, 17 March 1989, pp. 11233-36.)

unlimited profit if prices fall far enough, then *selling* a put can be quite a risky matter. In fact, when selling a put or call, profit is predetermined as the total value of the premium, while the loss risk is open-ended. Selling options is, then, a highly risky activity which is undertaken largely by a specialized group of traders or users. Unlike futures contracts, buyers of puts (e.g., livestock sellers or producers) do not offset buyers of calls (e.g., livestock users or packers). Speculators who are willing to sell either puts or calls are needed to keep options markets liquid.

As with other forms of hedging, options trading protects against particular forms of risk. As described above, they protect against price rises or falls. More complex strategies may be utilized to protect against price volatility in general. This may be seen most easily by considering the so-called *long straddle*. A long straddle is the simultaneous purchase of a put and call option for the same month and at the same strike price (Figure 14-12). With this example, profits will not be made unless the futures price falls below $.50/lb. or rises above $.55/lb. Thus, the options user is really betting that prices will move. He/she does not know the likely direction of movement but expects prices will not be stable. This use of options, then, is a hedging of the risk of price volatility.[7] Such complex strategies are more typically used by speculators but do have some legitimate roles in hedging.

Using options can provide great benefits but is specialized, like futures, and should not be undertaken without competent advice. If the future uncertainty is very great, then the true value will be high. Risk takers require a large premium to accept a large risk. At very elevated levels, users will find options inadvertently costly and select other risk-reducing mechanisms. Thus, options are not a panacea but rather provide an alternative which should be considered within the available mix.

USE BY PRODUCERS

Despite apparent advantages, the proportion of livestock producers (along with other producers) using futures and options is and has traditionally been small. A recent (1986 data) survey placed the number at five and 18 percent for futures with beef and pork producers, respectively. Similar figures for options were only two and four percent, but they were new at the time and may have risen subsequently (Smith 1989, Table 28). An "indexed" use of futures through forward cash contracting was large but

---

7. For a discussion of additional options strategies see Hauser and Eales 1987.

still limited: 12 and 23 percent for the two groups (Smith 1989, Table 27). The percentage of livestock hedged is greater as larger producers are heavier users, but the message is evident that livestock producers remain open to considerable price risk.

The reasons for that choice are less clear. However, use is positively correlated with education level and farm sales, and negatively with age. Thus there is clearly a significant educational factor in futures use (Makus, Carlson, and Krebill-Prather 1990, pp. 17-20). But knowledge is not the only issue; from one-third up to almost one-half of users and nonusers, respectively, indicated lack of confidence in the system as a moderate or serious concern (Makus, Carlson, and Krebill-Prather 1990, Tables 11 and 12). Multiple studies have generally not found a basis for concern that is limiting the use of hedging (see, e.g., Paul, Kahl, and Tomek 1981; Purcell and Hudson 1985; Collins and Irwin 1990). However, an anomaly remains, for smaller producers characterize themselves as more risk-adverse yet use futures and options least and are less willing to attend educational seminars (Smith 1989, Tables 12, 41, and 29). Given this situation, there is no real indication that the use of futures and options will expand notably over time.

SYNOPSIS

Futures and options are means of controlling (or more correctly, shifting) price risk. Futures are the contractual commitment to sell (a "short" position) or buy (a "long") a specified commodity for a pre-set price; options are the right, but not the obligation, to buy or sell a futures contract at the specified price. Futures can be advantageous because they fix a future price today, leaving the producer largely unaffected by rises or falls. While that is a benefit, with futures it also voids the benefit of favorable price swings (e.g., rises for sellers, declines for buyers). Options do not share that limitation since the cost of the option is the total cost, while the benefit is theoretically unlimited. However, the cost (total value) of options can be high in volatile markets, making them a less attractive choice.

Futures are not traditional markets in that product seldom changes hands. Rather, positions are typically voided before expiration by taking an offsetting position. That is, the holder of a sell contract can nullify his/her position by acquiring a buy contract for the same delivery date. Despite this, it is considered important to allow delivery so that futures and cash markets become one market at the expiration date. Most recently, hog contracts may be satisfied by cash payment and cattle can utilize a "certificate of delivery."

Futures contracts are sold on a margin, typically around ten percent. That allows substantial leveraging but means that a small unfavorable price move will wipe out the margin and trigger a "margin call." Options are sold at the full value; a proposal for marginalizing them has been made but not acted upon.

Studies have shown that hedging with futures — the process of taking an equal but opposite position on the futures and cash markets — leads to the same average price as without hedging but with reduced price variability. Higher prices are possible with selective hedging. Numerous alternatives are available but one should follow a procedure or strategy based on such indicators as a moving average. Similarly, cross hedging can be done for commodities without futures contracts, or margin hedging may be accomplished by hedging feed grains. These are all sophisticated activities that should be done in consultation with specialists.

Hedging actually changes the risk from the full price risk to the difference between the futures and cash price. This is known as the basis risk, and it must be projected by using historical relationships combined with additional current information. To hedge effectively, basis risk must be less than price risk (a frequent, but not universal, situation).

Despite the benefits of a carefully managed hedging/options program, these activities are little used by producers, including livestock producers. Use is associated with youth, education, and farm sales — all factors that are slow to change. Distrust of the markets is also relevant even though it has been frequently disputed in empirical investigation.

Study Questions

1. Livestock producers sometimes find themselves in a "Texas hedge," more commonly known as speculation. For a hog farrowing and finishing operation, describe: (a) a hedge and (b) speculation in the futures markets.
2. Considering risks in livestock and feed growing markets, as well as basis issues, describe how the maximum amount of risk can be shifted by using futures and/or options markets.
3. How is the choice between a futures contract or an option to be made?
4. Calculate the basis, over six months, for the red meat species of your choice in your part of the country. What is the necessary relationship

between basis and futures price risk in order to make hedging useful?
5. Define a hedging strategy. How does it differ from what might be called selective hedging activities?

REFERENCES

Chicago Board of Trade. *Commodity Trading Manual*. Department of the Chicago Board of Trade, 1989.

Chicago Mercantile Exchange. "Livestock Hedger's Workbook." 1977.

Collins, P. L. and S. H. Irwin. "The Reaction of Live Hog Futures Prices to USDA Hogs and Pigs Reports." *Am. J. Agr. Econ.* 72(1990):84-94.

Federal Register. 17 March 1989, pp. 11233-6.

Franzman, J. R. and J. D. Lehenbauer. "Hedging Feeder Cattle with the Aid of Moving Averages." Oklahoma Ag. Exp. Sta., Bul. 746, July 1979.

Franzman, J. R. and M. E. Shields (a). "Multiple Hedging Slaughter Cattle Using Moving Averages." Oklahoma Ag. Exp. Sta., Bul. B-753, January 1981.

Franzman, J. R. and M. E. Shields (b). "Managing Feedlot Price Risks: Fed Cattle, Feeder Cattle and Corn." Oklahoma Ag. Exp. Sta., Bul. B-759, October 1981.

Glaser, L. K. "Provisions of the Food Security Act of 1985." USDA, Econ. Res. Service, Ag. Info. Bul. 498, April 1986.

Hauser, R. J. and J. S. Eales. "Option Hedging Strategies." *North Central J. Agr. Econ.* 9(1987):123-34.

Hudson, M. A., T. A. Hieronymus, and S. R. Koontz. "Deliveries on the CME Live Cattle Contract: An Economic Assessment," *North Central J. Agr. Econ.* 10(1988):155-64.

Makus, L. D., J. E. Carlson, and R. Krebill-Prather. "An Evaluation of the Futures and Options Marketing Pilot Program." U. Idaho, Dept. Ag. Econ. and Rural Sociology, 1990.

Marshall, J. F., *Futures and Options Contracting: Theory and Practice*. Cincinnati, OH: South-West Publishing Co., 1989.

Miyat, C. D. and D. L. McLemore. "An Evaluation of Hedging Strategies for Backgrounding Feeder Cattle in Tennessee." Tennessee Ag. Exp. Sta., Bul. 607, February 1982.

Paul, A. B., K. H. Kahl, and W. T. Tomek. *Performance in Futures Markets: The Case of Potatoes*. USDA, Econ. and Stat. Service, Tech. Bul. 1636, January 1981.

Phillips, S. M. Testimony Before the Subcommittee on Government Information, Justice and Agriculture. Committee on Government Operations, May 8, 1986.

Price, R. V., S. P. Reed, and C. W. O'Connor. "A Descriptive Study of Live and Feeder Cattle Basis." Report Submitted to the National Cattleman's Foundation, January 1985.

Purcell, W. D. and M. A. Hudson. "The Economic Roles and Implications of

Trade in Livestock Futures." In *Futures Markets: Regulatory Issues*, ed. A. E. Peck. Washington, DC: Am. Enterprise Institute, 1985.

Querin, S. "Hedging Strategies Utilizing Technical Analysis: An Application to Corn in Western New York." Unpublished MS Thesis, Dept. of Agricultural Economics, Cornell University, August 1982.

Smith, R. D. "National Assessment of Producer Marketing Alternatives: Practices and Attitudes." Texas Ag. Exp. Service, April 1989.

Tendick, D. L. "Regulation of Futures Markets." In *The Barker and Hedging*. Chicago Mercantile Exchange, September 1975.

USDA, Economic Research Service. "Livestock and Poultry Update." March 23, 1990.

PART IV:
UNDERSTANDING MEAT MARKETING

Chapter 15

Meat Packing

The preceding chapters document the significant changes seen during recent years in the livestock *production* sector. No less affected over that period was the meat processing and distribution/retailing sectors. That group of firms was first affected by several rounds of technological change, along with an apparent shift in consumer taste for red meats. This chapter documents the adjustments made at the packer level and the impacts on livestock producers. Subsequent chapters document those changes/implications through other levels in the food distribution system.

TECHNOLOGICAL CHANGE, CONCENTRATION, AND SIZE ECONOMIES

In Chapter 3 we left the meat packing firms at the time of their "coming of age," dating to the Consent Decree of 1921. That Decree, while directed to specific actions and organizational arrangements, was clearly intended to reduce the dominance of the major firms (see Chapter 11). Indeed, at about that time those companies reached the near apex of their control over red meat supplies. In 1920, the top four controlled 49 percent of cattle and 44 percent of hog slaughter (but not the same four firms for both species). From that time, control over sales — referred to as market concentration by economists — declined to the 20 percent range for all cattle by the 1970s (Table 15-1). The decline was less the result of the Decree and more the response to both declining transport costs and mechanical refrigeration. These twin technological changes allowed packing to move nearer the sources of supply and, incidentally, to specialize in one species.

Beginning in the mid-1970s the trend in cattle slaughtering reversed, and increasing concentration was again apparent. Four-firm concentration rose to 31 percent by 1982. The leading firm that year, IBP, killed 16

TABLE 15-1. Concentration in Meat Packing by Livestock Type, Selected Years, 1920-1988[1] (Source: USDA, published in Nelson 1985, Table 1 and AMI 1989, p. 17)

Year	Boxed Beef	All Cattle	Steers and Heifers	Calves	Hogs	Sheep
1920		49.0		34.4	43.8	61.8
1930		48.5		45.5	37.5	68.8
1940		43.1		45.6	44.3	66.1
1950		36.4		35.4	40.9	63.6
1955		30.8		36.6	40.6	60.1
1960		23.5		29.0	34.9	54.7
1965		23.0		32.4	35.2	57.8
1970		21.3		23.8	31.5	53.1
1971		21.4		21.6	31.8	53.2
1972		22.3		21.9	31.6	54.7
1973		22.8		23.7	32.9	51.8
1974		20.9	26.0	23.5	32.7	55.7
1975		19.3	25.3	24.6	33.1	57.5
1976		19.6	25.1	25.3	32.3	53.6
1977		20.2	26.9	25.1	32.6	52.9
1978		22.9	29.6	28.2	34.4	56.4
1979	51	27.9	34.5	30.6	33.7	64.2
1980	53	28.4	35.7	30.7	33.6	55.9
1981	57	31.4	39.6	28.6	33.3	52.5
1982	59	32.0	41.4	27.8	35.8	43.6
1983	61	36.0	46.6	28.2	29.1	43.8
1984	62	37.2	49.5	28.8	35.0	48.9
1985	62	39.0	50.3	31.1	32.2	51.2
1986	68	42.3	55.1	26.5	32.5	54.3
1987	82	53.9	67.1	30.4	36.6	75.1
1988[2]/	NA	56.6	69.7	32.6	33.5	76.1

1/Percent share of commercial slaughter by four largest firms.
2/Preliminary
NA: Not Available

percent of the total federally inspected steers and heifers in 1980 (Cook 1981). This change, too, was partially technology driven by the change to boxed beef. Boxed beef refers to the reduction of a carcass to a number of primal and sub-primal cuts that can be placed in boxes (Figure 15-1). There are several real economies to this practice. Workers perform simple, repetitive tasks in a disassembly line, increasing labor productivity; additionally, more of the drop, bone, and fat is retained at the packing plant. This has the dual benefits of reducing weight and trucking costs for consumable meats, and making the retained by-products more valuable. For example, fat can often be kept clean enough for human, rather than industrial, use. Sanitation and shelf life are improved by the use of vacuum-packed cryovac bags. And the program is popular with retailers who can order only the carcass parts needed while reducing often high-priced unionized labor at the store level (Hall and MacBride 1980). One researcher places the attributable savings for boxing beef at 3.5 cents per pound in 1978 (Case and Co. 1978). Of course, these are system-wide savings as packer costs rise due to the greater inputs of labor and materials.

Safeway Stores, a major operator of supermarkets, is credited with inventing boxed beef in 1960. However, its use did not become commonplace until an aggressive new firm, IBP (or as it was first known, Iowa Beef Packing, Co.), built new plants to utilize the technology in the 1960s and 1970s. From that point, the use rose rapidly for beef, from 12 percent in 1971 to 28 percent in 1975 to 58 percent in 1982. These percentages refer to federal slaughter (Hall and MacBride 1980, Table 1; Nelson 1985, Table 9). (Federal slaughter accounts for about 95 percent of total steer and heifer slaughter.) This rapid change, and the ascendancy of IBP, were made possible by two factors other than the efficiencies of handling beef in boxes. These factors are size economies and labor costs.

Size Economies and Concentration

Size economies (often called scale economies) refer to the decline in average costs that is associated with increasing size. Costs may also rise for larger operations, but such cases are not often observed in practice. Size economies may apply separately at the plant and the firm level when more than one plant is operated. Here we shall focus on plant-level size economies only.

One suspects that size economies are important since the total number of packing plants has declined markedly over the past 15 years (Table 15-2). Yet, this masks the real trend, for large cattle plants (50,000+ head) fell from 153 in 1979 to 106 in 1988. Meanwhile, the *average*

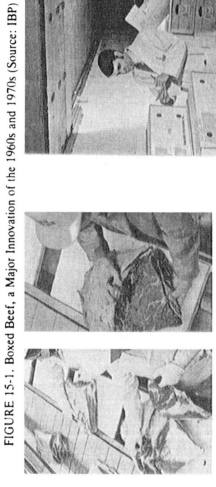

FIGURE 15-1. Boxed Beef, a Major Innovation of the 1960s and 1970s (Source: IBP)

TABLE 15-2. Commercial Livestock Slaughtering Establishments, January 1, 1972-1989 (Source: USDA, P&SA 1989, Table 26; AMI 1989, p. 16)

Year	Federally Inspected	Non-Federally Inspected	Total
1989	1,364	3,325	4,689
1988	1,387	3,453	4,840
1987	1,483	3,523	5,006
1986	1,544	3,701	5,245
1985	1,608	3,853	5,443
1984	1,666	3,982	5,558
1983	1,652	4,037	5,689
1982	1,688	4,648	5,736
1981	1,542	4,330	5,872
1980	1,627	4,399	6,026
1979	1,682	4,445	6,127
1978	1,701	4,434	6,135
1977	1,687	4,454	6,141
1976	1,741	4,514	6,255
1975	1,485	4,602	6,087
1974	1,437	4,440	5,877
1973	1,364	4,627	5,991
1972	984	5,172	6,156

number of head killed per plant in large plants (50,000 + cattle or 100,000 + hogs) rose from 141,800 and 521,400 in 1975, to 222,081 and 772,600 in 1983, to 295,472 and 1,216,567 in 1988, respectively—over a 200 percent increase (AMI 1984, p. 15; 1989, p. 18). Moreover, the number of plants in the largest size category—500,000 cattle and 1.0 million hogs annually—rose to 19 and 33, respectively, and constituted nearly 50 percent of total commercial cattle slaughter and three-quarters of hog slaughter in 1988. This was carried out in about 1.5 and 2.9 percent of plants by number, respectively (AMI 1989, p. 18). To be sure, these numbers are affected by numerous factors including inspection regulations in the mid-1960s (see Chapter 11), but some real size-related savings also seem to be operating.

Computing just what those economies are is difficult because the information is carefully held as proprietary. According to industry estimates for cattle, the lowest cost firm had costs in the $18 to 20 per head range in the early 1980s. Sergland (1985) estimated a minimum cost of $19.32 per head at 1,368,250 annual slaughter. Other efficient firms operated at up to $23 per head, or about 2.4 head per man-hour. These are "in-cooler" costs only, and were achieved in the largest existing plants with annual capacities of some 500,000 head (*Meat Industry* 1981). Other economies beyond the kill floor are associated with on-site waste water treatment and efficient use of by-products, including in-plant hide tanning and even the extraction of dental pulp from the teeth. The assembly costs for supplying animals to such large plants limit size economies. In the final analysis, actual costs must be determined on a site-specific basis.

Despite these seemingly overwhelming cost advantages of large plants, smaller plants can, and do, coexist through lower overhead in depreciated facilities, paying lower wage rates and livestock prices, or providing some specialized service that is not feasible for very large plants. As an example of the cost penalty, beef plants operating in the 1,500 to 6,000 head per year range (certainly minuscule by comparison with the industry giants) had estimated per-head costs of about $55 to $80 in 1979 (Figure 15-2). Seemingly, the industry can accommodate such relative high-cost operations only in specialized locales or market "niches." Size economies in boxing operations are reportedly greater, but smaller plants may sell carcasses to larger companies where they are broken down for wholesale distribution. As a result, four-firm concentration in box beef sales is higher yet, about 66 percent in 1982 (Nelson 1985, Table 10).

Mergers and Acquisitions

The continued operation of smaller plants aside, the recent cost pressures placed on the packing industry, especially with beef, have caused major disruptions including many firms closing up or selling out. Smaller firms tend to close, as evidenced by the declining numbers of packing plants (Table 15-2), while major ones change ownership. Major recent ownership changes are summarized in Table 15-3. As can be seen from this information, the recent trend in concentration (as opposed to plant numbers) in meat packing is largely attributable to mergers and acquisitions. Since 1987, the steer and heifer slaughter industry has been dominated by three firms — IBP, ConAgra, and Excel (Cargill) — the result of merger activity leading to an "unprecedented" rate of concentration increase (Marion 1988). Hog packing is slightly less concentrated. Since these involve multiple plant operations, they are generally unrelated to plant-level size economies. Many economists, in fact, believe that multiplant economies are modest compared to intra-plant economies (Scherer et al. 1975). Therefore, while increasing plant-level size economies during the 1970s and 1980s have been a major driving force behind rising concentration, other factors are also operating.

Cooperative Packers

Also disappearing during this period were many of the cooperatively owned packing plants. In 1975, cooperatives slaughtered .8 percent of all cattle and 2.3 percent of all hogs (Torgerson 1978, p. 16). Subsequent

FIGURE 15-2. Estimated Short- and Long-Run Average Cost Curves for Very Small Beef Packing Plants (Source: Durst and Kuehn 1979, Figure 1)

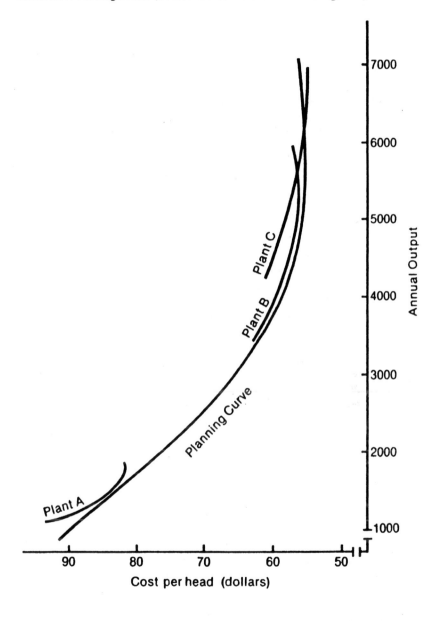

TABLE 15-3. Selected Ownership Changes in the United States Meat Packing Industry During the Late 1970s and 1980s (Source: USDA News, various issues)

1979	Cargill, Incorporated purchased MBPXL, renaming it Excel Corporation.
1980	Roth Packing Company employees received 60 percent ownership in the firm in exchange for wage and benefit concessions.
1981	Occidental Petroleum acquired Iowa Beef Processors, renaming it IBP.
1981	Swift Independent Packing Company (SIPCO) was spin-off from Esmark.
1981	LTV Corporation spin-off Wilson Foods.
1983	Armour Food Company was sold by Greyhound to ConAgra
1983	Wilson Foods, the then-leading pork packer filed for Chapter 11 bankruptcy, largely as means of removing itself from its main labor contract.
1987	ConAgra acquired F.A. Miller, Manfort and Swift Independent.
1987	Excel acquired Sterling Beef.

sales of packing plants by cooperatives have reduced this share somewhat. While there are many individual reasons for cooperative withdrawal from the sector, the general trend may be attributed to high capital investment, low operating profits, and difficulty sometimes experienced in limiting the amount of costly services provided. Cooperatives also tend to enter processing operations as a means of preserving a market outlet when a proprietary firm threatens bankruptcy. The hurried nature of the planning, along with other factors, can be detrimental to management for a considerable period.

LABOR ISSUES AND STRUCTURAL CHANGE

While much of the recent change in the structure of the beef packing industry is traceable to size economies and mergers/acquisitions, labor also played an overriding role. This was true because IBP, in particular, not only changed standard industry operating procedures in the widespread adoption of beef boxing technology, but further insisted on using non-unionized labor. Unions, following hard-fought campaigns, dominated labor supply in the larger plants. Firms adhered to a "master agreement" which assured that companies paid comparable wages, even if those were elevated compared to similar, non-unionized firms. In 1981, for example, Armour complained of labor costs 40 to 60 percent higher than those for such firms as IBP and MBPXL. With labor constituting the major non-product cost in the packing plant, or about 40 percent of the packer margin (AMI 1989, p. 46), IBP had a wild advantage over competitors tied to the master agreement. This period characterized IBP's greatest growth, but it could not last indefinitely. Following a protracted and often bitter strike at the Dakota City plant, IBP accepted unions in at least some of its plants. Concurrently, competitors found means of reducing their labor penalty. Many demanded and received several rounds of wage and work rule concessions from the union. Others achieved the same objective by closing and reopening under a different name or following a bankruptcy proceeding that voided the contract. By 1983, only an estimated one-third of all organized meat plant workers were receiving the "master agreement" rate (Cook 1981; Lublin 1983).

These changes took place in an era when packing plant labor productivity was rising 49 percent from 1960 to 1970, and an additional 20 percent from 1970 to 1980. Employee numbers, as a result, were in decline; a 20 percent loss of production workers was recorded from 1965 to 1988 (AMI 1989). In part, these productivity increases are attributable to the shift to larger plants.

As a direct result of these pressures on the unionized sector, beef packing labor wages in the mid-1980s were approximately equal for unionized and non-unionized beef plants at about $6.50 per hour, excluding fringes (McDowell and Lesser 1987). Across the meat packing industry, wages averaged $8.47 an hour in 1988 (AMI 1988, p. 48). In the final analysis, meat consumers are the major beneficiaries of this change in labor supply terms, and union members the major losers. Plant managers continued to seek wage and other concessions from labor, but with less success. In part, this request is based on the insistence that union work rules reduce labor productivity, but a study comparing kill floor productivity of unionized and non-unionized workers found no statistically significant difference between the two groups (McDowell and Lesser 1987).

Following the success in the beef business, IBP boasted loudly and repeatedly that hogs were next (Williams 1983). However, that was, and probably is, not to be. In part, technical reasons can explain the differences. Hogs are smaller and have a higher proportion of the cuts channelled to processed, as opposed to fresh, products. Hence the need for boxing is reduced. But, just as important, competitors were well aware of IBP tactics and operated in anticipation, rather than in response, to them. The 1980s, then, have seen union wage concessions and ownership changes which parallel the beef sector a decade earlier, but ownership has remained less concentrated than for beef (Table 15-1).

PERFORMANCE

Performance refers to the results of an industry's activities measured in terms of profits, margins, innovativeness, etc. (See Scherer 1980, Chapter 9.) Here we shall be concerned with performance defined loosely in terms of profitability. In brief, is the industry unduly profitable compared to the minimum returns required to keep it involved in these activities and to the detriment of producers and/or consumers?

The reported profitability of the meat packing sector of less than one percent of sales is certainly low when compared to other manufacturing enterprises (Table 15-4). For example, 1980 meat packing earnings as a percent of sales were .98 (and down to .30, before tax, in 1987) compared to a value of 4.87 percent for all manufacturing in 1980 (Nelson 1985, Table 13; AMI 1989, p. 46). But the sales dollar is not a useful numerair since the value added varies from sector to sector. For example, the livestock feeder who has an annual throughput of 50,000 head must have a net profit per sale dollar higher than the plant killing 500,000 head a year. A more reflective comparison is returns to equity, the investment of the owners (including stockholders) in the firm. Here, meat packing is more comparable to all manufacturing, 6.21 compared to 6.68, respectively, in 1980. An even better comparison is returns on net returns which in 1980 was 12.1 and 13.5 percent, respectively (Nelson 1985, Table 13).

There are two basic ways to empirically analyze the impacts of high packer concentration on producers and consumers. One is by examining industry selling prices and profits. Hjort (1980), for example, found no statistical significant "overcharges" attributable to the meat packing industry for the period 1972-75. (See also Connor 1980.) Using another approach for a longer time period, Hall et al. (1981) looked at the rate beef price changes were reflected from the farm gate to the wholesale level. Considering the period of 1960 to 1978 (a time characterized by strong inflationary trends, especially in food as an aftermath of the Russian wheat sale of 1972), they found 89 percent of a price change was passed through in the first month and virtually 100 percent by the second month. This type of study is always somewhat subjective as it requires a judgement about what is sufficiently rapid price adjustment for the industry to be considered competitive, but recognizing this, the results are generally consistent with a competitive beef sector. Strengthening this conclusion was the fact that no significant difference in rate of adjustment was found for price decreases compared to increases (1981, pp. 21-22). Hahn, writing in 1989, did find an asymmetrical response to price increases for both beef and pork. "Prices for beef and pork tend to increase faster than

TABLE 15-4. Four-Firm Concentration in the Ten Leading Steer and Heifer Slaughtering and Hog Slaughtering States, 1985 (Source: P&SA data reported in Ward 1988, Table 1.3)

Steers and Heifers		Hogs	
State	Four-Firm Concentration	State	Four-Firm Concentration
Kansas	88.4	Iowa	56.1
Texas	84.7	Illinois	86.4
Nebraska	72.3	Michigan	99.0
Colorado	99.9	Minnesota	99.9
Iowa	84.1	Nebraska	99.9
Illinois	89.6	Virginia	98.5
California	53.4	Indiana	99.0
Minnesota	99.1	Ohio	78.1
Washington	99.5	South Dakota	100.0
Idaho	98.5	Missouri	99.6

they decrease" (1989, p. iii). However, he rejects market power as a causal factor because ". . . the evidence here suggests that asymmetry works in producers' favor" (1989, p. 14).

Some of these studies are a little dated in the sense that concentration in meat packing has increased in the intervening time period, but there is no real evidence that the conclusion of a generally competitive industry would change if the studies were repeated today. In a more descriptive evaluation dating to the early 1980s, Nelson noted that the rapidly increasing concentration ratios in the sector were alarming to many, but, by most available standards of performance, the results were "consistent with aggressive competitive performance" (1985, p. 34). [See also Schnittker Associates (1980) for an industry-sponsored evaluation that reaches the same general conclusion.] One explanation for this is the general overcapacity that exists in the industry as a result of declining consumption. In a high fixed-cost, low-margin industry, firms are likely to compete keenly to maintain throughput.

A second approach is an examination of prices paid to producers. This form of analysis recognizes that procurement is a local activity; the normal supply region is said to be a 100-mile radius around a plant. Within that locally defined region, the concentration of buyers will often be far higher than at the national level as shown in Table 15-1. Table 15-4 is a listing of concentration on the state level in 1985, while Figure 15-3 shows averaged concentration for steer and heifer slaughter in 13 regional markets from 1972 to 1986. These values can be described as extremely high in most cases. Miller and Harris, and Quail, Marion, and Marquardt found that hog and fed cattle prices, respectively, were negatively related to the concentration of packer buyers at the state and regional levels (summarized in Connor et al. 1985, p. 238).

The impacts of individual plants are perhaps more evident in their absence, that is, when a local plant closes. In an evaluation of six plant closings from 1978 to 1981, Hayenga, Dieter, and Montoya (1986) found no significant price impacts for four hog slaughtering plants located in the Corn Belt. Two plants on the fringe of that area (Wisconsin and Oklahoma) did register declines, but only after a lag of a few weeks. The dependence of producers in lower concentration producing areas on individual plants echoes a 1965 analysis of Louisville, Kentucky, where prices fell 1.5 percent following the exit of a plant and the subsequent emergence of a dominant firm (Love and Shuffett 1965). Ward (1988, Chapter 7) provides an admiral summary of the recent analyses, the synopsis of which is reproduced in Table 15-5. Based on that review, he concludes, "Available evidence suggests that number of buyers is posi-

FIGURE 15-3. Average Four-Firm Concentration of Steer and Heifer Slaughter in 13 Regional Markets, 1971-1986 (Source: Marion 1988, Exhibit 4)

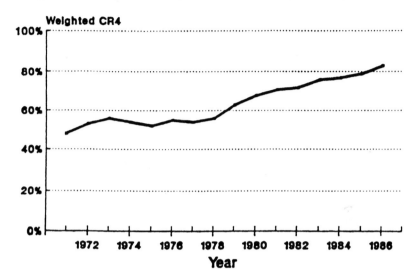

tively associated with prices paid for livestock. Adding a buyer tends to increase price and removing a buyer tends to lower price. Increasing concentration on high levels of concentration seems to negatively affect fed cattle price levels when measured on a state or regional basis"(1988, p. 168). Conceptually, the same applies to hog prices, but concentration has not been as great in that industry.

Marion (1988) provides one estimate of the magnitude of the impact caused by current concentration: 1/2 to 1 percent of the selling price. While small, it amounts in aggregate to ". . . at least $50 million annually that cattle feeders lost because of ineffective competition in cattle buying markets." As a remedy, Marion opposes any further mergers/acquisitions in beef and pork packing. Where concentration is already high, electronic markets are one means of attracting additional buyers, although efforts to date have not always been successful (see Chapter 13).

SYNOPSIS

Meat packing has changed in many critical ways from the earlier period as typified by the industry just prior to the Consent Decree of 1921. Largely due to technical changes, four-firm concentration fell to a low of below 20 percent for all cattle by mid-1970. Boxed beef, which was intro-

TABLE 15-5. Summary of Structure-Price Studies in Livestock Procurement and Meat Sales (Source: Ward 1988, Table 7.1)

Study	Empirical Objective	Outcome
MACRO STUDIES		
Parker and Connor, 1979	Estimate monopoly price overcharge for meat products	Meat Packing 0% Meat Processing 1.6-5.6%
MICRO STUDIES		
Love and Shuffett, 1965	Estimate price impact from an abrupt change in hog buyer structure at a terminal market	Significant price decline relative to two comparison markets
Aspelin and Engelman, 1966	Estimate price impact from a packer slaughtering its own fed cattle instead of buying cattle from a nearby terminal market	Significant price decline relative to other markets
Holder, 1979	Determine price impact from instituting a teleauction for slaughter lambs (i.e. increasing number of buyers)	Significant price increase relative to a more concentrated buying structure
Ward, 1981	Estimate relationship between fed cattle price and number of beef packers bidding on cattle	Significant price increase associated with increasing number of buyers in one of four markets; No significant impact in others
Menkhaus, et al., 1981	Estimate the relationship between fed cattle prices and beef packer concentration in selected states	Significant negative relationship across states for two separate years

378

Study	Purpose	Findings
Ward, 1982	Determine whether beef packers paid different prices for fed cattle; and examine market shares-prices relationship	Significant price difference between at least two buyers in six of twelve markets; no positive or negative relationship between buyer market shares and prices paid
Ward, 1983	Determine impact from porkpacking plant closing on prices at a nearby terminal market	Significant price decline immediately relative to three comparison markets; one year later, significant price decline relative to one comparison market
Ward, 1984	Determine price impact from number of lambpackers bidding at a slaughter lamb teleauction	Significant absolute and relative price increase a number of bidders increased
Rhodus, et al., 1985	Determine price impact from instituting an electronic market for slaughter hogs	Significant price increase relative to two comparison markets; Significant negative relationship between number of buyers and price
Ward, 1985	Determine beefpacker market shares-fed cattle prices relationship	Significant price difference between at least two beefpackers in all equations; Two positive and one negative market shares-prices relationship in three of four equations
Hayenga, et al. 1986	Determine price impacts from six porkpacking plant closings and two reopenings	Significant price declines for two to six weeks when plants closed in four of six markets relative to comparison markets; Significant price increase in one of two markets when plants reopened
Marion, et al., 1988	Estimate the relationship between fed cattle prices and beefpacker concentration in 13 regional markets	Significant negative relationship across regions over the period 1971 to 1980

duced about 1960, permitted the establishment of IBP (now the industry leader). It also led to many economies at the plant and, especially, at the distribution and retail levels. Efficient plant size grew accordingly to the point where, in 1988, 1.5 percent of plants slaughtered 49 percent of all cattle that year. Costs were further reduced by an IBP-led assault on the union master agreement so that union wages were frozen and, in some cases, even reduced.

Hog plants were subject to the same general economic forces but because of the degree of further processing – and hence the reduced need for boxing – they never reached the same level as beef.

A second level of change occurred in the 1980s when Excel (Cargill) and ConAgra corporations grew through acquisitions, pushing 1988 steer and heifer four-firm concentration up to nearly 70 percent. Similar activity occurred for hogs, but not to the same degree. This was, in part, due to the specialization in slaughter and further processing, meaning that some old-line firms like Oscar-Mayer spun off its slaughter operations.

The very high levels of steer and heifer concentration, particularly at the local level where procurement is done, has not been found to affect wholesale prices. However, producer prices are, to a small degree, negatively correlated to buyer concentration.

Study Questions

1. Describe the factors leading to the variation over time in the share of the largest packing firms.
2. Identify the factors that led to the rapid adoption of boxing operations for beef but not for pork.
3. What are the reasons for cooperatives' limited penetration into red meat packing? Are any factors likely to change in the foreseeable future?
4. Suppose the "performance" of the meat packing sector was judged as poor. What implications would that have for producers and consumers as groups, and for individual producers?

REFERENCES

American Meat Institute. "Meatfacts." 1982, 1983, 1984, and 1989.
Case and Co., Inc. *An Updated Look at the Beef Distribution System.* 1978.
Connor, J. M., R. T. Rogers, B. W. Marion, and W. F. Mueller. *The Food Manufacturing Industries: Structure, Strategies, Performance, and Policies.* Lexington, MA: Lexington Books, 1985.
Connor, J. M. "The U.S. Food and Manufacturing Industries: Market Structure,

Structural Change, and Economic Performance." USDA, Econ. Stat. and Coop. Service, AER 451, March 1980.

Cook, J. "Those Simple, Barefoot Boys from Iowa Beef." *Forbes*. June 22, 1981

Durst, R. L. and J. P. Kuehn. "The Feasibility of Establishing Low Volume Beef Slaughtering-Processing Plants in West Virginia." West Virginia Agricultural and Forestry Experiment Station, Bulletin 666, May 1979.

Hahn, W. F. "Asymmetric Price Interactions in Pork and Beef Markets." USDA, Economic Research Service, Tech. Bull. No. 1769, December 1989.

Hall, L. and M. MacBride. "Boxed Beef in the Marketing System: A Summary Appraisal." Cornell University, Department of Agricultural Economics, A. E. Res. 80-14, July 1980.

Hall, L. L., W. G. Tomek, N. L. Ruther, and S. S. Kyereme. "Case Studies in the Transmission of Farm Prices." Cornell University, Department of Agricultural Economics, A. E. Res. 81-12, August 1981.

Hayenga, M. L., R. E. Dieter, and C. Montoya. "Price Impacts Associated with the Closing of Hog Slaughtering Plants." *North Central J. Agr. Econ.* 8(1986):237-42.

Hjort, H. W. "Small Business Problems in the Marketing of Meat and Other Commodities: Part 7 — Monopoly Effects on Producers and Consumers." Testimony before the Committee on Small Business, House of Representatives. U.S. Government Printing Office, 1980.

Love, H. and D. Shuffett. "Short-Run Price Effects of a Structural Change in a Terminal Market for Hogs," *J. of Farm Econ.*, 47(1965):803-12.

Lublin, J. S. "Effort to Save Pay Scales in Meat Packing Brings Lewrie Anderson Many Spats, Not All with Firms," *The Wall Street Journal*, August 4, 1983, p. 46.

Marion, B. W. "Changes in the Structure of the Meat Packing Industries: Implication for Farmers and Consumers." Testimony before the Subcommittee on Monopolies and Commercial Law of the Committee on the Judiciary, House of Representatives. Serial No. 67, May 11, 1988.

McDowell, S. and W. Lesser. "The Effect of Unions on Productivity: An Analysis of the Cattle Kill Floor," *Agribusiness*, 3(1987):273-80.

Meat Industry. "Reader Opinions — What are Your Slaughter and Boxed Beef Costs." Oman Pub. Co., Mill Valley, CA, May 1981.

Nelson, K. E. "Issues and Developments in the U.S. Meat Packing Industry." USDA, Economic Research Service, National Economics Division, 1985.

Scherer, F. M. *Industrial Market Structure and Economics Performance*. Chicago: Rand McNally College Publishing Co., Second Edition, 1980.

Scherer, F. M., A. Beckenstein, E. Kaufer, and R. D. Murphy. *The Economics of Multi-Plant Operation: An International Comparisons Study*. Cambridge, MA: Harvard University Press, 1975.

Schnittker Associates. "An Economic Analysis of the Structure of the U.S. Meat Packing Industry." Washington, DC, November 11, 1980.

Sergland, C. J. "Cost Analysis of the Steer and Heifer Processing Industry and

Implications on Long Run Industry Structure." Unpublished PhD dissertation, Oklahoma State Univ., December 1985.

Torgerson, R. E. "The Future Role of Cooperatives in the Red Meats Industry." USDA, Econ. Stat. and Coop. Service, Mkt. Res. Rpt. 1089, 1978.

USDA, Office of Information. "News." Various issues.

Ward, E. E. *Meatpacking Competition and Pricing*. Blacksburg, VA: Research Inst. on Livestock Pricing, July 1988.

Williams, W. "An Upheaval in Meat Packing," *The New York Times*, June 20, 1983, pp. D1, D6.

Chapter 16

Processing and Distribution

Simply defined, firms involved in processing and distribution stand between the packer (Chapter 15), who supplies carcasses and boxed products, and the retailer-foodstore and foodservice operator alike (Chapters 17 and 18), where ultimate contact with the consumer is made. That said, there is nothing simple about this collection of firms. They serve a myriad of specialized activities from physical handling and processing to distribution to sales, a function carried out by any number of individual, specialized firms. Or the activities may be combined in one operation that slaughters, processes, and distributes. This sector, situated as it is between packers and retailers, has been affected by structural and technological changes at both ends of the system. In a general sense, the net trend has been toward fewer and larger firms, with most volume now handled directly by packers and supermarkets, leaving the specialized firms to serve the market niches. But the smaller operations do exist and, as a group, serve important functions for the sector. This chapter describes the activities of processors and distributors, and their implications for the livestock sector.

TYPES OF FIRMS

There are a number of ways to classify the types of firms in the sector. One classification is as follows (after Dietrich and Williams 1959):

- Packer Wholesalers: Sales offices associated with packers and centrally located.
- Packer Branch Offices: Field sales outlets operated by packers close to consumption points but not physically handling product.
- General Wholesalers: Independently owned and operated meat sales operations typically involved in the purchase and breaking down of carcasses.

- Fabricators or Purveyors: Specialize in the further preparation of fresh products such as single serving portions for restaurants and other institutional markets across the price spectrum.
- Processors: Independent firms specializing in the processing of fresh meats, especially pork, into prepared products.
- Brokers: Also known as agents or commission agents, the independent operators acting as sales agents on a commission basis without actually handling the product.
- Jobbers: Known alternatively as peddlers and truck jobbers, distribute on a small scale using trucks as the basis of the business.

Transportation is an important function provided by most of these classes of firms.

Here we shall make an initial break between processors and other firms. To a degree, this division is artificial since many packers are involved in processing while many processors sell and distribute their products. Nonetheless, in terms of their characteristics and implications, there is a clear distinction among processors (particularly large processors) and purveyors, wholesalers, and jobbers.

PROCESSING

Structure

The food manufacturing sector, **in total**, is one of the largest in the U.S., with 1988 sales of $342 billion, up 6.5 percent over the year. It also advertises heavily as a sector — totalling $8.5 billion in 1988 — as well as tends to be more profitable than other manufacturing (Figure 16-1) (USDA, ERS 1989, p. 9).

Food manufacturing, however, consists of 47 SIC industries of which two ("meatpacking" and "sausage and other prepared meats") are clearly related to the red meats sector. While recognizing that meat packers are often heavily involved in meat processing, the distinct processing sector (sausage and other prepared meats) is itself subdivided in three five digit SIC industries.[1]

---

1. Industries are identified by SIC numbers standing for Standard Industrial Classification. The smaller the number of digits, the more aggregate the measure (e.g., 20 refers to the food sector).

SIC	Description
20116	Pork, processed or canned
20117	Sausage and similar products
20118	Canned meats

Reported concentration in these industries remains modest and relatively stable over more than 30 years (Table 16-1). There are several reasons for this. One is that regional brands have continued to be important and foodservice, where branding is far less significant, is a major user of beef and pork (see Chapter 18) so that numerous smaller firms continue to survive (Marion 1988). Second — and this is more a measurement issue — sales are allocated to industries according to the preponderance of sales. Therefore, if a firm sold 75 percent beef and 25 percent pork, it would be classified as all beef. Among the large, diversified, food manufacturing companies, the allocation of products is even more complex so that SIC figures are not necessarily very representative.

Diversification is becoming more commonplace among the larger entities, especially with pork. Figure 16-2 is a partial listing of ownership of major meat processing firms by large food conglomerates. Marion (1988)

FIGURE 16-1. Net Profit Rate of Food Manufacturers and All Other Manufacturers, 1951-1955 to 1986-1987 (Source: Mueller 1988, Figure 1)

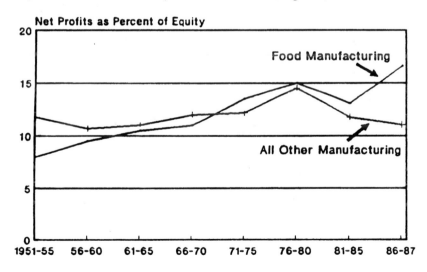

TABLE 16-1. Market Structure Data for Meat Processing Firms by SIC, 1954-1982 (Sources: Census Manufacturers, various years; Connor et al. 1985, Table B-2A)

SIC	Description	CR4						ADS
		1958	1963	1967	1972	1977	1982	
20116	Pork, processed and cured	37	33	30	34	31	34	.24
20117	Sausage and similar products	28	27	27	26	29	36	.62
20118	Canned meats	62	58	57	68	59	74	.99

Notes: CR4 – sales share of the leading four firms in industry
ADS – average advertising to sales ration for 1967, 1972 and 1977

FIGURE 16-2. Ownership of Meat Processing Operations in 1988, Selected Firms (Sources: AMI 1989; Marion 1988)

Sales Rank	Firm	Processing Operations Owned
2	ConAgra	Armour
4	Sara Lee Corp.	Kahns, Jimmy Dean, Hillshire, Brynans, 16 in total
10	Beatrice	Swift-Eckrich

refers to this as a "bifurcated" industry, with a growing separation of pork slaughter and processing, something he attributes to the higher profits in processing. Presently, old-line pork processors like Hormel and Oscar Mayer are selling off slaughter operations and concentrating on processing.

Costs, Margins, and Efficiency

The 1987 gross wholesale marketing margin, incorporating slaughter, transport, storage, and carcass-breaking, came to 36.4 cents per pound (trimmed for retail) for beef, and 42.7 cents per pound for pork, generally steady over a decade except for the costs of boxing beef. Measured as a percentage of the retail price, the wholesale margin is 15 percent and 23 percent for beef and pork, respectively (Table 16-2). Pork is higher in absolute terms because of the greater degree of processing, an activity that is classified as a wholesaling function. In percentage mark-up terms, pork is yet higher because of the lower average retail value of the products, while purchased wholesaling services (labor, transport, etc.) cost the same.

In total, transportation is a small cost, while the share for slaughtering operations has actually declined. This reflects the cost-cutting and efficiency-enhancing measures adopted by packers in recent years. Wages account for about one-third of the total (AMI 1989, p. 44).

Gross margins do not give a clear indication of the efficiency with which firms operate. Note, for example, the effect of the degree of processing and product value on the absolute and percentage margin. One factor that does provide insight is the number of ownership exchanges existing in the system between packer and retailer. If it is assumed packers and retailers are seldom the same firm, the minimum number of exchanges will be one, and this number will rise as the product exchanges hands more times. The implication is that more participants are less efficient as each must receive a margin, but, in reality, the larger number of participants may indicate specialization in handling which may be done more efficiently by a particular kind of firm. Recognizing the possible

TABLE 16-2. Processing and Distribution Costs for Choice Beef and Pork, 1978 and 1987 (Sources: Dunham 1980, Table 13; 1988, Table 22)

Activity	Year/Species			
	1978		1987	
	Beef	Pork	Beef	Pork
	-----cents per retail pound-----			
Slaughtering & Processing	2.5	28.6	3.6	26.8
Transportation	2.7	2.5	3.8	3.5
Wholesaling (including delivery)	10.8	8.5	16.0	12.4
Carcass breaking	--	--	13.0	--
Total	16.0	39.6	36.4	42.7
% Retail Price	10.4	27.6	15.0	22.7

ambiguities of the data, case studies of meat distribution patterns in six major southern metropolitan areas around 1980 showed the average number of exchanges for red meats was about 1.3 for beef and pork, with a range across the markets of 1.2 to 1.8 (Baker and Duewer 1983, p. 9). There are no similar earlier studies that would give an indication of trends, but clearly the number of firms handling red meats is approaching the practical minimum. As is evident, specialty firms handle a small portion of the total volume of meat.

General profitability studies of food manufacturing are not very meaningful because of the mix of products involved. The American Meat Institute, however, conducts its own survey and reports the profitability of firms identified as "processors" and "packer/processors." For 1987, the net, before tax, earnings of these two categories were, 3.90 and 4.40 percent of sales, respectively, compared to .55 for packers (AMI 1989, p. 44). Net returns on sales are not a clear measure of profitability for they ignore differences in equity and value added. But the difference between the packers [a highly capitalized industry found to be generally competitive in the mid-1970s (Connor et al. 1985, Table 7-8)] and the processors is so great that these figures do support the contention that processing is more remunerative than slaughtering, thereby explaining much of the structural change in the industry.

Another performance characteristic is the level of advertising to sales. Processed meat products are far higher than meat packing (where levels are near zero), but far less, at .25 to .99 percent, than some food items like pet foods (6.55%) and breakfast cereals (11.57%) (Connor et al. 1985, Table B-2A).

Connor (1981) has suggested that the number of minor changes to food products introduced annually can be both a cost imposed by the system and an anticompetitive tactic. This is a pertinent issue in food manufacturing in general, for over 9,000 new products were introduced in 1989 alone, representing a continuation of the rising trend (Gallo 1990, Figure 5). No breakout is available for meat products, but processed meat products (especially for microwave use) are one of the areas of development and one of the areas of increased sales of meat products (Table 16-3). Overall, the proliferation of prepared meat products is probably a benefit to the meat sector.

Branding of Meat Products

The great bulk of meat products move from the packer in fresh form to be sold as general, commodity-type products at supermarkets and restaurants of various types. This is in counterdistinction from many products

TABLE 16-3. Sales of Processed Meat Products, 1983-1986 (Source: Harris 1988, p. 35)

Item	Foodstore sales			
	1983	1984	1985	1986
		Pounds per capita		
Canned				
Tuna	2.69	2.81	3.03	3.11
Beef stew	0.45	0.46	0.48	0.46
Frozen				
Dinners	1.54	1.90	1.92	1.86
Poultry entree	1.48	1.56	1.63	1.62
Italian entree	0.95	1.19	1.41	1.43
Ground beef	0.87	0.90	1.01	1.03
Pot pies	0.86	0.95	0.99	1.03
Breaded fish	0.84	0.84	0.88	0.89
Meat entree	0.57	0.57	0.65	0.69
Mexican entree	0.34	0.37	0.44	0.51
Sausage dinner	0.30	0.36	0.45	0.47
Oriental entree	0.21	0.36	0.42	0.40
Refrigerated				
Frankfurters	3.78	3.66	3.98	3.87
Bacon	3.50	3.41	3.48	3.25
Sliced lunch meat	2.90	2.98	3.20	3.21
Breakfast sausage	1.48	1.42	1.45	1.54
Canned ham	0.56	0.43	0.46	0.41
Total	23.32	24.17	25.88	25.78

sold in the United States which carry a brand label, whether of a well-known international firm, a smaller regional brand, or a supermarket's own "private" label. For a long time, the branding of fresh meat products was considered unprofitable, but Frank Perdue changed much of that in recent decades by establishing recognition for "his" poultry products throughout much of the country. Armour has had a long-standing consumer franchise for its turkeys, but they are frozen, not strictly fresh.

At various levels, the red meat industry has attempted to achieve what Perdue did. Wholesaling of standardized products like fresh meat is a notoriously difficult place to establish brand identity. IBP, however, has had some success in getting trade recognition for its Cattle Pack (Reg) brand. It is more a marketing program perhaps than a product, highlighting the need to rely on services in part to differentiate these products. As Professor Brunk (1971) noted in an address to the Livestock Merchandising Institute, ". . . the more marketing services one adds to commodities he sells the more he transforms a common commodity into a unique and distinguishable product." Among these services can be added reliability of service and consistency of quality.

At retail, several fresh meat branding programs have been attempted using such designations as "Silver Platter Pork" by the P&C Food Company, a regional supermarket chain. Oscar Mayer, a leader in processed pork products, has a small fresh meat program, while Rath Packing Company experimented with a "Rath Pork Cuts" program. Pork appears to be a primary target for these efforts as packers attempt to smooth the fresh product demand to harmonize with the need to produce meat for processing on a more regularized schedule.

Where these efforts have been most successful (although total volume remains trivial) is for specialty beef products known as "natural" or "lite." It has been argued that the opportunity for branding most fresh meat products is limited because the retailer does the final preparation and packing, eliminating the supplier label. Moreover, some supermarkets like to maintain a low meat price as part of a low price image or to identify the store based on the quality of its meats. However, imaginative solutions are possible as with Perdue Farms which attaches identification tags to poultry parts packaged in the store as well as adding a harmless feed ingredient that makes its chickens a recognizable orange color.

One possible source of differentiation is consistency of quality, especially when the packer retains full control by providing prepackaged (tray ready) cuts (Commodity Econ. Div. 1988, p. 14). Conversely, supermarkets may wish to offer branded items as an alternative to store brands, as is now done with Perdue and Holly Farms chicken. Through whatever form, opportunities for branded red meat appear to be increasing, although the level it will reach in 5 to 10 years seems, for the present, limited.

Regulation

An additional performance component of the meat processing sector is the degree of regulation it is subject to. In all, some 74 regulations (of a total of 84 for the livestock sector from producer to consumer) in 12 categories were catalogued for 1979. (Also, see Burke and Dahl 1985.) These ranged from regulations applied to all plants, such as safety and unemployment, to more specialized forms, including labeling and restrictions on soya fortification. A Task Force of the Council for Agricultural Science and Technology (1979) concluded that only 33 of these had a net social benefit, while 25 had a negative impact. The conclusions were strictly judgemental, as opposed to analytical, and can be debated. However, the results are consistent with ongoing complaints by the industry that it is over-regulated, thereby contributing to higher consumer and lower producer prices than would be necessary in a more rational regulatory environment. Certainly, livestock producers are no strangers to the regulation issue, and many have strong opinions on the matters of pest control, antibiotic treatment of feed, and other matters. But careful, impartial analyses of the net *social* benefits of such regulations are difficult to conduct and are limited in number.

As an example of economic evaluation of a regulation, consider an earlier proposal by the Food Safety and Inspection Service that processed meat and poultry products be sold on a *net* weight basis, that is, excluding free liquids and fats and solids absorbed by packaging material. This regulation was intended to assist consumers by allowing a more direct comparison of prices based on the useful product weight. In addition, bulk buyers would have a more objective standard for comparing alternate suppliers. Smaller buyers were expected to benefit more due to the infeasibility of direct product testing.

In this instance, the costs of the proposed legislation were relatively easy to compute. These costs consisted of $59 to $116 million for industry to install and operate a mandated quality control system. Enforcement at the local level would cost at least $421,000 for new draining equipment, plus possible additional personnel. On a per pound of product basis, the total is about .05 cents per pound. Is it worth it? This is complicated to answer because there is little information on (a) any abuse of the existing regulations, and (b) consumers' use of the new information if provided (Handy, Sexauer, and Weingarter 1979). Indeed, only 20 percent of respondents in a 1977 study of food shopper opinions ranked "drained weight" as extremely important (Smith, Brown, and Weimer 1979). Perhaps consumers felt that the relative liquid levels could be judged by eye so that the system was self-policing, or that all products were the same.

At another level in the processing system is the controversy over mechanically deboned meats (MDM). Mechanical deboning permits the salvaging of more meat at a lower cost than is possible with hand methods, thus providing a significant potential benefit to producers and consumers since meat is used for human consumption rather than inedibles. Estimates place the quantities at about three percent of beef and hog carcasses (16 pounds for beef and 4 pounds for pork) which could be added to the 17 percent of pork products that are further processed and sold as sausage and related products (Field, quoted in Bullock and Ward 1981, p. 5). Regrettably, the process also introduces powdered bone and other "foreign" products into the meat, raising a health issue for some on restricted calcium diets and a content information issue for all. In response, the USDA in 1978 instituted a regulation specifying (a) the maximum content of MDM, (b) the labeling requirements concerning the presence of MDM and the content of powdered bone, and the range of protein, fat, and calcium content. That labeling requirement has, in the opinion of the industry, made the product unattractive so that the product was not produced in significant quantities. Bullock and Ward attributed a $500-million annual social cost to these regulations. The estimate does assume no harmful impacts for MDM consumption (1981, p. 2), but no assessment is made of the consumer's right/need to know in an easily observable manner the content of food products. Significantly, in 1981 the final labeling requirement was reduced to merely, "mechanically deboned (species)," a change which was affirmed by the U.S. Court of Appeals for the District of Columbia in 1984.

Currently under debate is the appropriate labeling of processed products, including meat. Presently, products must include a list of ingredients in descending order of content. Nutritional labeling is optional unless a nutritional claim is made. In this case, the claim must be substantiated on the label. (Note that nutritional claims are distinct from health claims.) Discussed at this time is the proper form and degree of label information, which is related to consumers' use of these data. The matter is complicated, as Padberg and Caswell (1990) remind us, because labels serve broader purposes than informing the shopper. They (a) define public values, (b) provide a forum for expert consensus, (c) influence product design, (d) provide a franchise to advertise, (e) provide a basis for public surveillance, and (f) offer a format for nutrition and food safety education. With this list of masters to serve, the labeling issue is sure to be with us for some time.

WHOLESALING AND DISTRIBUTION

This category of firms specializes in the distribution of items, and provides wholesaling/warehousing function for items that must be purchased in larger quantities than can be accommodated by individual customers.

Structure

Multiple product distributors — or so-called general line distributors — serve all or most of the needs of independent supermarkets, and even a few of the hyperstores. They remain large in number (nearly 25,000 firms recorded in 1986) but a large number of those are very small companies with less than $100,000 in assets. However, a handful of these firms handle the bulk of business. In 1986, 16 percent of firms controlled 88 percent of sector assets. Much of this concentration is a recent development attributable to mergers in the 1980s that left a few mammoth firms including Fleming and Super Value (USDA, ERS 1989, pp. 27-31). Again, divisions are not clear; many wholesalers operate their own stores as well as supplying foodservice, while some supermarket chains operate part-time as wholesalers and foodservice suppliers. Overall, however, this sector is quite concentrated, reducing the number of packer customers and supply lines to retail outlets.

Among the specialty meat wholesalers and distributors, the general trend to fewer but larger firms is also evident. In particular decline are packer branch offices as more of the sales functions are taken over at headquarters (Table 16-4). Wholesaler numbers have been in a slow decline since about 1970, perhaps partially due to the shrinkage of specialized meat outlets. Brokers act as local sales agents for firms, presenting new products, taking orders, and checking on delivery (Stafford and Grinnell 1982). Computerized ordering systems are reducing the role of the broker, to a degree, but they do continue to serve the function of a local presence for suppliers. Essentially a local activity, brokerage operations have been affiliating across states and even regions so that the number of separate firms is declining but the competition on the local level is little changed.

TABLE 16-4. Numbers of Meat Distribution Establishments by Classification and Sales, Selected Census Years, 1929-1982 (Source: U.S. Dept. of Commerce, Bureau Census, various years)

Year	ESTABLISHMENTS			SALES		
	Packer Branch Offices	Wholesalers	Brokers	Packer Branch Offices	Wholesalers	Brokers
	------ Number ------			------ Sales $M ------		
1929	1,157	2,225	---	1,923	690	---
1954	664	4,357	---	2,697	2,866	---
1969	616	5,041	163	2,811	7,395	853
1977	435	4,443	247	5,843	7,487	1,681
1982	326	4,218	245	9,349	26,547	2,688

Performance

General line wholesalers earned profits in line with those of supermarket chains during the 1980s. Small firms (below $1 million in assets) had higher profits than the largest group (over $250 million in assets) (11.6 vs. 7.5 percent return on equity in 1987) (USDA, ERS 1989, Table 23). The reasons for this are not quite clear, and after a few more years, may revert to a more expected pattern once the acquired firms are fully integrated.

The profit rates of specialty firms are roughly comparable to the meat packing industry. Measuring profits as a percentage of net worth, the specialty meat distribution sector earned (in 1982) 13.3 percent for firms with assets above $1 million compared to the aggregate of the meat packing industry of 11.7 percent that year. By comparison, all manufacturing earned a return of 9.1 percent for the same period (Epps 1986, Table 8; AMI 1984, Table A-1). One of the attributes of this industry is the apparent competition among the several classes of wholesalers. As Baker and Duewer note, ". . . the several different types of wholesalers are competing for the right to handle the meat. Some final outlets try to compete directly with wholesalers and buy directly from the packer" (1983, p. 7). If meat product brokers operate similarly to the more numerous grocery brokers, then costs and commissions are in the range of 3 to 3.5 percent of sales. Easy entry keeps this sector operating competitively even if firms tend to be too small to capture all size economies (Stafford and Grinnell 1982). This information, while admittedly limited and dated, does suggest that, in total, the meat distribution sector is performing in a reasonably efficient and competitive manner.

Wholesale Pricing

Until the advent of boxed products, most red meat was priced on a formula basis, that is, tied to some nationally publicized price. A typical arrangement would say "Yellow Sheet price for Thursday plus two cents a pound," referring to carcass weight. Further specifications would detail the quantity, delivery data, and location as well as a discount/premium schedule for quality. Some 70 percent of *carcass* beef is sold in this fashion (Hayenga and Schrader 1979, pp. 3-5). This approach had the benefit of ease plus providing a real assurance to sellers and buyers that their price would not be out of line with those paid by/to competitors. Serious questions have arisen periodically about the validity of "Yellow Sheet" and similar reported prices as reflections of national market clearing prices. Concerns focused particularly on the small number of observed prices underlying the reports and, hence, the potential for market participants to affect prices, even if for only a short time. Repeated evaluations have failed to substantiate any of these problems (see Chapter 7 for a review of this literature).

During the 1970s, as boxed products (notably for beef) began their domination of the market, this issue of "formula pricing" became muted. By 1978, most boxed product was priced by negotiation between packers and wholesalers and retailers. The apparent reason for this is the high value of some boxed sub-primals such that a small percentage error in the price would be costly in absolute dollars as well as competitive relationships with firms buying for a lower price (Hayenga 1978, p. 26). IBP continued to rely, to a degree, on formula pricing keyed to the "Yellow Sheet," perhaps because its "Cattle Pak" program involves selling most of the carcass rather than selected primals or sub-primals. In total, in 1979, 85 percent of all boxed beef was sold on a negotiated price basis (USDA, P&A 1982, Table 11).

Negotiated prices refers to both price arrived at through a negotiation process as well as the "offer-and-accept" method. This latter approach, strictly defined, omits the give and take between traders, restricting the role of the seller to accepting or rejecting the proposed sale. The acceptance method alleviates large buyers from some possible anti-trust legal problem while demanding less expertise of buyers, for the price negotiation proces is a substantial commitment (see Sarhan and Albanos 1985, pp. 15-16).

The situation with pork pricing is more mixed, a reflection of the product's division between fresh and processed markets. For fresh pork directed to retail outlets (loins, ribs, etc.), about 80 percent is sold on a negotiated price basis. The remainder is formula priced, especially by smaller buyers. However, when fresh *processing* products are included, the amount sold on a formula basis rises to 40 percent. That is, firms buying pork (bellies, hams, butts, etc.) for further processing tend to use formula methods. Processed products (about 40% of carcass weight) are specialized, many carry a brand identification like Armour or Hillshire Farms. These, like most retail food products, are sold on a listed price basis (Hayenga 1979).

One can make another exception for the very largest institutional buyers, of which McDonald's is a leading example. McDonald's generally procures from independent processors for hamburgers and, to some degree, sausage. The arrangements are typically cost-plus "handshake" agreements that allow the processor a specified margin, with the understanding that sales will not be made to competing firms. For more complex, but lower volume, products like Canadian bacon, prices are generally negotiated on a regular basis (Hayenga 1978 and 1979).

Electronic Marketing

Conceptually, the electronic marketing of meat products has the same benefits as its application for feeder and slaughter stock (see Chapter 13). Indeed, meat has certain characteristics that enhance its suitability for electronic trading over trading live animals: meat is routinely traded by description (USDA grades or packer grades), the number of buyers and sellers is more limited, and there are more frequent market participants compared to producers. This improves the feasibility of locating equipment in the office of each participant. Electronic trading would enhance the position of smaller participants for whom market information is more expensive to compile and new customers difficult to contact.

In a 1979 feasibility study, the USDA identified the following benefits to trading meat electronically (USDA, Ag. Mkt. Service 1979, pp. 38-39):

> Product Characteristics—high value, perishable necessitating rapid trading, long distance movement, frequent description trading large volume; Competition—improves information and coordination, expands customer/supplier base, enlarges trading area, equity.

Based on these considerations, electronic trading of meat was judged feasible using current technology (Figure 16-3). With this endorsement, the USDA funded a University of Illinois pilot project in 1980. CATS (Computer-Assisted Trading System) eventually operated for 22 weeks and traded 117 carloads of meat and meat products. The process was a sophisticated one, incorporating computers during the second phase systems for:

a. posting bids/offers by product type and location;
b. negotiating option in private between individuals;
c. data summary on bids/offers and trades (see Sarhan and Nelson 1981).

Yet in the end, the project failed to make the transition to private support due to the small volume traded. Partly this was an evolutionary matter—no one wished to trade because volume was low—and partly technological matters intervened, trading was actually slower than by telephone (30 minutes vs. 7 minutes, average, per transaction). But the failure was largely one of an inability to attract large volume buyers and, especially, packers (GAO 1987, p. 7). With no specific information, it is necessary to speculate that packers found the computerized process actually more

FIGURE 16-3. Once Seen as the Future of Trading Animals and Meats, Electronic Exchanges Have Had Difficulty Displacing the Tried-and-True Systems of Inspection and Telephone Contacts (Source: Equity Livestock Sales Cooperative)

costly than using the telephone. More significantly, opening the market to competitive price bidding could be a disadvantage to the seller in a favored position due to specialized services not easily describable in an anonymous exchange environment or simply due to the inertia of buyers. Without large packer commitments, these systems are inefficient, and the available evidence suggests that the large packers are not eager cooperators.

SYNOPSIS

The meat processing and distribution industry, while easy to conceptualize as those firms which stand between the packers and retailers, is indeed difficult to distinguish in the field. This is because there is much overlapping of activities; packers perform processing and distribution functions, while many processors distribute, etc. The message is that such a large sector as red meats permits great diversity of functions, from mammoth operations that combine numerous activities on a national and international scale to local operations concentrating on but one of the functions.

The meat processing sector is difficult to differentiate from food manufacturing as a whole, which is a huge ($342 billion) activity with numerous giant corporations. Food manufacturing, in general, is becoming more concentrated, has very high levels of advertising expenditures, and is more profitable than general manufacturing. The profit lure appears to be causing pork processors to cease slaughtering (a less profitable activity) and to concentrate on processing. Food manufacturers (especially the larger ones) are producing a number of new convenience meat products. This is a growth area in meat sales (but relatively small in terms of total volume). At the same time, branding of fresh red meats is limited largely to some specialty "natural" and "lite" items. Potential exists for growth in the next decade. Regulation, which is always substantial, can be expected to increase in the health- and safety-minded era, the recent concerns are about product labeling.

Wholesaling (often including distribution) is split between full (general) line firms and specialty operations. Both, like manufacturing, are increasing in concentration. Pricing is done largely through negotiations, while one effort to use electronic trading did not attract enough traders to be viable.

With the bulk of fresh meat moving directly from packers to supermarkets, the role of processors and wholesalers is not as important as in the past. Rising concentration in processing and wholesaling does mean fewer buyers, and possibly lower prices for packers, but that has not been documented. However, the loss of firms involves a shrinking of channels and niche markets which could reduce sales even if they are relatively costly alternatives.

Study Questions

1. Congressional hearings were held recently as but one expression of concern about rising concentration in food processing. What are the likely ramifications for livestock producers?
2. Identify the common factors that have led to fewer meat processors and wholesalers/distributors in recent years.
3. Discuss the several ways that wholesale meats are priced and the reasons for these different treatments.

REFERENCES

American Meat Institute. "Meatfacts." Various years.

Baker, A. J. and L. A. Duewer. "Meat Distribution Patterns in Six Souther Metro Areas." USDA, Economic Research Service, Agricultural Economics Rpt. 498, April 1983.

Brunk, M. ". . .The Myths Exposed." Address at the Livestock Marketing Congress — '71, Houston, Texas, June 16-18, 1971.

Bullock, J. B. and C. E. Ward. "Economic Impacts of Regulations on Mechanically Deboned Red Meats." Oklahoma State University, Agricultural Experiment Station, Report P-815, July 1981.

Burke, T. and D. C. Dahl. *Federal Regulation of the U.S. Food Marketing System*. University of Minnesota, Agriculture Experimental Station, Minnesota Economic Regulation Monograph, Misc. Rpt. Ad-MR-2338, 1985.

Connor, J. M., R. T. Rogers, B. W. Marion, and W. F. Mueller. *The Food Manufacturing Industries: Structure, Strategies, Performance, and Policies*. Lexington, MA: Lexington Books, 1985.

Connor, J. M. "Food Product Proliferation: A Market Structure Analysis," *Am. J. Agr. Econ.* 63(1981):607-17.

Council for Agricultural Science and Technology. "Impact of Government Regulations on the Beef Industry." Report No. 79, October 1979.

Dietrich, R. A. and W. F. Williams. "Red Meat Distribution in the Los Angeles Area." USDA, Agricultural Marketing Service, MRR-347, July 1959.

Dunham, D. "Food Cost Review, 1987." USDA, Economic Research Service, Agricultural Economics Rpt. 596, September 1988.

Dunham, D. "Developments in Marketing Spreads for Food Products." USDA, Economic, Statistics and Cooperative Service, Agricultural Economics Rpt. 449, March 1980.

Epps, W. B. "Specialty Grocery Wholesaling: Structure and Performance." USDA, Economic Research Service, Agricultural Economics Rpt. 547, March 1986.

General Accounting Office. "Electronic Marketing of Agricultural Commodities — An Evolutionary Trend." GAO/RCED 84-97, March 8, 1984.

Gallo, A. E. "The Food Marketing System in 1989." USDA, Economic Research Service, Ag. Information Bulletin No. 603, May 1990.

Handy, C.R., B. Sexauer, and L. Weingarter. "Assessment of Proposed Net Weight Labeling Regulations for Meat and Poultry Products." USDA, Economic, Statistics and Cooperative Service, Agricultural Economics Rpt. 443, December 1979.

Harris, M. "Spending on Meat, Poultry, Fish, and Shellfish." *National Food Review* 11(October-December 1988):33-36.

Hayenga, M. L. and L. F. Schrader. "Formula Pricing: A Comparative Analysis of Five Commodity Marketing Systems." In M. L. Hayenga (ed.), "Commodity Pricing Systems: Issues and Alternatives." U. Wisconsin, NC-117 WP-37, August 1979.

Hayenga M. L. "Vertical Coordination in the Beef Industry: Packer, Retailer and HRI Linkages." U. Wisconsin, NC-117 WP-22, October 1978.

Hayenga, M. L. "Pork Pricing Systems: The Importance and Economic Impact of Formula Pricing." U. Wisconsin, NC-117, WP-37, August 1979.

Marion, B. W. "Changes in the Structure of the Meat Packing Industries: Implication for Farmers and Consumers." Testimony before the Subcommittee on Monopolies and Commercial Law of the Committee on the Judiciary, House of Representatives, Serial No. 67, May 11, 1988.

Mueller, W. J. "Prepared Statement." Testimony before the Subcommittee on Monopolies and Commercial Law of the Committee on the Judiciary, House of Representatives, Serial No. 67, May 11, 1988.

Padberg, D. J. and J. A. Caswell. "Toward a More Comprehensive Theory of Food Labeling." Univ. Connecticut, NE-165 WP-19, May 1990.

Sarhan, M. E. and W. Albanos, Jr. "The U.S. Meat Industry: Components, Wholesale Pricing and Marketing Reporting." U. Illinois at Urbana-Champaign, Department of Agricultural Economics, AERR 198, October 1985.

Sarhan, M. E. and K. E. Nelson. "Electronic Meat Trading: An Alternative Exchange Mechanism to Wholesale Meat in the U.S." USDA, Economic and Statistic Service, Staff Report AGESS810428, April 1981.

Smith, R. B., J. A. Brown, and J. P. Weimer. "Consumer Attitude Toward Food Labeling and Other Shopping Aids." USDA, Economic, Statistics and Cooperative Service, Agricultural Economic Report 439, October 1979.

Stafford, T. H. and G. E. Grinnell. "Structure and Performance of Grocery Product Brokers." USDA, Economic Research Service, Agricultural Economic Report 490, September 1982.

USDA, Economic, Statistics and Cooperative Service, Agricultural Marketing Service. "The Feasibility of Electronic Marketing for the Wholesale Meat Trade." AMS-583, May 1979.

USDA, Economic Research Service. "Food Marketing Review, 1988." Ag. Econ. Rpt. No. 614, August 1989.

USDA, Packers and Stockyards Administration, "Résumé: Boxed Beef—Production, Pricing, and Distribution, 1979." Vol. IXIX, no. 10 (November 1982).

U.S. Dept. of Commerce, Bureau Census. "Census of Manufacturers: Concentration Ratios in Manufacturing." Washington, DC. Various years.

U.S. Dept. of Commerce, Bureau Census. Census of Wholesale Trade. Washington, DC. Various years.

Chapter 17

Retailing

The retailer is a key, all-important link between the producer and the consumer. Retailing takes on special significance just because it is the final or, more correctly, the first link to the consumer. It is here that preferences are recorded at the cash register. At this level, any error in measuring and transmitting these preferences will be magnified throughout the sector. Thus, all sector participants are critically concerned with how, and how well, the retailing component functions. In particular, producers and wholesalers must recognize that retailers may operate in ways which benefit them at some detriment to the producer. Note, for example, that retailers may decide for a time against reflecting lower wholesale prices at the retail level. Producer price reductions, then, do not move more volume, they just lead to higher retail margins.

The retail sector may be divided into sales for home consumption (notably supermarkets) and the providers of away-from-home meals consisting of restaurants and institutions (Chapter 18). The former dominates the latter—about 55 percent vs. 45 percent of total food expenditures in 1988—but away-from-home consumption has been growing at a rate of one percentage point per year in the later 1970s (Van Dress 1982; USDA, ERS 1989, Figure 1). More pertinent, each group and subclassifications within those groups, has its own attributes and lessons to teach the livestock sector.

RETAILING OF MEAT

Operations

Retailers break meat products down into portions of suitable size and cuts for individual or family use. For fresh products, that task involves: (a) projecting need; (b) ordering, receiving, and storing; (c) cutting into portion sizes and types acceptable to the local market; and (d) presenting the final products to consumers in a convenient and attractive fashion.

One aspect of the "attractive presentation" is, of course, appropriate pricing compared to competing products and stores. Prepackaged products — dominated by processed meat products such as hams, bacon, hot dogs, and luncheon meats — abbreviate the process by eliminating the need for step "c." Processed meats account for about one-third of meat of sales measured at the retail level (36% if canned meat is included) (AMI 1989, p. 36).

Until the dominance of boxed beef in the 1970s and subsequently, retailers would have bought meat in carcass form and broken it down at a central point or in the back room of the store (Figure 17-1). Subsequently, boxed beef has taken over in the bulk of retail outlets, largely eliminating the carcass breaking operation to the point where many stores abandoned the centralized facility and/or removed the "rail" necessary for handling carcasses. Such systems have been found to give the highest returns (but not necessarily the lowest cost, see O'Connor and Hammonds 1975), especially when cuts are selected as opposed to being purchased in carcass proportions. A second reason is the often lower labor costs prevailing in rural areas, where the packing is done, compared to the urbanized, heavily unionized environment that describes many large retail markets.

Increasingly, additional trimming and cutting activities are shifting to packers as part of the "tray-ready" systems. Here, packers trim and cut beef into retail cuts prior to vacuum packing, leaving only the physical

FIGURE 17-1. Centralized Supermarket Meat Breaking Operation, Now Largely an Outmoded Format Following Popularization of Boxed Meats (Source: Winn-Dixie, Jacksonville, Florida)

NEW ORLEANS Foreman Freeman Boudoin inspects new beef shipment. He is a 9-year associate.

task of wrapping for the retailer. This process accentuates the benefits found under the boxed beef system and has been estimated to be most profitable (Duewer 1985).

Structure of Food Retailing

Meat retailing outlets consist of supermarkets (both chain and "independent") grocery stores, convenience stores, and butcher shops. There is considerable overlap among these in terms of product line, especially among the first three groups, so that is most convenient to distinguish among them in terms of size. One classification system is as follows:

- Supermarkets: $1 million plus in annual sales by 1972 ($2.5 million in 1985 dollars), typically 20,000 sq. ft. and above in size. Chains have a minimum of 11 stores, typically served by a corporate-owned warehousing system. Independents are fewer in number and traditionally rely on wholesalers to warehouse and distribute product. A typical supermarket will stock 10,000 items as a minimum. In recent years, larger versions of supermarkets have appeared, known variously as superstores and hypermarkets, but the definitions are not very precise.
- Grocery Stores or Superette: smaller than supermarkets in terms of size and items stocked, but large enough to carry a full range of products including fresh meats and vegetables. Grocery stores tend to be individually owned and operated.
- Convenience Stores: are small stores in the range of 5,000 sq. ft. and less, and specializing in "fill in" items like milk and bread and snack foods. Fresh items, including meats, are uncommon. Convenience stores are increasingly chains, operating on the local, regional, and national levels.
- Meat/Butcher Shops: specialized meat stores operated by individuals. In order to enhance their competitive position with supermarkets, butcher shops emphasize some particular attribute like service, special products for ethnic clientele, or high quality meats for the well-heeled clientele.

Beginning with the establishment of the Great Atlantic and Pacific Tea Company after the Civil War, the chain movement began in earnest. However, it was not until the invention of the self-service food store in Memphis, during 1912 to 1920, that stores could exploit size economies. After this development and accelerating during the Depression, grocery store numbers began a sharp decline as they were replaced by fewer and larger supermarkets (Figure 17-2). During the mid-1980s, there were an estimated 166,000 grocery stores in the U.S. (USDA, ERS 1989, Table 7). From 1940 to about 1965, the number declined by half (Table 17-1). Convenience stores were first tracked as a separate class in 1975 with an estimated 13 percent of grocery store numbers, more than doubling in a decade. But while supermarkets are small in terms of numbers of stores, about one-tenth of total stores, their share of the eat-at-home food dollar approached three-quarters in 1987 (Figure 17-3). Clearly, selling for the home market is increasingly a matter of working with the supermarket industry.

Less evident to shoppers, but as significant in terms of the industry, has been the precipitous decline in the numbers of larger grocery wholesalers as a result of the merger wave of the 1980s. During 1985 alone, there were 64 mergers and acquisitions, some of which led to near-national-market-scope firms. The bulk of the 30,000-odd wholesalers remains very small (USDA, ERS 1989, Appendix Table 15). This means, among other

FIGURE 17-2. Despite Its Dated Appearance, This-1930s' Era A&P Store Embodies Most of the Concepts of Today's Self-Service Stores, Concepts That Date Back to 1914 [Note the prominence given to red meat, something else that has not changed.] (Source: Home Study Program, Cornell University)

TABLE 17-1. Foodstore Numbers, Selected Years 1940-1988 (Sources: Duewer 1985, Table 8; USDA, ERS 1989, Table 26)

Item and year	Affiliated	Independent Unaffiliated Number	Total	Chain	Grand Total
Stores:					
1940	108,750	296,250	405,000	41,350	446,350
1945	94,000	271,000	365,000	33,400	398,400
1950	122,000	253,000	375,000	25,700	400,700
1955	101,000	223,500	324,500	18,800	343,300

Year	Total	Supermarkets	Convenience stores	Superettes
1958	259,796	15,282	NA	NA
1963	244,838	21,167	NA	NA
1967	218,130	23,808	NA	NA
1972	194,346	27,231	NA	NA
1977	179,346	30,831	30,000	118,515
1982	168,041	26,640	38,700	102,701
1983[1]	167,615	26,821	40,400	100,394
1984[1]	167,186	26,947	42,950	97,289
1985[1]	166,755	27,266	45,400	94,089
1986[1]	166,322	26,995	47,000	92,327
1987[1]	165,887	26,522	50,000	89,365
1988[1]	165,453	NA	NA	NA

FIGURE 17-3. Foodstore Types: Shares of Establishments and Sales in 1987 (Source: USDA, ERS 1989, Figure 26)

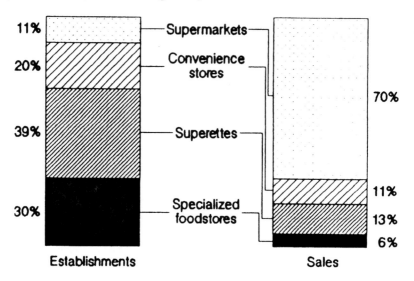

things, that the numbers of wholesalers who are available as customers has dropped by a like number. Wholesalers are private firms that perform the purchasing, merchandizing, and distribution functions for "independents," retail operations too small to carry out all these tasks in an efficient manner internally.

IMPORTANCE OF MEAT

Total meat sales contribute approximately one-fifth of total grocery store sales, a level which has remained remarkably constant into the 1980s but has shown some recent decline, perhaps because of increased nonfood sales in the new, large store formats (Table 17-2). Within this category, fresh beef is the most important, with a share in the 40 percent range that varies based on price fluctuations. Higher prices, as for beef between 1972 and 1981, tend to raise the share despite shifts to other meat products. Second in significance is processed meat products, encompassing cured meats and sausages.

The sales share of meat places it only second (admittedly a distant second) to general groceries, the large class of products encompassing

TABLE 17-2. Percent of Meat Sales in Grocery Stores by Type of Meat, Selected Years, 1972-1987 (Sources: Deuwer 1984, Table 7; AMI 1989, p. 36)

Item	1972	1976	1981	1987
		Million dollars		
Fresh meat, poultry, and provisions sales	20,084	29.109	46,920	51,657
		Percent		
Meat sales as a percentage of total store sales	21.5	21.1	21.2	17.5
Percentage of total meat sales	100	100	100	100
Beef, fresh	40	39	45	40
Lamb, fresh	3	2	2	2
Pork, fresh	8	9	9	8
Veal, fresh	3	3	1	1
Poultry	12	11	12	20
Provisions1/	34	36	31	29

1/Provisions are cured meats and sausage products.

canned foods, paper products, cereals, etc. Yet, with its relatively high value, the 20 percent of sales is done in only about ten percent of the total store space, even when the back-room cutting and wrapping areas are included (Harwell, Friedman, and Feig 1980, p. 27).

PRICING AND MERCHANDISING

The characteristics of meat and their role in the store make the pricing and selling of these products complex issues, with the results of those decisions having broad ramifications for the livestock sector.

Pricing Procedures

Most grocery store items are sold as received, with the store serving the primary function of display. Fresh meats are different since they typically arrive in boxed form as primals or sub-primals and require further preparation prior to sale. The first adjustment is for the trim loss of fat and bone. Second, the labor input must be considered and, finally, the prices set for different cuts of varying value to yield the target margin.

As a preliminary step, stores must determine the amount of salable product net of trim loss. This can only be determined accurately by doing an in-store "cutting test." The test involves weighing the product prior to and following cutting, with the trimmed products weighed as part of the post-cut weight. A sample cutting test result is shown in Figure 17-4. As shown there, not all the product can be accounted for after cutting, the difference is shrink and cutting loss. This, too, is an expense of retailing that must be reflected in the final product price. In recent years, microcomputer programs have appeared which simplify the computation of the test results. Some are integrated with the scales to automate the process. But whether the process is done by hand or automated, retailers pay close attention to cutting test results, and use them as one factor for selecting among packer-suppliers.

Step two involves setting the target retail value for all the cuts coming from the primal or sub-primal. This may be done by using the following formula:

$$\text{Total retail value} = \frac{\text{Cost}}{100 - \text{target gross margin}}$$

The chuck from Figure 17-4 cost $101.78. If the total markup, or gross margin, were 25 percent, then the retail price for the combined cuts is:

Total retail price = $101.78 (1.00 − .25) = $101.78/.75 = $135.71.

FIGURE 17-4. Sample Cutting Test for Beef Boneless Chuck (Source: IBP)

IOWA BEEF PROCESSORS, INC.
BEEF CUTTING TEST FORM

Date __10-2-78__

Store Company __ABC MARKETS__

Person Doing Test __JOHN DOE__

Grade __CHOICE__

Primal Being Tested __BNLS. C-P CHUCK__

Weight __75.02__ Cost/Cwt __101.78__

Estimate of Cutability

High (Average) Low

Org Wt __75.40__
B & C __.38__
Naked Wt __74.80__
Shrink __.22__
Net Wt __75.02__

Indicate on the carcass outline the cutting method used

Cut	Weight (In Decimals)	Pounds Per 100	Test Selling Price	Sales Value	Test Selling Price	Sales Value
BONELESS SHOULDER STK.	6.50	8.66	2.09	18.10		
BONELESS SHOULDER RST.	5.30	7.06	1.99	14.05		
CHUCK TENDER STEAK	1.79	2.39	2.19	5.23		
BONELESS CHUCK ROAST	18.57	24.75	1.79	44.30		
CHUCK EYE STEAK	1.49	1.99	2.19	4.36		
CUBED STEAK	2.17	2.89	1.99	5.75		
BONELESS SHORT RIBS	2.08	2.77	1.49	4.13		
LONDON BROIL	2.14	2.85	2.09	5.96		
BONELESS LEAN STEW	4.48	5.97	1.79	10.69		
TRIMMINGS 81/19	7.53	10.04	1.39	13.96		
TRIMMINGS 73/27	14.06	18.74	1.19	22.30		
SALEABLE YIELD		(88.11)				
FAT	8.54	11.38				
BONE	-0-	-0-				
SHRINK	.22	.29				
CUT LOSS	.15	.22				
Totals	75.02	100.0				

Convert all cuts into pounds-per-hundred. Do this by dividing the weight of each cut by the original weight of the wholesale cut and multiply the answer by 100.

To determine the percentage margin, divide the DPH margin by the Realized Sales Value and multiply the answer by 100. The result is the percentage margin for that particular wholesale cut.

Projected Sales Value __148.83__

Less Marketing Loss __2% 2.98__

Realized Sales Value __145.85__

Less Cost __101.78__

DPH Margin __44.07__

or

Percent Margin __30.22__

E2

Form 1510 – Rev 7 23 74

The decision remaining, most difficult of all, is how to price the individual products to achieve this target while maintaining a reasonable balance in demand among the products. Both judgement and competitive conditions will be of help here.

Setting the Gross Margin

Gross margins averaged across all items in supermarkets are now approaching 25 percent, up from the low 20s earlier in the decade (McLaughlin and Hawkes 1986, p. 9). In contrast, gross margins for all meats have declined slightly: 20.3 percent in 1986, 20.9 percent in 1980, and 22.4 percent in 1969 (*Chain Store Age, Progressive Grocer,* various years). This decline is partly attributable to the increased usage of boxed beef, resulting in less store labor used. Moreover, the wholesale price is higher, so a given percentage markup yields a higher absolute dollar margin.

If *gross* margins are comparable for grocery and meat items, the *net* margins are not (where net refers to after variable costs are deducted). Figures on costs by store department are not published, but it is clear the profits realized from meats, at least fresh meats, are lower than for many other items. Meats must support the costs of meat cutters (skilled workers earning union wages in some areas), spoilage, refrigerated display equipment, and energy costs. Indeed, many meat department managers price to break even as opposed to making a profit for the store. Stores do this because they believe that consumers form much of their store selection decisions based on the price and quality of the fresh meat offerings (Duewer 1984, p. 18). Alternative strategies may be followed, incorporating high quality, low, regular prices or selected, deep, price cut specials. Each chain has its own approach, but the net effect is that the meat department is heavily featured in food store advertisements.

Surveys of shoppers' store selection criteria have indeed shown the meat department to be a notable determining factor. In one survey run annually by the Food Marketing Institute, a trade association, meat quality was placed eighth in importance to store selection (1986, Table 8). Store-level studies have also shown meat sales to be quite responsive to price differentials among competing stores. While *aggregate* retail demand elasticity for beef has been estimated in the $-.64$ to -1.0 range (see Chapters 2 and 7), store level demand elasticity for individual cuts can be quite high, up to 139 (in absolute value). Moreover, the interproduct elasticity (cross elasticity of demand) can be equally high, suggesting that item selection at the store level is heavily influenced by relative prices. To confound the situation further, considerable variability was found among stores and weeks; the temporal factor was closely related to paydays (Marion and Walker 1978). These empirical findings tend to support the qualitative judgements of meat merchandisers about consumer's choice preferences.

Processed meat items are treated quite differently; they are priced more in line with general grocery items. Sales of these products through in-store delicatessen departments is a growing trend, with an estimated 61 percent of supermarkets in 1988 having such departments (USDA, ERS 1989, p. 33). Deli departments are managed with average gross margins of 48 percent, high due to the substantial labor inputs. Net margins, based on a small sample, are roughly four times the store average (McLaughlin, German, and Uetz 1986).

The heavy promotion of fresh meat in retail food stores has a mixed impact on the livestock sector. On the one hand, it maintains a lower price and higher profile for these products, which helps stimulate demand. Demand for competing products, namely poultry, is likewise stimulated, with the net result uncertain. At the same time, the intermittent use of specials, often at or below cost, makes slaughter prices a less reliable indicator of retail demand. And with slaughter prices a major determinant of future supply, the irregular retail pricing of fresh meats undoubtedly contributes to the instability of production and prices. Even in the short term, specialization makes it more difficult to limit demand on scarce products by raising wholesale prices and, alternately, to stimulate sales by lowering prices. Yet, supermarkets are the sector's major customer and, as such, the remainder of sector participants must conform to their wishes, not vice versa.

Merchandising

Quality constitutes the other dimension along with price in determining consumer value. Quality, though, is not limited to the characteristics of the product itself, but also involves the preparation and presentation of the cuts as a composite. The process begins with proper cutting, considering local preferences and portion size needs. Standard cuts exist but may need to be modified to maximize local market potential (Figure 17-5).

When making merchandising decisions about cuts, it has been found particularly valuable to pay close attention to impulse sales and counter-seasonal sales. Seasonality affects such products as blade roasts, which typically decline in sales during the warmer months. Grinding these cuts is always a possibility, but a lower value one. Alternately, more steaks can be produced from the chuck to fit better with seasonality of demand. During the summer, pork may move more easily as chops or boneless cutlets

FIGURE 17-5. Standard Cuts from Cattle Carcasses [Sometimes creativity in portioning and presenting the cuts is required to get the highest profits from the meat department.] (Source: National Livestock and Meat Board)

than as roasts. According to the standard wisdom in meat departments, impulse sales are largely related to appearance, appearance of the cut and of the entire department. Eye appeal is improved by quality control, good lighting, and effective displays. Some merchandisers place the impulse items first in the customer flow pattern, assuming it is more tempting to select an item when it does not involve replacing another one.

Merchandising the *department* takes a somewhat different approach. There, the emphasis is on selection of cuts and sizes (Figure 17-6). A broad selection reduces the need for custom cutting and the likelihood of the customer walking away unsatisfied. There are two basic ways to order the cuts, once selected. One is placement by species, the other by cooking method. The species grouping is based on the assumption that consumers first decide on beef, pork, or lamb, and on the exact cut, second. Ordering by cooking method operates as if the shopper wishes to, say, barbecue, then selects a type of meat from among beef steaks, pork chops, or lamb kabobs, to mention a few possibilities. The ordering-by-cooking type does allow for more substitution, thereby placing less pressure on keeping an item in stock. Sometimes, by using shelved display cases, merchandisers can combine the two approaches. But while there are technological solutions to some marketing choices and some general principles to follow, merchandising remains very much an art. This explains how so many different approaches can be simultaneously successful. (For further information on meat merchandising, see Harwell, Friedman, and Feig, 1980, Chapter 9.)

PERFORMANCE

Performance refers to the impact of ownership and operational practices on the prices, profits, and other results of those activities (see Scherer 1980, Chapter 9). Food retailing receives particular scrutiny because of its close interface with a large portion of the population. The result is that a number of studies have been conducted over the years. These studies generally begin with the hypothesis that concentration—the proportion of sales controlled by, say, the largest four, eight, or 20 firms—is positively associated with higher profits and prices.

Concentration is measured in relation to the appropriate market area. For food stores, this is quite localized, a diameter of several miles around the store. However, data are collected on an SMSA (or Standard Metropolitan Statistical Area) as defined by the Census Bureau. With smaller cities like Texarkana, the SMSA is a reasonable approximation of the market area of a food store. However, for Los Angeles and New York, to

FIGURE 17-6. A Full Range of Cuts and Sizes Combined with an Overall Attractive Appearance Are Keys to Profitable Self-Service Meat Departments: (A) Fresh; (B) Processed (Source: E. McLaughlin, Cornell University)

(A)

(B)

identify the extremes, the SMSA clearly overstates the market area and, hence, understates the effective concentration. In 1977, 277 SMSAs were defined and contained nearly three-quarters of the U.S. population.

Using 1982 data, the most recent available, the average four-firm concentration ratio was 58.3 percent (USDA, ERS 1989, Table 29). This figure may be interpreted as indicating that the largest four firms, on average, are responsible for over 50 percent of sales in each area. To give an idea of the range in 1982, Denver had a four-firm concentration ratio of 83.7 percent. At the low end was Huntington-Ashland, West Virginia, with a figure of 29.8 percent (data in Kaufman and Handy 1989, Table 1). In general, the smaller markets have higher concentration levels, attributable in part to the number of efficiently sized stores that can be accommodated, although, as noted, these figures are not strictly comparable with the largest SMSAs. In 1982, the smallest SMSAs had average four-firm concentrations of 63.6 percent (USDA, ERS 1989, Table 29).

The relatively high current concentration levels are a recent and continuing phenomenon. The highest concentration group (above 60 percent for the top four firms) was found in only six percent of markets in 1954, but increased to nearly 40 percent by 1977 (Table 17-3). Components in the trend are increasing store size economies, changing store formats (favoring larger, general merchandise stores) the declining role of independent grocery stores, and the merger movement among supermarkets and wholesalers (see USDA, ERS 1989, pp. 33-41). Cotterill estimates, using industry data for 94 SMSAs, that average concentration increased by 4.8 percent primarily due to mergers and acquisitions (1988, see also Tables 1, 2, and 3).

Concentration levels receive this much attention because they have been found to be positively correlated with profits and prices. In a recent review of the literature, Marion and the NC-117 Committee concluded, "the bulk of the evidence indicates that market concentration and a retailer's position in the market have a strong influence over that firm's prices and profits" (1986, p. 324). Perhaps the most detailed of these studies, conducted for the U.S. Legislature, warrants further attention (Marion et al. 1979). Due to the association with the Joint Economic Committee, the researchers were able to request data not publicly available, in particular, quarterly sales and profit data for 17 large chains for the period 1970-1973.

The results of several detailed models showed that *both* concentration and the relative share of the largest firm in each market had statistically significant effects on net profits. For example, the models predict that, as

TABLE 17-3. Changes in Four-Firm Concentration in Grocery Retailing, Selected Years 1954-1985 (Sources: Parker 1986, Table 3-5; USDA, ERS 1987, Table 13)

Concentration	1954	1958	1963	1967	1972	1977	1982	1985
>60%	5.8%	16.7%	18.8%	19.7%	24.3%	39.0%	NA	NA
50-57	25.0%	27.0%	33.0%	36.2%	31.2%	28.0%	NA	NA
40-49	38.9%	39.1%	31.7%	30.1%	35.4%	23.8%	NA	NA
<40	30.3%	16.7%	16.5%	14.0%	9.1%	9.4%	NA	NA
Top 4 chains	NA	21.7%	20.0%	19.0%	17.5%	17.4%	16.1%	19.0%

Top 4 chain data based on all SMSAs. Concentration by level based on 196 comparable SMSAs.

the four-firm concentration rises from 40 to 50 and the share of the largest firm in each case goes from 25 percent to 50 percent, profits as a proportion of sales increase by about four and a half times (1979, Table 1 and p. 432). The relationship of food prices to concentration showed a similar pattern. Again, both concentration and the relative share of the largest firm had separate and positive effects on prices (1979, Table 3). This portion of the analysis has, however, been criticized for the small price sample used: 94 items, one month, three chains for a total of 39 observations (1979, p. 421).

In a study more directly related to meat, Hall, Schmitz, and Cothern (1979) examined the relationship between retail beef prices and supermarket concentration at the SMSA level. Technically, the analysis is focused on the packer retail *margin*, but since the study assumes all stores pay the "Yellow Sheet" price for choice carcasses, the only factor introducing variation in the margin is the retail price. As a source of price data, the Bureau of Labor Statistics series (the same used for computing the inflation rate) is used. The authors conclude, "the degree of concentration existing in a market does appear to be an important factor affecting the price-cost marketing margin in a particular region" (1979, p. 299). More recently, Cotterill (1988) estimates that concentration increases, due to actual and proposed mergers in 1986, would cause consumers in the effected 16 SMSAs to pay $.5 million more, annually, for food.

Kaufman and Handy (1989) reached quite different conclusions; in particular, that concentration and market share were not determinants of price levels after adjusting for other factors. The different results are, in their judgement, due to a far larger sample of products and randomized data collection procedure in 28 SMSAs. Also, with the rise of superstores and hypermarkets in the 1980s it is apparent that supermarket competition dynamics are in a state of flux, but how much of that is reflected in the 1982 data used in the analysis is unclear. The Kaufman and Handy (1989) approach is different in several other aspects from the studies reported above, and a detailed comparison has yet to appear in the literature. Until that time, some caution should be applied to accepting results that are contrary to many previously accepted results.

Thus, the evidence strongly, but not universally, supports the hypothesis that concentration leads to higher retail prices for meat as well as other food products. This should be of some concern to the livestock sector, especially during a period of rising concentration. But the impact is probably not very strong, for, although higher prices reduce consumption to a degree, elasticity estimates indicate the overall effect is not great.

SYNOPSIS

The food retailer, and especially the supermarket, is the major contact point between the livestock producer and the consumer, and is responsible for 55 percent of food expenditures in 1988. The sector is a complex one, with numerous store formats of which supermarkets are the most important in terms of sales, handling three-quarters of total food sales through one-tenth of total food store numbers in the late 1980s. Supermarkets are defined as having more than $2.5 million in sales in 1985, although much larger stores—known as superstores and hyperstores—are increasing in number while total food store numbers decline.

Meat contributes one-fifth of total grocery store sales, of which fresh meats have a 40-percent share. However, meat has an importance to the store beyond its sales share, even though net meat department profits are generally low. Care is given to the handling and merchandising of fresh meat products. For beef, boxed product has largely replaced carcass purchases due to greater efficiency and merchandise flexibility. Prices are generally negotiated between retailers and packers.

The 1980s showed considerable merger activity in food retailing as in many other areas. Higher concentration at the local market or SMSA (Standard Metropolitan Statistical Area) level has been associated with higher meat prices as with other items, although that relationship has come under some recent reappraisal. The implication for livestock producers is probably limited.

Study Questions

1. Trace the activities of supermarkets in the procurement, preparation, and selling of fresh red meats.
2. What are the steps involved in pricing fresh meat in the supermarket?
3. How do supermarket pricing practices sometimes contribute to the price instability of the livestock sector?

REFERENCES

American Meat Institute. "Meat Facts." 1989.

Chain Store Age, Progressive Grocer. "Annual Report of the Grocery Industry." April, various years.

Cotterill, R. W. "Mergers and Concentration in Food Retailing: Implications for Performance and Merger Policy." Testimony before Subcommittee on Monopolies and Commercial Law of the Committee on the Judiciary, House of Representatives, Serial No. 67, May 11, 1988.

Duewer, L. A. "Costs of Retail Beef — Handling Systems: A Modeling Approach." USDA, Economic Research Service, Technical Bulletin 1704, June 1985.

Duewer, L. A. "Changing Trends in the Red Meat Distribution System." USDA, Economic Research Service, Agricultural Economics Report 509, February 1984.

Food Marketing Institute. "Trends: Consumer Attitudes and the Supermarket." Washington, DC, 1986.

Hall, L., A. Schmitz, and J. Cothern. "Beef Wholesale-Retail Marketing Margins and Concentration." *Economica* 46(1979):295-300.

Harwell, E. M., H. Friedman, and S. Feig. *Meat Management and Operations*. New York: Lebhar-Friedman Books, 1980.

Kaufman, P. R. and C. R. Handy. "Supermarket Prices and Price Differences." USDA, Economic Research Service, Tech. Bull. No. 1776, December 1989.

Marion, B. W. and the NC-117 Committee. *The Organization and Performance of the U.S. Food System*. Lexington, MA: Lexington Books, 1986.

Marion, B. W., W. F. Mueller, R. W. Cotterill, F. E. Geithman, and J. R. Schmelzer. "The Price and Profit Performance of Leading Food Chains," *Am. J. Agr. Econ.* 61(1979):420-33.

Marion, B. W. and F. E. Walker. "Short-Run Predictive Models for Retail Meat Sales." *Am. J. Agr. Econ.* 60(1978):667-73.

McLaughlin, E. W., and G. A. German, and M. P. Uetz. "The Economics of the Supermarket Delicatessen." Cornell University, Department of Agricultural Economics, Agricultural Economics Research 86-23 September 1986.

McLaughlin, E. W. and G. F. Hawkes. "Operating Results of Food Chains, 1985-86." Cornell University, College of Agriculture and Life Sciences, November 1986.

O'Connor, C. W. and T. M. Hammonds. "Measurement of the Economic Efficiency of Central-Fabrication-versus- Carcass Meat Handling System." *Am. J. Agr. Econ.*, 57(1975):665-75.

Parker, R. C. "Concentration, Integration and Diversification in the Grocery Relating Industry." Federal Trade Commission, Bureau of Economics, March 1986.

Scherer, F. M. *Industrial Market Structure and Economics Performance*. Chicago: Rand McNally College Publishing Co., Second Edition, 1980.

USDA, Economic Research Service, "Food Marketing Review, 1988. Ag. Econ. Rpt. No. 614, August 1989.

USDA, Economic Research Service, *National Food Review*, 1987 Yearbook NFR-37, Summer 1987.

Van Dress, M. G. "The Food Service Industry." USDA, Economic Research Service, Statistic Bulletin 690, September 1982.

Chapter 18

Away-from-Home Consumption

By 1988, Americans spent 45 cents of every food dollar on food prepared outside the home (Figure 18-1). This amount represents a rise of about one percentage point per year, although growth may be ending (Dunham 1988, Table 15; Van Dress 1982; Gallo 1990), and is the result of a number of demographic changes including increases in two-job couples and the number of individuals living or eating in institutions. In 1981, expenditures on these meals consisted of 55 percent for beef and 30 percent for pork, but this overstates the importance in quantity terms since the margin for restaurant meals is high. Using the farm value share as more reflective of the quantity consumed, the figure that year was 27 percent for beef and 20 percent for pork (Duewer 1984, p. 19).

The diversity of reasons for eating away from home is reflected in the diversity of firms serving this need. This chapter describes the away-from-home food industry and its significance for the livestock sector.

STRUCTURE OF THE SECTOR

In 1986, there were more than 700,000 establishments serving away-from-home food in the United States. The range of types, from hot dog stands to elegant restaurants to school cafeterias to cafés, are easier to describe as subclasses. The first subdivision may be made between so-called commercial and noncommercial sectors. The noncommercial establishments are found in the military, businesses, and schools—anywhere food is provided as a service but not as a profit center in its own right. The split is nearly one to four with commercial, for-profit establishments. The commercial sector may be further subdivided into restaurants and fast-food outlets predominantly, with several others contributing smaller shares (Figure 18-2).

The commercial component of the food service sector (as firms providing away-from-home food service are often termed) may be divided more

FIGURE 18-1. Shares of Food Sales Accounted for by Food Service Sector, Selected Years, 1963-1989 (Source: Gallo 1990, Figure 2)

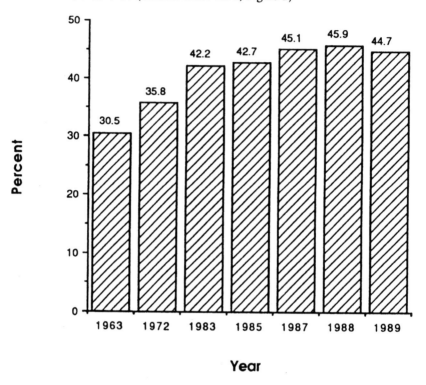

finely into those with a food service specialization (about 75 percent of the total) and those providing food and snacks as a sideline (25 percent), such as department stores and bowling alleys. As a final designation, commercial restaurants may be classified as fast food and other. Fast-food operations require special attention because of their rapid growth compared to the sector as a whole. From 1977 to 1985, fast-food outlets grew by 24 percent compared to an overall sector growth of just two percent (Table 18-1). For McDonald's, it is estimated a new outlet opens somewhere in the world every 15 hours or so, with a spectacular opening in Moscow in 1990.

With a share of 46 percent of sales in 1987, franchised outlets dominate the commercial sector. This share, in fact, exactly doubled from 1970 to 1984 (USDA, ERS 1987, p. 41; 1989, pp. 43-44), and includes both company-owned and franchised outlets. The leading firm, McDonald's,

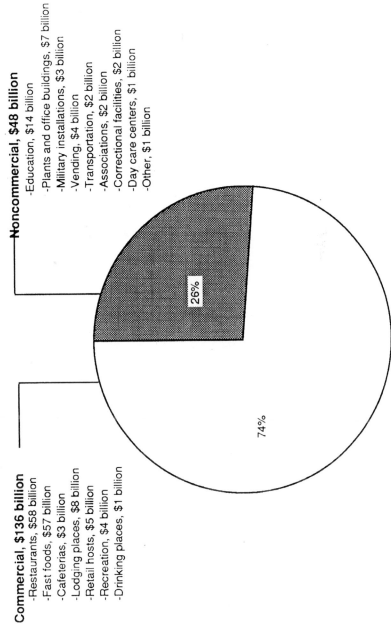

FIGURE 18-2. Away-from-Home Food Sales by Sector, 1987 (Source: USDA, ERS 1989, Figure 29)

Commercial, $136 billion
- Restaurants, $58 billion
- Fast foods, $57 billion
- Cafeterias, $3 billion
- Lodging places, $8 billion
- Retail hosts, $5 billion
- Recreation, $4 billion
- Drinking places, $1 billion

Noncommercial, $48 billion
- Education, $14 billion
- Plants and office buildings, $7 billion
- Military installations, $3 billion
- Vending, $4 billion
- Transportation, $2 billion
- Associations, $2 billion
- Correctional facilities, $2 billion
- Day care centers, $1 billion
- Other, $1 billion

26%

74%

TABLE 18-1. Number of Food Service Establishments, 1977 and 1985 (Source: USDA, ERS 1987, Table 19)

Industry segment	1977	1985[1]	Percentage change
	Number	Number	Percent
Commercial feeding	401,502	409,869	2.1
Separate eating places	229,892	255,699	11.2
Restaurants, lunchrooms	118,896	125,502	5.6
Fast food outlets	100,493	124,809	24.2
Cafeterias	7,001	5,388	-23.0
Lodging places	25,931	22,613	-12.8
Retail hosts	60,652	56,005	7.7
Recreation, entertainment	33,619	34,910	3.8
Separate drinking places	51,408	40,642	-20.9
Noncommercial feeding	223,005	301,962	35.4
Education	97,325	95,775	-1.6
Elementary, secondary	91,300	89,400	-2.1
Colleges, universities	3,095	3,299	6.6
Other education	2,930	3,076	4.9
Plants, office buildings	15,187	15,963	5.1
Hospitals	7,099	6,835	-3.7
Care facilities	21,117	29,711	40.1
Vending	3,737	3,535	-5.4
Military services	3,971	3,270	-17.6
Troop feeding	1,435	1,290	-10.1
Clubs, exchanges	2,536	1,980	-21.9
Transportation	799	640	-19.9
Associations	18,966	19,450	2.5
Correctional facilities	6,907	7,204	4.3
Child daycare	18,967	88,410	366.1
Elderly feeding programs	11,173	14,068	25.9
Other	17,757	17,101	-3.7
Total	624,507	711,831	13.9

[1]/Preliminary figures.

had about an eight percent share of commercial sales in 1987, while the top three commanded a 13-percent share. The latest 25 restaurant companies, as a group, had a 31-percent share of total commercial food service sales (USDA, ERS 1989, Appendix Table 21).

ROLE OF MEAT

As previously noted, food service accounted for 27 percent and 20 percent of beef and pork sales, respectively, in 1979. Meat dominates menu selections in all but a small portion of these outlets (Figure 18-3). In total, meat constitutes about 11 percent of the *volume* of products sold, but a

FIGURE 18-3. Distribution of Food Service Menu Specialties, 1979 (Source: Van Dress 1982, Figure 1)

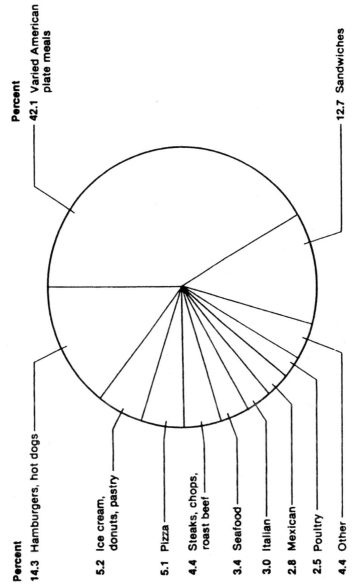

Percent

42.1 Varied American plate meals

12.7 Sandwiches

Percent

14.3 Hamburgers, hot dogs

5.2 Ice cream, donuts, pastry

5.1 Pizza

4.4 Steaks, chops, roast beef

3.4 Seafood

3.0 Italian

2.8 Mexican

2.5 Poultry

4.4 Other

*Figures may not add to 100% because of rounding.

427

larger portion of value due to the relatively high price of meats (Figure 18-4). Ground beef and veal itself accounted for 3.3 percent of total volume sales in 1979 (Van Dress 1982, p. 21). "Other meats" (essentially pork) has shown a share growth over beef. This would be attributable, in part, to fast-food restaurants offering breakfast menus where pork would be the featured meat product. The addition of breakfast and dinner entrees to the menus of fast-food outlets is a reflection of the decline in the growth rates of firms. While once they led the increase of ground beef sales over all other types of meat, recent menu diversification has led to greater expansions in other meats, including poultry. Yet, food service remains a highly important and visible outlet for red meats.

PERFORMANCE

The performance of the food service sector is more difficult to characterize than grocery retailing because of the differences in products and services provided. Nonetheless, some general insights may be gained by examining various aspects of these firms' operations.

After food, labor is the major cost of food service establishments, accounting for nearly 25 percent of the check amount (Van Dress 1982, p. 28). In the decade following 1977, labor productivity, measured as sales per labor hour, has actually declined ten percent. The reasons are not fully clear but seem related to longer hours, more varied menus, and reduced labor supplies, especially among teenagers who are very important for fast food outlets (USDA, ERS 1989, Appendix Table 40). Despite this, the return on stockholders' equity (a preferred means of measuring profitability across firms and industries) was 15.7 percent (weighted average) for 1985-1986 for the 90 top publicly held companies (reported in USDA, ERS 1987, p. 46). This is a favorable rate of return and suggests that, in the face of falling labor productivity, returns were maintained by higher prices. By 1989, however, restaurant prices rose by less than retail grocery prices, possibly an indication of increased pressure from supermarkets and convenience stores which are now heavily involved in competition with restaurants and fast-food outlets for sales (Gallo 1990, Figure 3).

One of the ways some food service firms are able to maintain high returns is through the use of substantial advertising. As a group, the sector spent in excess of $1 billion in 1987, with McDonald's alone spending one-quarter of this (USDA, ERS 1989, p. 46; Gallo 1990, Figure 6).

While the food service sector retains a very large group of small, independent operations, the clear and ongoing trend is to fewer and larger companies. Such firms operate as corporate-owned or franchiser-owned enterprises, or a combination of the two. For the livestock sector, this

FIGURE 18-4. Distribution of Major Commodities Sold by the Food Service Sector, 1979 (Source: Van Dress 1982, Figure 3)

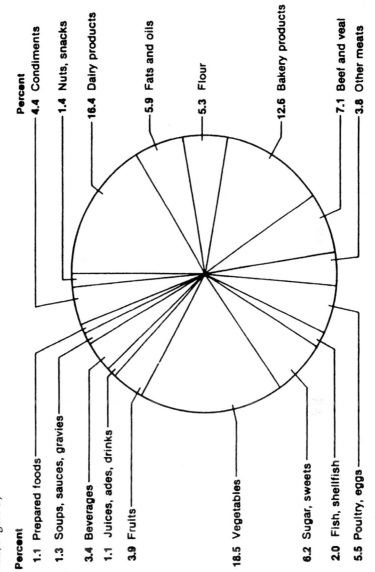

Percent

1.1 Prepared foods
1.3 Soups, sauces, gravies
3.4 Beverages
1.1 Juices, ades, drinks
3.9 Fruits
18.5 Vegetables
6.2 Sugar, sweets
2.0 Fish, shellfish
5.5 Poultry, eggs

Percent

4.4 Condiments
1.4 Nuts, snacks
16.4 Dairy products
5.9 Fats and oils
5.3 Flour
12.6 Bakery products
7.1 Beef and veal
3.8 Other meats

*Figures may not add to 100% because of rounding.

429

implies a higher concentration of customers along with greater homogeneity of the requirements of these customers. This trend parallels that which has occurred with food retailing, but can lead to some loss in the diversity and flexibility of the sector.

SYNOPSIS

Food service sales for away-from-home consumption have been claiming a larger share of food dollars, but may be leveling off at 45 cents per food dollar spent. The sector is composed of about 25 percent of noncommercial sales (hospitals and schools) and 75 percent of commercial restaurant and fast-food sales. Franchised outlets, especially in the fast-food business, dominate the commercial component.

Meat, especially beef, is a major menu item, with all meat constituting 11 percent of the volume of product sold (but a higher proportion of value). Overall, food service and some chains in particular, remains very profitable but declining labor productivity, while wages rise due to security and increased competition from supermarkets and convenience stores, is putting pressure on prices and margins. Increasingly, firms are franchises, which means fewer independent buyers for livestock products.

Study Questions

1. What factors have contributed to the rise in away-from-home food consumption?
2. What is the importance of this industry to the cattle, hog, and sheep sectors?
3. How do the greater service component and higher price of away-from-home meat consumption affect total meat demand and substitution among the red meats and with poultry and fish?

REFERENCES

Duewer, L. A. "Changing Trends in the Red Meat Distribution System." USDA, Economic Research Service, Agricultural Economics Report 509, February 1984.

Dunham, D. "Food Cost Review, 1987." USDA, Economic Research Service, Agricultural Economics Report 596, September 1988.

Gallo, A. E. "The Food Marketing System in 1989." USDA, Economic Research Service, Ag. Info. Bull. No. 603, May 1990.

USDA, Economic Research Service. "Food Marketing Review." Various years.

Van Dress, M. G. "The Food Service Industry." USDA, Economic Research Service, Statistic Bulletin 690, September 1982.

Chapter 19

A View to the Future

As the livestock sector looks back to a dynamic, if sometimes turbulent, past, it must also look ahead. The future — what glimpses are available from the poor vantage point of the present — appears just as or more uncertain than for similar periods in the past. Some of this approaching change is generic to society and the demographic and lifestyle changes underway. Some is generic to agriculture as the world becomes more integrated through trade, while rising agricultural productivity, political considerations, and economic factors leave large surpluses on world markets. Much of this surplus can be moved into consumption only at ruinously low prices. And some change is reserved for the unique combination of products and producers that define the red meat sector. In this chapter, the major components of change in the red meats sector apparent at this time are reviewed, and their likely impact on livestock production and prices are investigated. The aspects selected for evaluation are (a) production technology, and (b) consumption, including the effects of health/safety issues, promotional activities, and new products.

Some of the changes already underway are undeniably radical and can well lead to a livestock sector measurably different in 10 or 20 years than the one familiar today. However, the fundamental attractiveness of red meat to consumers is not in question. What is changing is consumers' relative valuation of these products and the way red meats can best be presented to consumers. By understanding and controlling these forces, the sector can shape the future rather than merely acclimate to change imposed from the outside.

PRODUCTION

Over much of the 1980s, red meat producers have seen their market share, and profitability, erode to the benefit of poultry (see Chapter 6). While the complex of causal factors is indeed intricate, relative cost is a major component. Cost, in turn, is largely derived from feed conversion efficiencies: two pounds per pound of gain for poultry, but up to eight for beef. Feed costs are about 45 percent of cash cattle feeding expenses (Shapouri et al. 1990, Figure 3). In short, red meats must be more cost competitive if they are not to become low-volume luxury goods. Gains have, and continue to be, made through traditional breeding and management improvements, and distribution efficiencies contribute to the overall price attractiveness of this product. But something more dynamic is needed, and that something is becoming available in the form of biotechnology.

Biotechnology and Livestock Production

Mankind has comprised biotechnology for thousands of years, if biotechnology is defined broadly as the manipulation of naturally occurring organisms for our benefit. This manipulation became more direct when animal scientists gained the ability to fertilize eggs outside the womb and implant those eggs in a surrogate mother (sometimes a mother of a different species). Modern biotechnology may, however, be traced to the breakthrough in 1972 when the first successful transfer of genes from one form of life to another was reported. Early on, these techniques were directed to simple organisms like viruses and bacteria, giving them the ability to produce a range of products (primarily human and animal pharmaceuticals). (For an overview of early developments, see OTA 1984.) More recently, genetic manipulation techniques and, as significantly, the growing understanding of animal growth and regulatory systems have advanced to the degree where genetic engineering is possible on complex animals including hogs, sheep, and cattle.

During the heyday of biotechnology start-up companies in the late 1970s and early 1980s, the potential of this field was depicted as limitless. Dairy cows as large as elephants were presented as feasible in the foreseeable future (*Business Week* 1982). Reality arrived later in the decade, leading to sharply reduced (but possibly still overly ambitious) projections for the remainder of the decade. Current expectations of commercialization dates for applications to complex animals are as follows (see Hansel 1986):

- Embryo transplants: currently available
- Hormone injections: 1990s
- Genetically manipulated animals — single genes: 2000s
- Genetically manipulated animals — multiple genes: later in next century

Each of these developments will be discussed in detail.

Embryo Transplants

Early animal breeding efforts consisted of mating the best performing animals while restricting the opportunity of poorer stock to pass along their inferior genetic material (see Chapter 3). Refinements in the practice of "traditional" animal breeding consisted, very basically, of systematically cataloging the performance attributes of individual animals (feed conversion efficiency, reproductive facility, handling ease, disease resistance, etc.) and experimenting to determine the degree to which these traits are inheritable. This information allows breeders to select sires and dams of a desired progeny, although the low inheritability of many desirable traits makes the process probable at best. With the inheritability rates as low as five percent for cattle litter size and pig's birth weights, it is no wonder animal breeding is described as often as an art as a science (data in Acker 1983, Table 18-1).

When success is achieved, the multiplication rate is limited by the reproductive potential of the species. For males, this limit can be eased through the use of artificial insemination, allowing a single bull to sire thousands of calves. The commercial feasibility of this practice, though, is limited in production herds, with only some two to four percent of beef cattle and hogs artificially inseminated annually (Gilliam 1984; Van Arsdall and Nelson 1984). Females are biologically more limited; they generally have a single calf or fewer than 20 pigs annually.

Embryo transplants (artificial inembryonation) lifts the biological reproductive limitation on the female in a conceptually similar way to artificial insemination practices with sperm. With transplants, the genetically superior cow conceives (using sperm from a genetically superior male), but rather than carrying the fetus to maturity, it is removed and implanted in a genetically inferior brood animal. In this way, a cow may produce dozens of calves annually rather than one. The sow can produce a hundred or more pigs. Assisting the process are recent advances in non-surgical embryo recovery, superovulation (multiple egg production), embryo splitting, embryo freezing, and regulation of ovulation (Hansel 1986; Donahue 1985; Leibo 1986). Practical procedures for embryo sexing and in vitro fertilization are under intensive development (Brackett et al. 1982; Anderson 1986).

A non-surgical implant method for cattle was recently developed. This **development has brought the cost of the procedure down from the $2,000-$3,000 range to as low as $200-$300 when the farmer on large operations provides the donor and recipient and performs the initial injections. Of course, costs are higher when recipient animals are purchased and when transportation is involved. Hogs must still be treated surgically, a costly process, so that the roughly 150,000 annual transplants performed in North America are almost exclusively confined to dairy and beef cattle (Foote 1987, Cornell University, personal communication). Other procedures, either in or near commercial use, include sexing of sperm, dividing embryos, and cloning, each of which promises to speed the development and dissemination of superior genes (Foote 1990).

Hormone Introduction

Since the 1930s, it has been known that the injection of growth hormone stimulates milk production in bovines (Miller, Martial, and Baxter 1980). That knowledge was of little practical value at the time due to the high cost of extracting the hormone from pituitary glands. That restraint was lifted in the 1980s when a bacterium was engineered to produce the identical product at a cost estimated to be as low as 8.5-18.6 cents per daily dose (Kalter et al. 1985). Subsequent experiments with the manufactured hormone led to yield increases of 10 to 40 percent, averaging up to 25 percent over the entire lactation (Bauman et al. 1985). With an estimated high return to early adopters and conditions favorable for rapid adoption, one view is that the dairy industry will soon be revolutionized following the commercial introduction of this product (Kalter et al. 1985). Higher production per cow would, however, reduce the required herd size at current milk consumption levels, leading to a possible reduction in dairy farm numbers in New York State of about 15 percent, or 30 percent if the milk price support system is disbanded (Magrath and Tauer 1986). Other analysts are less sanguine, identifying reasons for slower adoption and less sectoral impact (Buttel and Geisler 1989; Larson and Kuchler 1990; Kuchler, McClelland, and Offutt 1989).

While the use of growth hormone, also known as somatotropin, for the stimulation of milk production is anticipated to be the earliest truly dramatic impact of biotechnology on animal agriculture, its use is by no means limited to lactating dairy cows. Research has shown that the hormone has a growth-enhancing effect when introduced into young cattle, hogs, and sheep. To date, results with poultry have been largely unsuccessful.

Efficiency enhancements for dairy stock come from the dilution of the maintenance feed requirement, about 30 percent of the total. Since animals treated with somatotropin during the growth period grow faster, there is also an increase in feed efficiency for meat-type animals. The major efficiency improvement is due to the partitioning of feed utilization to lean tissue and away from fat accumulation. Since lean accretion is made at a cost of 1-2 kcal net energy per unit of gain, while fat is produced at a cost of 6-9 kcal per unit of gain, there is a major increase in feed efficiency for animals treated with somatotropin during the growth stage (Kalter and Milligan 1986).

Experimental results indicate feed efficiency improvements for hogs of 10 to 20 percent over the entire feeding cycle, and up to 30 to 50 percent during the last two months of the fattening cycle (Meltzer 1987; Etherton et al. 1986; Boyd et al. 1986; Kuchler, McClelland, and Offutt 1989). Lamb feed efficiency has been in the same range as pigs (summarized in Meltzer 1987). Results for beef cattle are more preliminary, but indicate efficiency gains in the range of 20 percent (Fabry et al. 1985). Achievable results in production herds, where the same degree of control applied in experiments is not possible, are likely to be notably less.

While the somatotropin treatment increases the efficiency of feed energy utilization, it also increases the protein requirement in the feed (Boyd et al. 1986). Adjusting for the higher cost of dietary protein implied when raising the hog fattening ration from 14 percent to 20 percent protein means the cost of hog fattening is reduced by an estimated 6 to 11 percent, depending on the assumed increase in feed utilization efficiency. This figure excludes the cost of the hormone. For cattle, the estimated production cost reduction ranges from 5 to 8.5 percent (Kalter and Milligan 1986, Table 1).

For hogs, the expected feed efficiency gains place this species close to the level achievable with poultry, which will help to make red meats more cost competitive with the white meats. However, in the absence of increases in red meat consumption due to lower relative cost or other factors, the number of feeding facilities can be expected to decline as each is able to finish more head per year. Similarly, with more rapid gain, the management component becomes increasingly important. Better managers will tend to expand their operations, while poorer ones will be forced out. Over time, these effects are anticipated to lead to larger and more concentrated feeding operations, everything else held constant. For example, Lemieux and Richardson (1989), in a simulation of hog feeding for 1988-1992, found the large-scale producers receive the greatest benefits, but smaller scale operations benefited also. The production of feeder animals will not be affected directly unless hormone use is found to stimulate growth at earlier stages in the growth cycle.

The reduction in fat deposition has a direct impact on the carcass quality. In some trials, hog backfat has been reduced by 70 percent, and similar results are expected for cattle (Bauman and McCutcheon 1986; Boyd et al. 1986). This dramatic change in the surface fat content will reduce waste and trimming costs for the packer, purveyor, and retailer (Figure 19-1). It may further enhance the attractiveness of red meats to a health-conscious society, leading to increased sales (see below).

The timing of commercial introduction of growth hormone remains very uncertain. Bovine growth hormone approval for use during lactation has been under test for several years, during which the sale of milk from experimental trials has been allowed. Yet, final approval has not yet been granted due to considerations of animal health and the ongoing uncertainty over even such basic parameters as how many generations must be tested. Further delaying the procedure is the intense scrutiny placed on the testing and approval of biotechnology products by some "watchdog" groups. Altogether, approval appears possible every year.

Once approval is granted, a matter of commercial practicality remains. At this time, to be effective, the compound must be injected into the body daily. This procedure is highly impractical for commercial herds. An implant is under development and, if commercially successful, will remove much of the practical complexities.

Genetically-Based Hormone Production

While growth hormone may be introduced externally, there are significant additional benefits to carrying the trait in the animal's genes. This would lift the cost and complexity of controlling periodic treatments from the manager. The introduction of an enhanced ability to produce growth hormone is currently possible with modern genetic engineering techniques. Genes conferring hormone production in the red meat species have been identified and cloned and, on an experimental basis, inserted into goats, sheep, and hogs (Michalska et al. 1986; Hammer et al. 1985; Hammer et al. 1986).

Several factors limit the practicality of these efforts. For characteristics like growth hormone, which are controlled by single genes, the principal deterrent is the low success rate. Inoculation of eggs with the transmuted genes has a success rate on the order of .2 percent. Even when the treatment is successful, researchers have been unable to control the secretion rate, leading to highly abnormal animals. Commercialization of these developments is some time off, occurring no sooner than 2000 (Hansel 1986). Even that date may be optimistic, as one leading researcher in the area describes current research as ". . . still at the Orville Wright stage . . ." (quoted in Miller 1987).

Complex traits, like mature weight and number born, are controlled by

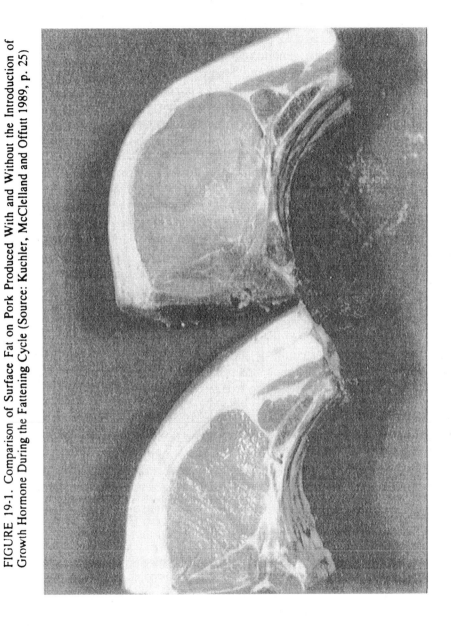

FIGURE 19-1. Comparison of Surface Fat on Pork Produced With and Without the Introduction of Growth Hormone During the Fattening Cycle (Source: Kuchler, McClelland and Offutt 1989, p. 25)

437

multiple genes. Introducing changes in these genes requires an understanding of the controlling genes and their interaction, a far more complex task. Marketable advances in these areas will not be seen until well into the next century.

Research on the genetic manipulation of animals was recently given a boost by the U.S. Patent and Trademark Office decision to allow patents for higher (multi-celled) animals (*Ex parte Allen*, April 3, 1987).[1] A patent grants the holder temporary monopoly rights, increasing the opportunity to profit from the invention. This enhances the incentive to invest in research in this area. Animal patents, though, are not without their complexities. Chief among these is the matter of control over the progeny of patented animals. At this point, appearances are that the patent holder will have rights to progeny that carry the patented trait, for example, enhanced growth hormone production. The number of offspring that inherit the trait varies from half, for single gene-based characteristics when only one percent carries the trait, to very small, for highly complex traits like milk production and birth weights (see Acker 1983, Table 18-1). Moreover, the right to a royalty on a calf or pig is distinct from the practicality of enforcing that right, especially in light of certain opposition by livestock producers as well as animal rights groups.[2] Lesser (1990) identifies the collection of royalties on offspring as one of the most complex and controversial issues of the commercial use of patented livestock. A costly collection system could negate much of the value of the efficiency improvements, so that some streamlined system must be identified. Milligan and Lesser (1990) recognize two possible areas of bioengineering in livestock: those that enhance production efficiency (such as elevated somatotropin production) and those that change the meat or carcass characteristics. The former are likely to be made widely available and can lower costs, while the latter may remain more tightly controlled and provide a way for firms to differentiate their products. Overall, the role of patents in transgenic livestock is a complex one that will evolve over the coming years. The procedures for inducing the changes are, however, becoming

---

1. The first and, to date, only higher animal patent in the world was granted for a laboratory mouse in 1988. The U.S. Patent Office is believed to have at least 20 applications under review but few, if any, at this point are likely to be appropriate for commercial livestock producers.

2. Brody (1990) distinguishes between animal welfare — e.g., the right to minimal suffering — and animal rights, the inalienable right of animals to certain less tangible privileges such as some self-determination.

more efficient and predictable so that an increased level of products can be expected in the future (Van Brunt 1990).

FAT, HEALTH, AND SAFETY

The rising prosperity since World War II, combined with unprecedented medical advances, have led to a rising life expectancy for Americans. A male born in 1985 may now expect to live 71 years, and a female, 78 years (Census Bureau 1986, No. 105). Even the century mark is no longer a rarity. Acting to hold back this trend, especially for men, is the high rate of heart attacks, now numbering over one million annually and about half of which are fatal. With heart disease a leading national killer, there has been no shortage of research on how to lessen the danger. And much of the research focuses on the role of cholesterol.

Cholesterol is a naturally occurring and necessary component of blood, contributing to such functions as the production of cell walls. About two-thirds of blood cholesterol is produced by the liver, with the remainder derived from the diet. The role of diet becomes important because excessively high levels of blood serum cholesterol, or more particularly the damaging or LDL (low-density lipoprotein) variant, tends to lead to the buildup of fibrous "plaque" in the arteries. The resultant restriction of blood flow, in turn, can cause heart attacks and stroke. Among the components of diet, animal fat consumption is associated with arteriosclerosis and heart disease. At the national level, heart trouble is a leading killer in North America and Europe where animal fat consumption is high, but it is largely nonexistent in Japan and other nations where seafood is the major protein source.

One response is to reduce animal fat consumption, and Americans have been doing just that. Once considered synonymous with the good life, the juicy beef steak is now viewed by many as a possible health threat. The American Heart Association recommends no more than 300 mg of cholesterol per day for men and 225 for women. That is roughly equivalent to a single egg or nine ounces of veal. In 1977, the U.S. Department of Agriculture suggested adults eat but 2-3 ounces of lean meat, fish, or poultry daily, but that was for a very low-calorie diet. Clearly, many Americans are exceeding this recommended level, and many worry about it. A recommended diet contains 30 percent fat, one-third saturated, while Americans in 1989 were eating 13 percent saturated fats and 37 percent total fats (USDA data reported in Hellmich 1989). Nearly three-quarters of consumers surveyed in 1989 had heard of health problems associated with too much fat intake, yet one-third did not consider it important to avoid too much fat. One possible explanation is that 78 percent of respondents felt there were so many recommendations they did not know what to believe (reported in Hellmich 1989). However, this may be changing as, in 1990, low-fat content was the primary nutritional concern for 46 percent of a sample done by the Food Marketing Institute. This is a 17-percent jump from 1989's figures (AMI 1989).

The livestock industry response, for a long period, was one of denial. This position was based on the lack of scientific proof of the causal relationship between animal fat consumption and heart disease. National-level studies are merely correlative and it may properly be argued that other national differences are causing the observed health factors. More specifically for the industry, there was no evidence linking the *reduction* in saturated fat consumption to the cholesterol level in the blood. Earlier studies had shown that lower cholesterol levels reduced the risk of heart attack, but those levels were achieved through the use of drugs (Wallis 1984). Complicating the analysis are individual differences in the genetic response to dietary fat combined with lifestyle differences in exercise, smoking and alcohol consumption.

In 1987, a report in the *Journal of the American Medical Association* documented the relationship between diet and blood-cholesterol and arteriosclerosis (Blankenborn et al. 1987). Moreover, in a small number of cases, the reduced intake of animal fats led to actual reductions in arterial plaque. New drugs can reduce cholesterol levels further, but dietary changes remain an essential, if not always sufficient, component (Begley 1987).

These findings will not end the controversy, and indeed a recent review of evidence by Moore (1989) casts further doubt on the "conventional wisdom." According to his assessment, the dangers of high blood-cholesterol levels have frequently been exaggerated. Most notably, recommendations have been based on the experience of middle-aged men with very high cholesterol levels, while the situation for women and the elderly is little understood (see also Kolata 1989). And even the relationship between dietary change and heart attack risk is tenuous for middle-aged men. Nonetheless, the contention by many in the meat industry that there is not ". . . a shred of evidence . . ." linking meat consumption with heart disease would seem to be invalid (see, e.g., Fleming 1982; National Livestock and Meat Board 1983). According to one long-time industry observer, "The time has passed for the beef industry to be defensive about its product and keep saying, 'It's 100 percent healthy.' It's not" (Cross 1987).

Just what to say remains a complex matter, in part because consumers' perceptions of health issues frequently differ from the medical "facts." For example, a consumer attitudes survey in the late 1970s showed that a large portion of the public did not distinguish between the consumption of animal and vegetable fats (Branson and Rosson 1981, Table 9). In 1989, 73 percent of surveyed consumers never heard of monounsaturated fats (Hellmich 1989). This, despite the ample evidence that arteriosclerosis is advanced by the saturated fats present in all animal-based products but lacking in most vegetable oils. In another case, the trimming of surface fat led to the perception of lower cholesterol content in beef, although the actual impact on the consumed portion is negligible (Cross 1986). Yet, the trend to lower animal fat consumption by a large segment of the population is apparent, and the industry is responding at an increasingly rapid rate to those changing tastes. The early 1990s saw the end of the use of animal fats in fryers in major fast-food chains. Change is evident in both the products being sold and what the industry is saying about them.

The industry is saying plenty. Using funds generated from the producer "check-off" programs (see below), an information program has been developed to provide another side to a conventional perception of the health effects of red meat consumption. In the case of beef, physicians, dietitians, parents of preschool children and young athletes, among other groups, are being targeted. A theme is, "beef is healthful and should be included in a varied and balanced low-fat diet" (Meat Board 1990, p. 7).

Production Innovation

With the focus on fat-related issues, a primary product innovation of the industry has been to reduce the level of fat in marketed cuts. This desire, in part, underlies the recent and continuing changes in grade standards to lower fat levels (see Chapter 12). The parallel reason is the concomitant reduction in the feeding period and, hence, cost and price. Also apparent is the changing emphasis toward leaner breeds such as Brahman and Charolais and away from Angus. This change represents a risk since fat content is positively correlated with flavor, at least in beef (see Chapter 12), but consumer surveys have shown the probable existence of a significant market segment interested in lean beef (Branson et al. 1986). In one actual test comparing two beef grades corresponding roughly to USDA "select" and "choice," total meat sales were enhanced by providing this consumer alternative. The survey of Giant Food Inc.'s "Giant Lean" program showed a strong preference by of a class of relatively heavy beef consumers for lean beef. The lean product was sold at a 10-20-cent-per-pound discount under the choice beef, but survey respondence suggested that it was perceived as equal in price and quality to choice (A. M. I. 1987).

What cannot be kept off in the production process is being removed prior to sale. Major beef packers are trimming surface fat to one-fourth of an inch. Surveys show that consumers rate otherwise identical beef cuts higher when the surface fat covering is removed. In response, some supermarket operators are trimming the cuts even further. ShopRite, a Northeast chain of independents, has instituted a "Trim 'n' Lean" program with a maximum 1/8-inch fat covering, subsequently reduced to zero. The 1/8-inch level has now become the industry standard (Figure 19-2). In total, some 50 to 60 million pounds of fat were *not* sold in 1986 as a result of these trimming programs (Cross 1987). As these programs continue to grow in popularity, the actual trimming will likely shift back to the packer level where it can often be done for less cost and can provide a higher use value for the trimmed fat.[3]

But even with trimming, the basic beef product remains essentially the same as that available throughout the century. During that period, the needs and wants of the consumer diverged greatly, particularly as family size declined, the population aged, and women joined the work force in great numbers. With these increasingly prosperous families, time, not money, becomes the limiting factor in food preparation. Meats like the hamburger are no longer convenience foods since other sectors package

FIGURE 19-2. External Fat Trim on Beef Has Declined [The program to remove all but 1/8-inch of surface fat has led to greater customer satisfaction and increased sales.] (Source: AMI 1989, p. 30)

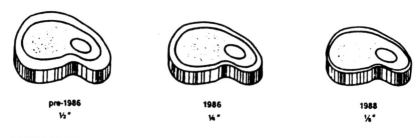

pre-1986   1986   1988
½"   ¼"   ⅛"

---

3. Increased supply and unchanged demand will, and is, depressing prices.

their products in new and tasty forms ready to be microwaved and table-ready in minutes. In the words of Manly Molpus, President of the American Meat Institute, ". . . the marketplace changed and we did not change with it."

The position of the livestock sector, especially those involved with beef, regarding new products is changing rapidly. Much of the actual innovation is coming from food processors like Campbell Soup Company and Armour which have incorporated red meats in popular new prepared products like "Le Menu" and "Dinner Classics" (high-quality, prepared dinners specially adapted to microwave ovens). Per capita consumption of frozen meat entrees rose 20 percent from 1983 to 1986, while many other meat items were stagnant at best (Harris 1988, p. 35). Sector participants, most notably packers, are taking an increasingly greater role in the product innovation process. Excel, a recent leader in this area, applied the same technology used for boxed beef to preserve individual beef cuts for up to seven days in the refrigerator. This process removes the need to freeze fresh meat, a procedure viewed with disfavor by consumers. Excel is further testing the prospects of vacuum packed *cooked* beef, a microwave-ready product. With this treatment, a steak can be ready in two minutes.

Much of the product innovation activity is being directed toward beef. Pork shares the same inherent product characteristics as beef, but this sector had remained more up to date than beef. A quarter of a century ago, the sector recognized the trend to leaner meat and adjusted the breeding and feeding of hogs accordingly (see Chapter 1). Processing of pork products, too, has had a higher priority. With about 40 percent of the carcass subject to additional processing, pork packers have remained in closer contact with consumers. The result has been pre-cooked sausages (decades ago) as well as closely trimmed canned hams, a new twist on a traditional product. Indeed, with traditional packers like Oscar Mayer withdrawing from slaughtering operations, the sector is moving to specialization in commodity (carcass) processing and further product preparation (see Chapter 16). The beef sector, at this time, has no further processors with the strong customer franchise of an Oscar Mayer, Armour, or Rath.

Safety

Food safety means the absence of contaminants with effects ranging from diarrhea to cancer. There are numerous sources, including microbial and environmental contaminants; inadvertent contamination by cost-reducing products such as pesticides, antibiotics, and other drug residues; and additives intended to enhance the flavor, shelf life, or other attributes of a product (Roberts and van Ravenswaay 1989a). Here the attention shall be on drug residues, even though microbial contamination is recognized as a greater health threat (Roberts and van Ravenswaay 1989a).

The responsible agency, the Food Safety and Inspection Service (FSIS) of the USDA, samples carcasses to estimate the average violation rate. If the rate exceeds one percent, a detection and enforcement program goes into effect, one aspect of which is the recent swine ID program (see Chapter 11). This approach has led to reductions in residue levels of sulfa drugs and sulfamethazine but, according to Roberts and van Ravenswaay (1989b, p. 7), "adequate, timely detection methods do not exist for approximately 70 percent of the animal drug residues in meat." Enforcement is difficult, but a 1990 California law permits packers to sue violators for treble damages for condemned carcasses, plus the collection of attorneys' fees and civil penalties. Supported by the P&SA, this kind of law could be adopted in other states (AMI 1990).

Of perhaps equal importance to consumers is the possible residues of synthetic growth hormones which, when properly administered in both dose and method, pose no threat to human health according to scientists from both the WHO and FDA. Since FSIS's National Residue Program was begun in 1978, no misuses were revealed (Kenney and Fallert 1989). Lingering concerns, nonetheless, have led to the development of a new beef product referred to as "natural" or "organic" with a typical 20-50 percent retail price premium but, as yet, small scale sales (Johnson et al. 1989). With no formal monitoring program underway or even a standard definition of what these terms mean, consumers must accept, as a matter of faith, the complete absence of pharmaceuticals and/or synthetic hormones in these products. The National Research Council has recommended a standardized definition of "natural" be adopted as it applies to meat (see Glaser 1989, p. 28).

While the apparent effect of meat safety issues on sales is limited at this time, it is likely to grow in the future. Observers of the beef sector have labeled the 1990s the "decade of safety."

PROMOTIONAL ACTIVITIES

Fresh meats are indeed one of the last commodity groups to receive brand names at the retail level. Until recently, fresh meats have also received one of the lowest levels of advertising expenditure measured as a proportion of sales. In 1974-1976, total advertising expenditures through major channels came to .6 percent of sales. This can be contrasted with levels of 3.2 percent for all food and up to 9.9 percent for breakfast cereals (Connor et al. 1985, Table 3-6). The principal reasons for low advertising expenditures for meats is their commodity characteristic; all producers receive the benefit of advertising, so no single firm has the incentive to make the investment.

The obvious beneficiaries are *all* producers, but they had not been organized to collect and disseminate the necessary funds. Beef producers twice (in 1977 and 1980) voted down in referendum a "check-off program" to support research and promotional programs. Technically, these check-offs are known as marketing orders (for background see Ward, Thompson, and Armbruster 1983). Wool and lamb producers have had a program in effect since the passage of the National Wool Act of 1954, but the funds involved have been modest — less than $2 million in 1980 (reported in Ward, Thompson, and Armbruster 1983, Table 4-1). One of the limitations of this kind of program, operated under the auspices of the Agricultural Marketing Service of the USDA, is allowing producers to request a refund of collections made during the marketing process. Contributions are then quasi-voluntary, leading to at least the perception of an inequitable distribution of costs and benefits. This perception tends to reduce overall support for the program.

The situation for beef and hog producers changed with the passage of the Food Security Act of 1985 (P.L. 99-198), the 1985 "Farm Bill." This Act contains mandatory promotional orders for beef and pork known as the "Beef Promotion and Research Act of 1985" and the "Pork Promotion, Research and Consumer Information Act of 1985" respectively. Both Acts required a vote of support of a simple majority of voting producers within approximately two years of enacting the legislation.

The Acts are administered by promotional boards made up of the affected groups, producers, and importers. Representation is by state or region and is based on the number of head marketed. Each seller must pay $1 per head of cattle, or the equivalent for products (including imports), and between .25 and .5 percent of the market value of hogs and pork products (Table 19-1). The half-percent value is the ceiling amount that can be reached by maximum .1 percent annual increments beginning from the base .25 percent. Refunds are not permitted following the acceptance of these programs.

The initial use of the funds thereby generated are effectively summarized in the 1985 Meat Board "Consumer Marketing Plan":

TABLE 19-1. Major Aspects of the Beef and Pork Promotional Programs (Sources: Adapted from Glaser 1986, Table 13; Lipton 1988; Lipton 1989)

Item	Beef	Pork
Implementation	Mandatory	Mandatory
Persons affected	Beef producers and importers	Pack producers and importers
Administrative Organization	Cattlemen's Beef Promotion and Research Board	National Pork Producers Delegate Body
	Beef Promotion Operating Committee	National Pork Board
Assessment rate	$1/head cattle or equivalent for beef and beef products	.25-.5% of market value of hogs and pork
Referendum date	May 1988	September 1988
Approval required for continuation	Majority of those voting	Majority of those voting
Check-off Receipts, 1989 (est.)	$89.2 million	$26.7 million

- To increase customer understanding and appreciation for red meat's role in a balanced and varied diet.
- To increase positive price/value perception among consumers about red meats.
- To enhance the development of new value-added meat products and encourage adoption of new merchandising systems that increase consumer demand for red meats. (1985, p. 2)

The goals for pork are similar. Most funds are channelled through several contractors including the National Pork Producers Council (NPPC), the National Livestock and Meat Board's Pork Industry Group (PIG), and the Beef Industry Council of the National Livestock and Meat Board. For pork, legislation requires that, in 1989, approximately one-fifth of funds collected be returned to state Councils. The state Beef Boards collect the check-off funds, forwarding half to the national board. Of the half retained, about 50 percent is contributed to the national programs and the remainder is used for in-state education and information programs.

The bulk of the funds in 1989—approaching three-quarters—were spent on promotion, a general term for demand-enhancing activities directed to consumers (advertising) as well as food service and retail merchandising. Other expenditure categories include research, producer communication, and consumer and producer information. Administrative expenses are modest, less than five percent (Figure 19-3). Legislation prohibits the use of check-off dollars for lobbying. The programs are described briefly below.

Promotion

Advertising funds have been used for TV, radio, and print advertisements, among other non-media campaigns. Beef programs have focused on the "Real Food for Real People" message, using such celebrities as James Garner (Figure 19-4). More recently "regular" people from around the country have been featured. The intent is to emphasize the contemporary nature and broad appeal of beef. The Pork Producers Council has focused on a "The Other White Meat" theme which emphasizes the low fat content of pork along with its substitutability for chicken, in particular, in a range of dishes.

Promotion is more than consumer advertising. Programs are also directed to retailers and food service operators to encourage the featuring of red meat and to maintain a level of excitement about the product. Nor are the funds restricted to the U.S. In 1989, monies were spent in Japan and Europe to promote the consumption of U.S.-produced red meats.

FIGURE 19-3. Beef Board and National Pork Producers Council Expenditures by Category in 1989 (Sources: Beef Board 1989 (l); National Pork Producers Council 1989 (r))

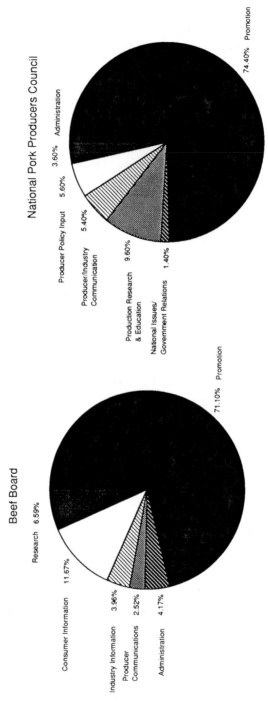

FIGURE 19-4. Part of the Beef Board's "Real Food for Real People" Advertising Themes (Source: Beef Board 1988)

"Nice thing about sirloin. Makes the whole brain happy."

"I've devised my own way to cook perfect steaks.... I call it my test steak. ...., it's done a little early...and no one knows, 'cause I eat the evidence."

Information

Informational programs may have material on the nutritional value of red meats, and their place in a well-balanced diet, with suggestions on new uses for these meats. Materials are presented directed to consumers and channeled through "opinion leaders" like athletes, coaches, and health professionals.

Research

Research programs are many and directed to increasing sales as well as reducing production costs. A sampling of sales-increasing projects are:

- Beef consumption patterns — at home and away from home;
- A study of [supermarket meat] case space allocation;
- In-state pilot nutrition information program on beef, pork, and lamb;
- Fat and caloric content of pork trimmed of fat before or after cooking;
- Study of producing an acceptable low-fat, ground beef patty.

These, and similar programs, support the position that the "information we disseminate will be supported by facts and science" (NLMB 1989). Sales-increasing projects can better be described under the headings of marketing research, nutrition research, and product development, each of which received about one-third of available funds in 1989.

Production-related research has focused on disease prevention and control, especially for pseudorabies. Pork producers are designing voluntary swine guidelines for humane, quality animal care in confinement operations.

PROGRAM EFFECTIVENESS

Limited systematic evaluation has been conducted on the impacts of the advertiser, information, and research funds collected and spent under the check-off programs. This limited analysis may be attributed to both the recentness of the efforts — which means there is only a small data base available — and the absence of a mandate for evaluation in the authorizing legislation.

Ward (1989) has attempted a preliminary estimate of promotion's impact on sales by statistically modelling beef demand for 1979-1986, just prior to the period of program expenditures. This model is then used to project demand for 1987-1989 and compare these projections to actual demand in that period. The implicit assumption is that the difference between the two sets of figures is attributable to program activities. Estimates were made at the retail, box beef, and live weight levels; the results shown graphically in Figure 19-5. This figure can be interpreted by noting that, for the period of promotional expenditures beginning in 1987, actual

FIGURE 19-5. Comparisons of Forecast Residuals and Beef Industry Program Expenditures Starting with 82:2.

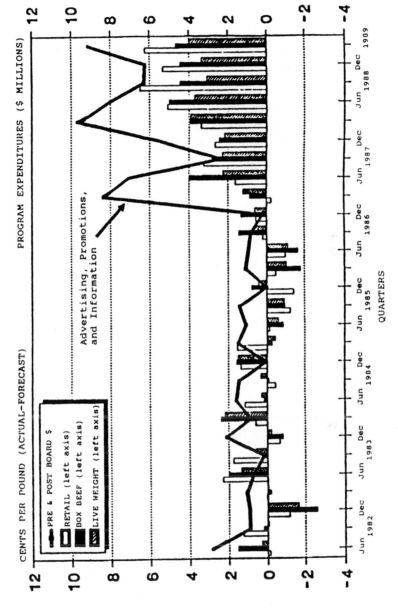

451

prices are higher than would be expected otherwise.[4] Ward estimates that the average quarterly expenditures of $8 million led to prices of 1 to 2.8 cents per pound higher than they would have been otherwise (1989, pp. 17-19).

Nutritional and other consumer information is more difficult to relate to product sales. It is therefore helpful to examine the record for other products.

In an extensive 1985 literature review led by David Johnson, 303 individual studies, covering a wide range of products, were integrated and accessed. The analysis concluded that persons of all age groups who participated in nutritional educational programs were more knowledgeable about nutrition than non-participants. Moreover, 58 percent of participants made changes in purchasing and consumption decisions that reflected their better understanding of nutritional issues.

At a commodity-specific level, the National Dairy Council has, for some time, supported nutritional education programs in primary and secondary schools, and evaluated the impacts of those programs. In one study of a high-school-level program, students were found to have significantly enhanced their nutritional sophistication. Furthermore, dairy product consumption increased 36 percent (reported in Tauer and Forker 1987, p. 211). The nutritional image of dairy products, especially with regard to cholesterol levels, shares many parallels with that of red meats. These experiences with dairy industry programs strongly suggest that nutritional education about meat will improve consumers' factual knowledge and purchases.

Some indirect evidence exists for beef and pork. Consumers who are aware of industry advertising are 59 percent more likely to believe beef fits into their lifestyle, and 34 percent more likely to say beef can fit dietary guidelines for health and diet. For pork, dietitians were monitored over a multi-year information campaign. Almost 25 percent of those surveyed have a more positive attitude about pork because of leaner animals and leaner products (Meat Board 1990). Thus, these informational programs appear to successfully present a counter-position to the simplistic anti-cholesterol message which is so prevalent in the media and advertising today.

---

4. That is, the residuals shown in the figure are all positive when the expected pattern would be a random distribution, positive and negative, around the zero line.

SYNOPSIS

Several major changes are anticipated in the 1990s and beyond which will have major implications for the future of the red meat sector. Here, three are examined: biotechnology and production, health and safety, and promotional activities.

Biotechnology has already had an impact on red meat production through embryo transplants. These, however, remain costly, especially for hogs, so the influence is slow. Potentially more substantial is the use of growth hormones (somatotropin) to stimulate lean gain compared to fat deposition. The result is a product that is more attractive to consumers at a lower cost, an estimated 6-11 and 5-8.5 percent lower for hogs and cattle, respectively. An injected or implanted form of this product is near commercial readiness but regulatory approval has been slow. In the longer term, an inserted gene can produce internally elevated levels of somatotropin. However, that advance is some time off, as are changes in characteristics controlled by multiple genes.

Health and safety issues, especially cholesterol, have been perplexing for consumers due to the varied nature of recommendations. In general, consumers are advised to lower their intake of fats and especially saturated (largely) animal fats. The benefits of dietary change are, however, being questioned, especially for women and the elderly.

Promotional activities advanced under 1985 laws allowing "checkoff" funds to be collected from cattle and hog producers (and importers). By 1990, about 70 percent of these monies were used to promote the consumption of red meats. Print and especially TV advertising is used. Preliminary results suggest that, for beef, the programs led to prices 1-2.8 cents per pound higher than in the absence of the programs. Other monies are spent on, among other things, consumer information, including nutritional information and recipes.

Study Questions

1. What does the net effect of biotechnology on red meat production appear to be at this time?
2. Cholesterol used to be so easy to hate. What recent information is making it difficult for consumers to evaluate the implications of cholesterol in the diet?
3. Cholesterol is but one component of the potential health risks of red meats. What are the others, and how might they affect the sector?

4. What is the industry doing to counter the negative health image of red meats? How effective is it?

REFERENCES

Acker, D. *Animal Science and Industry.* Englewood Cliffs, NJ: Prentice-Hall, Third Edition, 1983.

American Meat Institute. "California Law Hits Suppliers' Pocketbooks for Residue Violations," *Newsletter* May 25, 1990, pp. 3-4.

American Meat Institute. "Meat Facts." 1989.

American Meat Institute. "AMI Study Examines 'Giant Lean' Beef Program," *AMI Newsletter* March 6, 1987, p. 3.

Anderson, G. B. "Identification of Sex in Mammalian Embryos." In *Genetic Engineering of Animals,* ed. J. Warren Evans and Alexander Hollaender. New York: Plenum Press, 1986, pp. 243-250, Basic Life Sciences Vol. 37.

Bauman, D. E., P. S. Eppard, M. J. DeGeeter, and G. M. Lanza. "Responses of High Producing Dairy Cows to Long-Term Treatment with Pituitary- and Recombinant-Somatotropin," *Journal of Dairy Science,* 68(1985).

Bauman, D. E. and S. N. McCutcheon. "The Effects of Growth Hormone and Prolactin on Metabolism." In *Proceedings, VI International Symposium on Ruminant Physiology: Control of Digestion and Metabolism in Ruminants,* ed. L. P. Milligan, W. L. Grovum and A. Dobson. Englewood Cliffs, NJ: Prentice-Hall, 1986, Chapter 23, pp. 436-455.

Beef Board. "Annual Report." Englewood, CO, 1988 and 1989.

Begley, S. "Hope for Clogged Arteries" *Newsweek* September 14, 1987, p. 74.

Blankenborn, D. H., et al. "Beneficial Effects of Combined Colestipol-Niacin Therapy on Coronary Atherosclerosis and Coronary Venous Bypass Grafts," *J. American Medical Association* 257(1987):3233-3240.

Boyd, R. D., D. E. Bauman, D. H. Beermann, A. F. Neergaard, L. Souza, and W. R. Butler. (Abstract) "Titration of the Porcine Growth Hormone Dose Which Maximizes Growth Performance and Lean Deposition in Swine." Paper presented at the Annual Meeting of the American Society of Animal Science, Kansas State University, July 29-August 1, 1986.

Brackett, B. G., D. Bousquet, M. L. Boice, W. J. Donawick, J. F. Evans, and M. A. Dressel. "Normal Development Following in vitro Fertilization in the Cow," *Biol. Reprod.* 27 (1982):147-158.

Branson, R. E., H. R. Cross, J. W. Sauell, G. C. Smith, and R. A. Edwards. "Marketing Implications from the Natural Consumer Beef Study," *Western J. of Agri. Econ.,* 11(1986):82-91.

Branson, R. E. and P. Rosson. "Consumer Attitudes Regarding Leanness in Beef, Animal Fats, and Beef Grading Systems." Texas Agricultural Experiment Station, Texas A&M University, MP-1495, September 1981.

Brody, B. "Evaluation of the Ethical Arguments Commonly Raised Against the Patenting of Transgenic Animals." In *Animal Patents: The Legal, Economic and Social Issues,* ed. W. Lesser. New York: Stockton Press, 1990.

Business Week. "The Livestock Industry's Genetic Revolution." June 21, 1982, pp. 124, 126, 130, 132.

Buttel, F. H. and C. C. Geisler. "The Social Impacts of Bovine Somatotropin: Emerging Issues." Chapter 7 in *Biotechnology and the New Agricultural Revolution,* ed. J. J. Molnar and H. Kinnucan. Boulder, CO: Westview Press, 1989.

Connor, J. M., R. T. Rogers, B. W. Marion, and W. F. Mueller. *The Food Manufacturing Industries: Structure, Strategies, Performance, and Policies.* Lexington, MA: Lexington Books, 1985.

Cross, H. R., quoted in "It's Leaner Not Meaner," *The Ithaca Journal* March 26, 1987, p. 14B.

Cross, H. R. "The Change Factors." Keynote talk at Livestock Industry Institute Forum, "Risk and the Emerging Livestock and Meat Industry." June 18, 1986, Seattle, WA.

Donahue, S. E. "A Technique for Bisection of Embryos to Produce Identical Twins." In *Genetic Engineering of Animals,* eds. J. W. Evans and A. Hollaender. New York: Plenum Press, 1986, p. 163-173 (Basic Life Sciences, Vol. 37).

Etherton, T. D., C. M. Evock, C. S. Chung, P. E. Walton, M. N. Sillence, K. A. Magri, and R. E. Ivy. "Stimulation of Pig Growth Performance by Long-Term Treatment with Pituitary Porcine Growth Hormone (pGH) and a Recombinant pGH," *J. of Animal Science* 63(1986, Suppl. 1):219.

Fabry, J., L. Ruelle, V. Claes, and E. Ettaib. "Efficacity of Exogenous Bovine Growth Hormone for Increased Weight Gains, Feed Efficiency and Carcass Quality in Beef Heifers," *J. Animal Science* 61(1985, Suppl. 1):261-262.

Fleming, B. Opinion Page. *National Hog Farmer* November 15, 1982, p. 13.

Foote, R. "The Technology and Costs of Deposits." In *Animal Patents: The Legal, Economic and Social Issues,* ed. W. Lesser. New York: Stockton Press, 1990.

General Accounting Office. "Electronic Marketing of Agricultural Commodities—An Evolutionary Trend." GAO/RCED 84-97, March 8, 1984.

Gilliam, H. C. Jr. "The U.S. Beef Cow-Calf Industry." USDA, Economic Research Service, Agricultural Economic Report 515 (September 1984).

Glaser, L. K. "Options for Making Animal Products Part of a Leaner Diet," *National Food Review* 12(January-March 1989):27-30.

Glaser, L. K. "Provisions of the Food Security Act of 1985." USDA, Economic Research Service, Ag. Information Bull. 498, April 1986.

Hammer, R. E., V. B. Pursel, C. E. Rexroad, Jr., R. J. Wall, D. J. Bolt, K. M. Ebert, R. H. Palmiter, and R. L. Brinster. "Production of Transgenic Rabbits, Sheep, and Pigs by Microinjection," *Nature* 318(1985):680-683.

Hammer, R. E., V. B. Pursel, C. E. Rexrood, Jr., R. J. Wall, D. J. Bolt, R. H. Palmiter, and R. L. Brinster. "Genetic Engineering of Mamallian Embryos," *J. Animal Science* 63(1986):269-78.

Hansel, W. "Animal Agriculture for the Year 2000 and Beyond." William Henry Hatch Memorial Lecture, November 10, 1986, Phoenix, AZ.

Harris, M. "Spending on Meat, Poultry, Fish and Shellfish," *National Food Review* 11(October-December 1988):33-36.

Hellmich, N. "Eating Habits Reflect Dietary Confusion," *USA Today* 1989, p. 1.

Johnson, D. G., J. M. Connor, T. Josling, A. Schmitz, and G. E. Schuh. *Competitive Issues in the Beef Sector: Can Beef Compete in the 1990s?* Minneapolis, MN: The H. H. Humphrey Inst. Public Affairs, Univ. Minnesota, September 1989.

Johnson, D. "Nutrition Education: A Model for Effectiveness, A Synthesis of Research," *J. Nutrition Ed.* 17 (June 1985, Supplement).

Kalter, R. J. and R. A. Milligan. "Emerging Agricultural Technologies: Economic and Policy Implications for Animal Production." Paper presented at the Symposium on Food Animal Research, Lexington, KY, November 2-4, 1986.

Kalter, R. J., R. A. Milligan, W. Lesser, W. Magrath, L. Tauer, and D. Bauman. "Biotechnology and the Dairy Industry: Production Costs, Commercial Potential, and the Economic Impacts." A. E. Res. 85-20, Department of Agricultural Economics, Cornell University, December 1985.

Kenney, J. and D. Fallert. "Livestock Hormones in the United States," *National Food Review* 12(July-September 1989):21-24.

Kuchler, F., J. McClelland, and S. E. Offutt. "Regulating Food Safety: The Case of Animal Growth Hormones," *National Food Review* 12(July-September 1989):25-30.

Kolata, G. "Major Study Aims to Learn Who Should Lower Cholesterol," *The New York Times* September 26, 1989, pp. C1, C11.

Larson, B. A. and F. Kuchler."The Simple Analytics of Technology Adoption: Bovine Growth Hormone and the Dairy Industry," *North Central J. Agr. Econ.* 2(1990):109-24.

Leibo, S. P. "Cryobiology: Preservation of Mammalian Embryos." In *Genetic Engineering of Animals*, Eds. J. W. Evans and A. Hollaender. New York: Plenum Press, 1986, p. 251-272 (Basic Life Sciences, Vol. 37).

Lemieux, C. M. and J. W. Richardson."Economic Impacts of Porcine Somatotropin on Midwest Hog Producers," *North Central J. Agr. Econ.* 11(1989):171-81.

Lesser, W. "Implications for Breeders." In *Animal Patents: The Legal, Economic and Social Issues*, ed. W. Lesser. New York: Stockton Press, 1990.

Lipton, K. L. "USDA Actions," *National Food Review* 12(July-September 1988):44-46.

Lipton, K. L. "USDA Actions, "*National Food Review* 11(July-September 1988):36-38.

Magrath, W. B. and L. W. Tauer."The Economic Impact of bGH on the New York State Dairy Section: Comparative Static Results," *Northeastern J. Agr. and Resource Econ.* 15(1986):6-13.

Meat Board, National Livestock and Meat Board. *Annual Report*. Winter 1990.

Meat Board."Consumer Marketing Plan." 1985.

Meltzer, M. I. Repartitioning Agents in Livestock: Ecomonic Impact of Porcine Growth Hormone. Master's thesis, Cornell University, 1987.

Michalska, A., P. Vize, R. J. Ashman, B. A. Stone, P. Quinn, J. R. E. Wells, and R. F. Seamark. "Expression of Porcine Growth Hormone cDNA in Transgenic Pigs." Proc. 18th Animal Conference. Aust. Society for Reprod. Biol. Brisbane, 1986, P. 13 (abst).

Miller, C. "Growth Hormone Genes Bring Super Pigs Closer to Market," *Genetic Eng. News*, 1987, p. 7 (May).

Miller, W. L., J. A. Martial, and J. D. Baxter, "Molecular Cloning of DNA Complementary to Bovine Growth Hormone in RNA," *J. Biological Chem.* 255:7521(1980).

Milligan, R. and W. Lesser. "Implications for Agriculture." In *Animal Patents: The Legal, Economic and Social Issues,* ed. W. Lesser. New York: Stockton Press, 1990.

Moore, T. J. *Heart Failure.* New York: Random House, Inc., 1989.

National Livestock and Meat Board. "Investments in Research, 1987-1990." Chicago, 1989.

National Livestock and Meat Board. "Exploring the Unknown: Meat, Diet and Health." Chicago, 1983.

National Pork Producers Council. "Annual Report." Des Moines, IA, 1989.

Roberts, T. and E. van Ravenswaay. "The Economics of Safeguarding the U.S. Food Supply." USDA, Economic Research Service, Ag. Info. Bulletin No. 566, July 1989(a).

Roberts, T. and E. van Ravenswaay. "The Economics of Food Safety," *National Food Review* 12(July-September 1989 (b)):1-8.

Shapouri, H., R. Bowe, T. Crawford, and W. Jessee, "Costs of Producing U.S. Livestock, 1972-87." USDA, Economic Research Service, Ag. Econ. Rpt. No. 632, April 1990.

Tauer, J. R. and O. D. Forker. *Dairy Promotions in the United States, 1979-1986.* Department of Agricultural Economics, Cornell University, A. E. Res 87-5, June 1987.

U.S. Congress. Office of Technology Assessment. *Commercial Biotechnology, An International Analysis.* Washington, DC, 1984.

U.S. Department of Commerce, Bureau of Census. *Statistical Abstract of the U.S., 1987.* Washington, DC, December 1986.

Van Arsdall, R. N. and K. E. Nelson. "U.S. Hog Industry." USDA, Economic Research Service, Agr. Econ. Rpt. 511, June 1984.

Van Brunt, J. "Transgenics Primed for Research," *Bio/Technology* 8(August 1990):725-28.

Wallis, C. "Hold the Eggs and Butter," *Time* March 26, 1984, pp. 56, 58-63.

Ward, R. W., S. R. Thompson, and W. J. Armbruster. "Advertising, Promotion and Research." Chapter 4 in *Federal Marketing Programs in Agriculture: Issues and Options,* eds. W. J. Armbruster, D. R. Henderson, and R. D. Knutson, Danville, IL: Interstate Printers and Publishers, 1983.

Ward, R. W. "Economic Evaluation of Beef Promotion and Information Programs." Study Conducted for the National Cattlemen's Association. September 1989.

Subject Index

Additives and residues, 239,240, 443,444

Auctions, 38,176,177
  auction market, 9,10,212,223, 283,286,292,293,311
  electronic, 314
    live animals, 313,317,318,319, 320
  meat, 398,399

Away-from-home meat
    consumption, 5,117,403, 423-430
  importance of/for meat, 123-128, 148,423,426-428
  performance, 428-430

Basis (Futures)
  calculation, 356,362
  concept, 343,355,362

Biotechnology, 432-439,453
  growth hormones (somatotropin), 434-436,444,453

Boxed beef, 175,244,367,377-380, 397,404-405,412,443

Branded meat products, 385, 389-392,397,400,444

Buying
  buying stations, 214,215,220,285, 294-297,298,319
  direct-to-consumers, 34,107,238, 312-313,319
  grade and yield, 244,301-304,319
  order buyers, 214,215-217,218, 220-221,243,283,298,318
  packer buyers, 183,212-214,221, 243,290,296,297,299,300, 318,376

public markets, 34,177,178,184, 218,219,228,229,231,232, 243,244,247,283,285, 286-294,295,319,330

Buying stations, 214-215,220,285, 294-297,298,319

Cattle
  cycles, 51,65,70,97,192-198,201, 202,203,206
  feeding, 38,58,60,65,70,74,160, 192,195,203,301,304,305, 357,432,435
  grades, 255,256-271
    quality, 256,257,262,264,267
    yield, 256-257,258,262-264, 265,268
  international trade, 77,79,81,82, 83
  seasonal trade, 158,159,160

Chicago Union Stock Yard, 41

Commingling, 292,294

Commission agent(s), 41,215,286, 289,291,384

Comparative advantage, 58,85,175

Consent Decree of 1920, 47,241, 244,365,377

Consumer surplus, 176,177

Consumption, 113-148
  beef and veal
    U.S., 60,119,123,128,312
    world, 83,132,135,137,138, 141,143,146,147,148,202, 426,430,443
  lamb
    U.S., 119,123,312
    world, 135,138,141,148

459

Author Index

The New Encyclopedia Britannica, 48
The Washington Post, 236,248
Thompson, J. W., 30,32,34,48
Thompson, S. R., 445,457
Tilley, D. S., 180,182,186
Tomek, W. G., 11,13,23,27,129, 138,146,150,178,186,361, 363,381
Torgerson, R. E., 322
Tryfos, P., 150
Tryphonopoulos, N., 150

U.S. Congress, Office of Technology Assessment, 250,251,253,254,432,457
U.S. Department of Commerce, 76, 111,457
U.S. Department of Commerce, Bureau Census, 70,74,76,78, 80,82,111,113,150,402
U.S. General Accounting Office, 322,398,401,455
Uetz, M. P., 128,150,413,421
United Nations, 76,111,113,150
United Nations, Food and Agriculture Organization, 110,186
United States Congress, Office of Technology Assessment, 282
Unnevehr, L. J., 149
USDA, 1,8,29,30,48,59,60,62,63, 64,65,75,76,77,79,82,111, 128,150,154,155,156,157, 186,239,248,263,366,372
USDA, Agricultural Marketing Service, 174,175,186,250, 258,260,265,267,269,270, 271,273,274,275,276,280, 282,398
USDA, Economic Research Service, 1,5,13,24,60,61,76,97,111, 150,155,160,161,162,163, 164,165,166,167,171,172,

186,192,193,198,207,220, 221,233,238,248,298,312, 322,326,364,384,394,396, 402,403,406,408,413,417, 421,424,426,428,430
USDA, Economic, Statistics and Cooperative Service, 150,402
USDA, Extension Service, 194,207
USDA, Food Safety and Inspection Service, 237,238,239,240, 248
USDA, Foreign Agricultural Service, 77,83,95,99,101, 109,111
USDA, National Agricultural Statistics Service, 13,74,76, 161,186
USDA, News Center, 256,282
USDA, News Division, 255,282
USDA, Office of Governmental and Public Affairs, 96,111
USDA, Office of Information, 382
USDA, Packers and Stockyards Administration, 217,229, 233,243,248,286,298,299, 301,304,305,316,369,397, 402
USDA, Statistical Reporting Service, 221,227,229,233
USDA, Statistical Research Service, 63,76
Usman, M., 185

Van Arsdall, R. N., 322,433,457
Van Brunt, J., 439,457
Van Dress, M. G., 403,421,423, 428,430
van Ravenswaay, E. 444,457
Venadikian, H. M., 108,111
Vertin, J. P., 147,149
Virginia Tech., Cooperative Extension Service, 313,322
Vize, P., 456